Pieces of Ringeck:
The Definitive Edition of the Gloss of Sigmund Ainring

Pieces of ringeck

The Definitive Edition of the Gloss of Sigmund Ainring

MICHAEL CHIDESTER

DIERK HAGEDORN

HEMA Bookshelf

Illustrations from the *Cluny fight book* (P) © RMN-Grand Palais/Art Resource, NY.

Le illustrazioni del *Peter von Danzig fight book* (PD) sono riprodotte per concessione della Biblioteca dell'Accademia Nazionale dei Lincei e Corsiniana.

All other art in this book is believed to be licensed for free use (as indicated in the captions) or to be in the public domain.

HEMA Bookshelf endeavors to respect the rights of all copyright holders in a manner consistent with its educational mission. If you believe any material has been included or described in this publication improperly, please contact us so we can make revisions to future editions.

Published by HEMA Bookshelf, LLC.

47 High Street #443
Medford, MA 02155
United States of America
www.hemabookshelf.com

Version 1.2, 2025.

ISBN 978-1-953683-41-0 (prestige hardcover)
ISBN 978-1-953683-42-7 (softcover)
ISBN 978-1-953683-47-2 (standard hardcover)

Library of Congress Control Number: 2024951142

Designed by Michael Chidester. The illustration on the front cover is a detail from GF: 2^{v}.

Typeset in Libertinus Serif Display and Libertinus Sans, used under the Open Font License http://libertine-fonts.org/

An environmentally-friendly book printed and bound in England by bookvault.

Table of Contents

Johannes Liechtenauer, who was the grand master of our art,
travelled widely and learned the art of fencing in many lands.

He allowed his teachings to be written in obscure and cryptic words
in a long poem called the Record.

Liechtenauer had a fellowship with sixteen masters.

Of those sixteen, seven left teachings that survive to our time.

Of those seven, two left commentaries on Liechtenauer's teachings:
Peter von Danzig and Sigmund Ainring.

This book is about Sigmund Ainring.

FOREWORD

TEA KEW

When I teach at events, I'm often asked why I use Ringeck as my primary gloss. This is a great question, and it's easy to give a "learned" answer—my favourite is his longsword gloss is the only one which is directly attributed to an author on Kal's list of the *Gesellschaft*. But that's not really why I focus on Ringeck.

Eleven years ago, I walked into a cheap school hall in Cambridge for my first ever class and learned about Ringeck for the first time. Cambridge HEMA in those days was a very new club. We had half a dozen members, some cheap synthetic swords and a few fencing masks. But we had a treatise to work from, in the form of KEITH FARRELL's 2011 translation of Ringeck. With the release of the Wiktenauer Compendiums a couple of years later, I realised that I would need a heuristic to address situations where the different glosses disagreed with each other. It seemed simplest to pick the text I was already familiar with—Ringeck—and give that primacy any time there's a disagreement. So it all comes down to simple chance that day in October.

I first had the privilege of meeting MICHAEL himself at WMAW in 2017; on my first HEMA trip outside of the UK. WMAW, HEMAC Dijon and other less tournament-oriented events are a fantastic way to build connections and share ideas. Without regular travel, it's very easy to settle into the comfortable default behaviour of your club or local scene. This can mask issues when your local area is particularly weak at certain actions or tactics, leading to blind spots in your understanding and interpretations. If I can give you one piece of advice as a newer fencer or coach—travel!

Of course in 2020, travel became very much impossible. Instead, MIKE SMALLRIDGE organised *Fencing by the Book*, a podcast where a panel of early Liechtenauer specialists did a line-by-line discussion of every part of the unarmoured longsword gloss. That ended up taking us over a year of weekly podcasts to complete, with contributions on language, history and nearly every other topic which might be related to the *Zettel* and its meaning. If you're interested and have enough time (or enough difficulty sleeping), I hear it still holds up well.

In this book, we have the culmination of all the scholarship that's been done on Ringeck until now. The manuscript discussion (Chapter 4) is particularly valuable and provides a comprehensive reference to what we know about these manuscripts. For any sort of academic or research question I have going forward, it will be the first thing I pull off the shelf. I am very grateful that MICHAEL has taken the time and focus to produce it as part of HEMA Bookshelf—and I'll do my part to try and force a second edition in the near future!

Preface

Michael Chidester

In March of 2022, as I was despairing over the endless delays in the Meyer project, I contemplated what kind of project I could start that would fill in the gap and keep me busy until the translation was done. The answer came to me in the shower one day: Sigmund Ringeck.

My first encounter with Sigmund ain Ringeck's text was David Lindholm's 2003 long sword translation, which I bought in 2005 and found utterly baffling (all I knew at the time was what my club taught, which was Liechtenauer-*inspired* at best). When I took over as head instructor at my club in 2007, I revisited it in my efforts to develop a curriculum that really was based on Liechtenauer, and I also bought Christian Henry Tobler's excellent 2001 translation and interpretation. Both of these books shared a problem, though: they were based on only one manuscript—the Dresden manuscript (**SR**)—and not even the best one for the master.

As my shower thoughts turned to Ainringck, it occurred to me that my friend Dierk Hagedorn had spearheaded an ambitious publication of the Augsburg Lew manuscript (**L**) in 2017, and had joined forces with Christian to publish a solid edition of the Rome Danzig manuscript (**PD**) in 2021. It was only a matter of time until someone (maybe Dierk, maybe not) published a lavish Ainringck edition, but while **L** is an adequate representative of that gloss and **PD** is an exemplary copy of its glosses, there's no comparable manuscript we could look to for Ainringck. All we have are larger and smaller *fragments* of text which we try to piece together as best we can, but if someone were to put out a high-quality edition of the Glasgow fight book (**GF**), for example—or worse, **SR** again—that would risk becoming the 'default' Ainringck text in most peoples' minds for at least a decade and all the important variations would fall by the wayside.

Instead, I thought it would be cool to do an edition focused on building a text by putting all the pieces of the Ainring puzzle together, and even do something no one had tried in HEMA before—publish all the individual versions of a gloss in separate translation, in addition to a clean eclectic translation that merged them all together.

I reached out to Dierk with this idea, and he didn't take much convincing to jump on board. He authored new transcriptions of everything, including manuscripts he'd done before and others he hadn't touched yet. While I did most of the writing for this book, it's peppered with research provided by Dierk, both from his prior publications (see the Bibliography) and new research that no one's seen yet—some of which he was still doing even in late 2024 as this book was on the verge of publication.

Two and a half years after my first Patreon post, the book is complete—and what a book it is. For the first time, the mounted fencing from GF and the armored fencing from the Pirckheimer fight book (WP) appear in print, in both German transcription and English translation. Illustrations from GF and the Cluny fight book (P) are published here for the first time, and Dierk and I have packed a ton of original research on the manuscripts into a codicological study of the sources which I expect to be a valuable resource for all researchers going forward. We've also written some introductory chapters that will hopefully be a helpful foundation for students who are just dipping a toe into manual study for the first time.

This book will sit on my shelf proudly next to *Jude Lew* and *Peter von Danzig*, and I hope it will help advance the study of Liechtenauer for a new generation of fencers.

This project has stretched over the better part of three years, and as with any project spanning so much time, a large cast of supporters came together to make it possible.

First and foremost, my partner Kendra Brown not only not only reminded me to cook food and go outside periodically during this project, but also reviewed drafts of various parts of the book and offered tons of feedback that improved the finished product immensely.

Jack Baker, Stephen Cheney, Michael-Forest Meservy, and Brandon F. Whitehill reviewed drafts of the translation, and Brian Puckett and Christopher VanSlambrouck reviewed drafts of the other chapters. Their feedback was all very helpful.

The translation was only possible due to many long conversations about terminology and translation style with Rebecca Garber, Christian Trosclair, Cory Winslow, and my fellow panelists on the *Fencing by the Book* podcast: Kendra Brown, Stephen Cheney, Jessica Finley, Johanna Hopfgartner, Tea Kew, and Mike Smallridge.

Natasha and Christian Darcé of *Purple Heart Armoury* funded the license fees for the illustrations used in Chapter 3, and without their gracious support this would have been a very different book.

Finally, my supporters on Patreon helped me keep enthusiasm and momentum going as I worked though this project. I was going to list you all, but I'm told that might be a privacy problem. You know who you are, and thank you!

Introduction: Why I *used to* Despise the Sigmund Ringeck Manuscript

Dierk Hagedorn

It is not that I completely detest the manuscript SR, attributed to a certain fencing master Sigmund Ringeck, but I do think it is fairly overrated, and I have been quite vocal about this in the past. I held a number of talks or lectures about the matter, on the occasion of one event or another conference.

So, why do I dislike it so much? The manuscript is extraordinarily sloppily written; it does not credit authors where credit is due; it lacks images; and lastly, there never was a person called Sigmund Ringeck.

1. The Fame of the Manuscript—Rightfully Deserved, or Not So Much?

The subject of this discussion is a small volume, written on paper by an obviously hasty hand, bound in a soft parchment cover called a 'limp binding'. It is a collection of several sections that describe armoured and unarmoured combat in various disciplines; and it belongs to a genre of pragmatic literature called *Fechtbücher* ("fight books").

The manuscript is in large parts based on the teachings of a certain master Johannes Liechtenauer, who is a rather elusive figure since no biographical data whatsoever exist. We can only assume from the various datings of renditions of his lore—which only survives in quite cryptic verses that rhyme more or less convincingly—that he lived sometime around the turn of the fourteenth to the fifteenth century.

Liechtenauer's verses were commented on—in so-called 'glosses'—by a number of disciples. Some of these glosses have spawned a fair number of further manuscripts that can be divided into specific family groups, depending on the similarity of one source to another. Back then, in the fifteenth century, the only way to possess a book was to copy it by hand. Even after the invention of the printing press in 1454, a great number of books were written by hand. Depending on editorial intent or diligence, those copies were more or less true to their template—absolute equality, however, was hardly ever achieved (or intended).

The main text sections that make up manuscript SR, such as fencing with the long sword and armoured and mounted combat, have been copied individually numerous times, but mostly independently and never in conjunction in the same structure. The long sword section, for instance, was used extensively at least two other times (in GF and JMR), and to a lesser extent, with only brief excerpts, in another four sources (**AR**, **HM**, **HS**, and **P**). All these are the subjects of this edition.

Fight books in general do not necessarily attract an enormous audience, so they are *per se* only *comparatively* famous, at best. The interest in them was never excessive, but a small and steady flow of publications was issued throughout time. Some prolific early publications came out at the turn of the twentieth century and featured the works of Albrecht Dürer[1] or fencing master Hans Talhoffer.[2] Those were representative picture and coffee table books, testaments of the pinnacle of the printing technology of the time. Talhoffer's fight books[3] are notorious for their utter scarcity of text; Dürer's fight book is in some parts much more verbose, but lacks in other parts any text whatsoever that would accompany the illustrations. Furthermore, the edition of his manuscript by FRIEDRICH DÖRNHÖFFER neglects the text-only sections that constitute almost half of its contents.

Thus, it is a curious and unprecedented circumstance that in 1965, MARTIN WIERSCHIN published his PhD thesis about a fight book that contains nothing but text—not one single image explains the vast amount of technical descriptions.[4] WIERSCHIN's edition of the "Sigmund Ringeck" manuscript reproduces the text as a slightly modernised transcription of the original Early New High German; there is no easy-access transference into a modern, more approachable language. This item of the specialised literature was not exactly a crowd pleaser or a bestseller; nevertheless, it gained some recognition in the niche when at the turn of the last century more and more people practised what today is known as HEMA, 'Historical European Martial Arts'. Back then, in the 1980s and 1990s, source material was scarce, the internet was still in development, and acquiring valuable source material was a time-consuming and costly endeavour; early practitioners of the lost arts had to turn to whatever material could be unearthed, and WIERSCHIN's book was available. Thus, it spawned two major follow-up editions that aimed, as special how-to books, at the historical fencing community.

Thirty-six years after WIERSCHIN's edition, CHRISTIAN HENRY TOBLER came forth with an entire interpretation of the manuscript in black-

[1] DÖRNHÖFFER 1910.

[2] HERGSELL 1887, HERGSELL 1889A and HERGSELL 1889B.

[3] There are five manuscripts that can be considered as archetypes from his lifetime, namely HTB, HTG, HTK, HTKo, and HTM. HTG, HTK, and HTM in particular have spawned several copies, some of which have been created up to the 19th century. The contents of these manuscripts varies considerably, and while certain aspects are kept throughout, each source has its unique collocation. HTG and HTKo offer longer text sections but the illustrated material is accompanied only by short texts or captions, as it is the habit of the other manuscripts throughout.

[4] WIERSCHIN 1965.

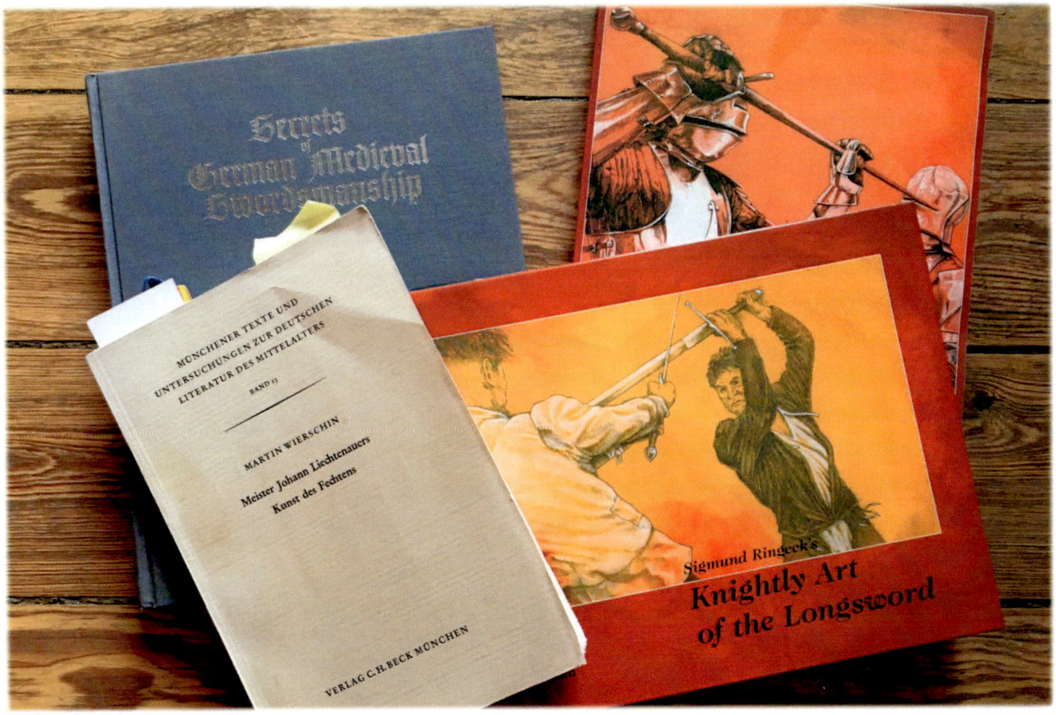

Figure 1: Several editions of the so-called 'Sigmund Ringeck' manuscript.

and-white images.[5] The book offered the complete text of the manuscript translated into the English language, though only the cryptic verses by Master Johannes Liechtenauer were also echoed as a transcription. The heavy hardcover book has gained quite some popularity and was a solid starting point for many followers of the historical fighting arts.

A strong competitor appeared two years later when DAVID LINDHOLM and PETER SVÄRD published another interpretation of "Ringeck's" text.[6] Instead of photos, their version relied on clear and precise illustrations that, as an extra benefit, were accompanied by footwork diagrams on the lower margin of each page. However, they offered only the long sword section. Considering the predominant interest particularly in this weapon, their choice was understandable. Another three years later, however, they finally filled the gap and edited the rest of the manuscript in a similar way: illustrations were maintained as the explanatory means, although they were executed much more coarsely; also, the footwork marginalia were dropped.[7]

[5] TOBLER 2001.
[6] LINDHOLM & SVÄRD 2003.
[7] LINDHOLM & SVÄRD 2006.

2. Master Sigmund, what was your name again?

The name of the alleged author of the glosses on Master Johannes Liechtenauer's verses about how to fence with the long sword is traded in a surprising variety of spellings. Master Sigmund appears in seven known sources (Fig. 2), and while his first name is the same throughout, the opposite is true for his last name.

Only two sources, PKM and PKB, use the same spelling: *Amring*. Another version in PKW adds a diaeresis over the letter 'a', thus creating an 'ä' which would change the pronunciation to sound like *Emring*—as we encounter the spelling of the name in GF. Writing a tittle was not overarchingly common in the 15th century which makes it occasionally daunting to distinguish between the German *dem* ("the", dative) and *dein* ("your"). So adding a coincidental tittle to *Emring* may have resulted in *Einring*, as in JMR. A potential next step would consist of not only spelling the syllable *ein* as *ain* (which would sound absolutely identical in German—like "ine" in "line") but also tearing apart the name and adding the suffix *-eck* with completely unclear origin.

The final rendition of Master Sigmund's last name in manuscript HM as *Schining*, *Schninig*, or even *Schming*—the position and/or existence of tittles in this case is not unambiguous—comes out of nowhere and remains a mystery.

3. Patron, master, and apprentice

Dating the manuscript is not possible without doubt; WIERSCHIN dated it—without much evidence—to the first half of the fifteenth century;[8] later examinations of the watermarks hint at a time of creation at the beginning of the 16th century.[9] The text itself refers to a certain Albrecht, *Pfalzgraf bei Rhein* ("Count Palatine of the Rhine") and *Herzog in Bayern* ("Duke of Bavaria"), which may give us a cue. Master Sigmund was his fencing master,[10] but inconveniently, there were four Albrechts in that position, namely:

Albrecht I (1336–1404), duke from 1347 to 1404,
Albrecht II (1368–1397), duke from 1387 to 1397,
Albrecht III (the Pious; 1401–1460), duke from 1438 to 1460,
Albrecht IV (the Wise; 1447–1508), duke from 1465 to 1508.

[8] WIERSCHIN 1965: 31.
[9] HOFFMANN 2022: 319–323.
[10] SR: 11^r.

There are several assumptions to be taken into consideration. Master Sigmund is mentioned as a member of 'Liechtenauer's company' (*Gesellschaft Liechtenauers*) in three manuscripts authored by Master Paulus

Figure 2: The many names of Master Sigmund.

Maister sigmund amring (PKM: 2^{r}, late 15th cen.).

Maister Sigmund åmring (PKB: 3^{v}, late 15th cen.).

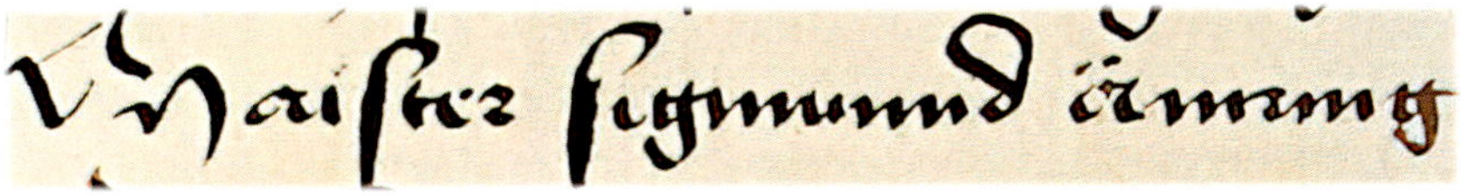

Maister sigmunnd ämring (PKW: 9^{r}, late 15th cen.)

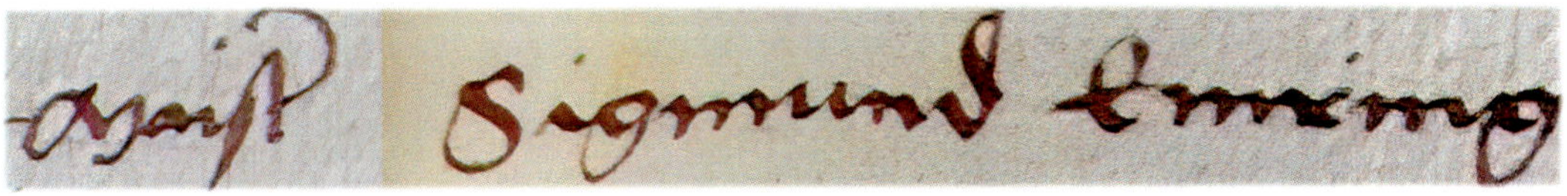

Maister Sigmund Emring (GF: 22^{r}, 1508).

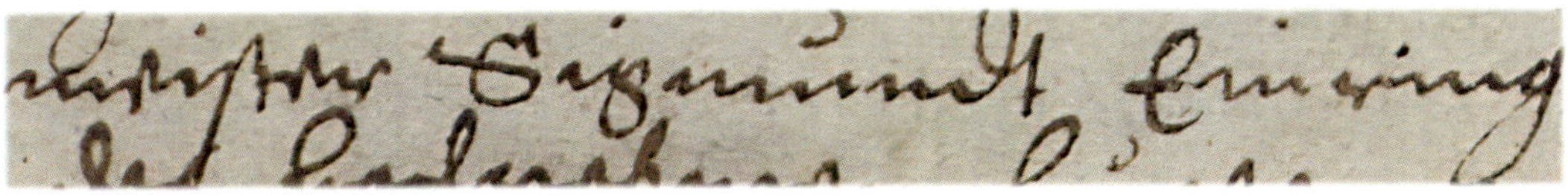

meister Sigmundt Einring (JMR: 6^{r}, 1570).

maister Sigmund ain ringeck (SR: 11^{r}, after 1500).

maister Sigmund Schining or *Schninig* (HM: 21^{r}, 1539).

Kal.[11] However, it remains unclear what kind of fellowship exactly it was. It may have been an informal community of fencers or a union of war-hardened mercenaries under the leadership of Liechtenauer.

Since Liechtenauer's life dates are unknown, we cannot use them as a basis to draw any conclusions about those of his potential pupil or comrade-in-arms Sigmund. In manuscript PD, reliably dated to 1452, Liechtenauer is mentioned as deceased with the formula "God have mercy on him". Another manuscript from Nuremberg, N, gives the impression that Liechtenauer was still alive at the time of writing, as he is repeatedly referred to in the present tense. However, the traditional dating of the manuscript to 1389 has recently been critically questioned, and a creation date in the fifteenth century is under discussion.[12]

If manuscript N is correctly dated to around 1389, then Liechtenauer was a fourteenth century master and Sigmund's patron was probably Albrecht I, who reigned from 1353 to 1404. If, on the other hand, Liechtenauer was a master of the early fifteenth century and had—hypothetically—assembled his company to fight, for example, in the Hussite Wars of the 1420s and 30s, then Sigmund's patron could have been Albrecht III, who held the title from 1438 to 1460. Albrecht IV claimed the title in 1460 and could therefore also have been Sigmund's patron; however, this would mean that our master was probably not a direct disciple of Liechtenauer, but a later heir to the tradition. However, Albrecht IV lived until 1508 and so the manuscripts of Dresden (SR) and Glasgow (GF) were, at least partially, written during his reign.

4. Sigmund and the others

When we single out at the long sword teachings that are based on Master Johannes Liechtenauer's teachings for a more detailed glance and compare particularly the glosses, we realise that while a certain similarity throughout the corpus of manuscripts can be observed, there are decisive differences too. It is hardly possible to determine direct relations between two manuscripts since it is not entirely clear which manuscript may have served as a template for one or more copies. One reason is the lack of reliable datings (some manuscripts are dated internally, others can only approximately be dated by the means of watermarks or palaeographical

[11] Fencing master Paul Kal is the initiator of a series of illustrated manuscripts among which PKB, PKM, and PKW are provided with a short introduction in which the names of the members of Liechtenauer's company are mentioned.

[12] Ondřej Vodička suggests a date of creation in the first third of the fifteenth century (Vodička 2019: 88), whereas Eric Burkart follows Lotte Kurras in dating it *around* 1389 (Burkart 2016: 451, p. 453; Kurras 1980: 15–17).

analyses); another one is that many authors copied repeatedly from older sources and occasionally intermingled them with passages of their own invention, so that a full family tree is next to impossible to establish.

It is furthermore unlikely that the mere un-glossed verses have spawned only one single branch of glosses, thus explaining more than one expanded and glossed-on version of the same source material.

A comparative look at specimens of three of the most prominent tradition lines (Master Sigmund, Peter von Danzig, Lew the Jew) reveals that occasionally, the similarity is striking, occasionally, however, not so much. In the following, three representative manuscripts of the respective lineages describe a technique called *Durchwechseln* ("changing through"). It becomes clear that SR (and other witnesses close to this source) is less extensive than the other two, which are much more similar to each other. The following translations by three different translators highlight the complexity of how to transfer the original text into modern English; many choices must be made which can result in diverse interpretations, even if the original source material is not too dissimilar in its phrasing.

SR: 41^r

> *In the approach, when you cut in or thrust toward him, if he wants to bind on your sword (with a cut or with parrying), then let your point sneak through under his sword and stab in greatly toward the other side with it. Thus you find his exposure quickly.* [13]

PD: 31^r

> *When you come to him in the* Zufechten, *then strike strongly above to him. If he then strikes against you to your sword and not your body, then let the point slip through under his sword before he can bind against your sword, and thrust to his other side to the chest. If he becomes aware of the thrust and quickly goes after with his sword to parry, then change through again; and always do this when he goes after your sword to parry.* [14]

13 *Wenn du im in dem zufechten ein hawest oder zu stichst will er dir denn mitt ainem hawe oder mitt veseczen an daß schwert binden so lauß den ort vnder sinem schwert durch wischen Unnd stich im do mitt sere ain eÿn zu der anderen sÿtten so fündest du an im schier de blöß.* See Chapter 5, section 71.

14 *Wenn dw mit dem zw uechten zw Im kumpst So haw im oben starck ein hawt er denn wider gegen dir zw dem swert vnd nicht zu dem leib So lass den ort mit dem haw vnden durch sein swert wischen ee wenn er dir an das swert pindt vnd stich Im zw anderen seitten zw der prust wirt er denn des stichs gewar vnd vert mit dem swert dem stich pald nach mit vor satzung So wechsel aber durch vnd das thue albeg wenn er dir mit vor seczen nach dem swert vert.* Translation by CHRISTIAN HENRY TOBLER from HAGEDORN & TOBLER 2021: 127.

L: 39^{r}

If you approach him in the Zufechten, *strike above long to his head. When he strikes against you towards the sword and not to your body, let your point drop with the strike and change through before he binds at your sword and thrust to the other side. If he notices the thrust and follows it with the sword with a displacement, then again change through to the other side. Do this always, on both sides, when he is going against your sword with a displacement.*[15]

5. I WOULD LIKE TO SEE SOME PICTURES, PLEASE

The two manuscripts that—in the long sword section—share the closest similarity to **SR**, frequently refer to images, or even show them. **JMR**, the personal copy written by fencing master Joachim Meyer, frequently mentions images but refrains from showing them. He presumably copied from an illustrated version of the lore but found it unnecessary to maintain the images—or he lacked time or skill.

GF, on the other hand, accompanies the majority of techniques with an individual illustration. Unfortunately, the first quires of the codex are missing. The compiler responsible for **SR**, however, deleted any notion of images throughout except for one single incident, where he writes *alß am nesten gemelt ist* ("as it is painted hereafter"). Nevertheless, possibly due to the overall carelessness of the execution of the manuscript, he writes *gemelt* instead of *gemalt* ("painted"), which could be read instead as *gemeldet* ("announced"); thus the exchange of one single vowel may be a hint at an otherwise complete and thorough extermination of any reference to images.

The entire paragraph of **SR**: 30^{v}–31^{r} reads as follows (the line breaks are mine in order to better understand the structure of the technique):

Glosa |Das ist wan dü Im mit der e^{r}sten verfÿrunge zu sine^{r} rechten sytten
zu dem kepff geschlagen haūst,
alß am nesten gemelt ist
|So schlach bald wyderūmb Im zu der rechten sytten zu dem kopff

[15] *Wann du mit dem zufechten zu Im kompst So haw Im oben lanck ein zum kopffe Hewet er dann gegen dir zum swert vnd nicht zum leibe So las den ort mit dem haw vnden durch wischen ee er dir an das swert pindet vnd stich In zu der andern seitten Würdt er dann des stichs gewar vnd fert mit dem swert den stich bald nach mit versatzung So wechsel aber durch zu der andern seitten vnd das treibe albegen wann er dir nach dem swert fert mit der versatzung zu beiden seitten etc.* Translation by FALKO FRITZ and THOMAS REHM, taken from HAGEDORN 2017: 169.

vnd far mit der kurtzen schnÿden mit auß gecrũtzten armen ṽber
sin schwe^{r}t
vnd spring Imlincke |das ist auff dein lincken sÿtten
vnd schnyd In mit der langen schnÿden durch das ma^{u}l ~

The same passage for comparison in GF: 5^{r}, in which the notion of a picture is repeated at the end of the passage:

Glosa. Das ist wen du in mit der ersten verfũrung zu seiner rechten
seitten
zu dem kopf geschlagen hast,
als vor am nachstn gemalt stet
so nym damit den schnit vndten in sein armm,
vnd schlag pald widerumb im zu der lincken seitten zu dem kopf,
vnd far mit der kurtzn schneid auß gekreutznt armen vber sein
schwert
vnd spring im linck das ist auf sein lincke seitt,
vnd schneid in mit der langen schneid durch das maul,
als da gemalt stet.

Figure 3: Illustrations from GF: 4^{v} & 5^{r} which are referenced in the text.

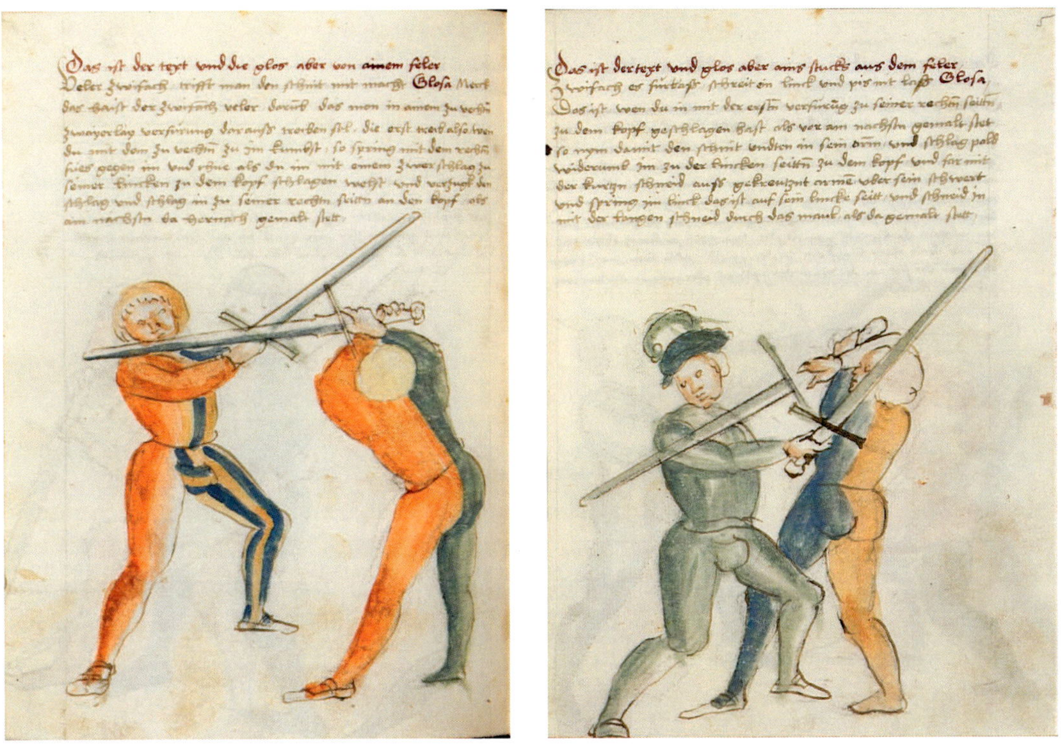

6. GET A NIB—THAT COULD BE A TAD TIDIER

The overall impression of the book is that it is cursory at best. Its small format is filled with a fluid yet hasty script that maintains the same impression throughout the pages. Several corrections can be made out, and also—in comparison with other sources—omissions. Possibly the manuscript was nothing more than a personal copy that didn't need to fulfill any representative purposes. Nevertheless, the scribe was not shy in applying rubrication—that is a second, red colour—throughout, thus adding a more elaborate touch.

For a brief overview, Figure 4 includes some examples of scribal hands from selected manuscripts. All of them describe the same technique—the *Schielhaw*—the exact phrasing, however, differs from one manuscript to the other.

7. TO LOSE ONE NAME MAY BE REGARDED AS A MISFORTUNE; TO LOSE MANY LOOKS LIKE CARELESSNESS

Manuscript **SR** mentions the names of only two authors, namely Johannes Liechtenauer and Sigmund ain Ringeck. Both appear in the first section of the manuscript that deals with swordplay out of armour.[16] Sigmund is referred to as the person who made good use of Liechtenauer's verses in order to enrich them with explanations—glosses—and thus make them more accessible and comprehensible. This leads to the impression that the entire text of the manuscript was authored by Sigmund ain Ringeck as no further names are mentioned. It is by no means unheard of that medieval or Renaissance copyists or compilers omitted the names of the original authors of their respective editions. One example from the realm of fight books is the notorious council servant Paul Hektor Mayr. He issued three massive double-volumes in the middle of the 16th century, and he refrains from mentioning the name of any fencing authority altogether—including that of the elsewise popular and revered Master Liechtenauer.

However, it is noteworthy that other compilers of the same material included in manuscript **SR** were more friendly to their authorial predecessors.

Manuscript **GF**, for instance, mentions several names—*Johannes Liechtenawer*: long sword verses, mounted and armoured combat techniques (ff. 22^r, 75^r, 83^r); *Sigmund Emring*: explanation of the long sword verses (f. 22^r); *Martein Syber*: long sword (f. 24^v); *Andres Jude, Jobs von der Nyssen, Nicklas Prewsen, Hans Döbringer*: additional fencing instructions

[16] SR: 3^r, 10^v, 11^r.

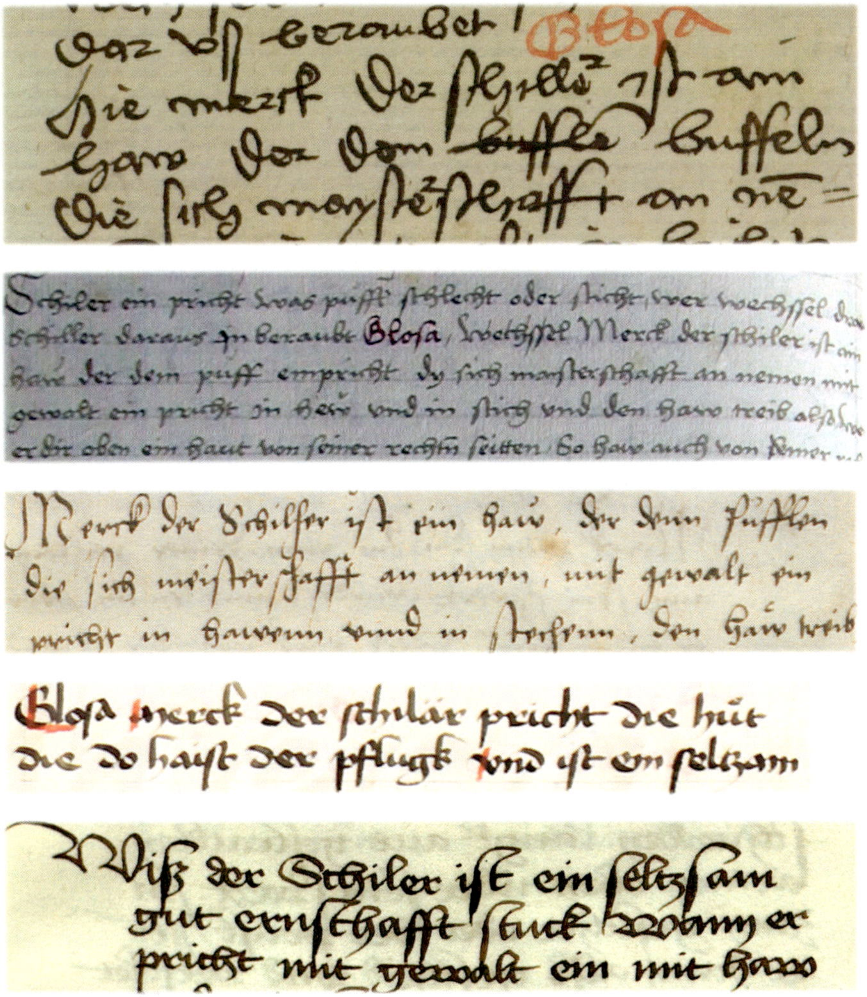

Figure 4: Scribal hands used in (top to bottom) SR (f. 29^{r}), GF (f. 5^{v}), JMR (f. 11^{v}), PD (f. 23^{r}), and L (f. 22^{r}).

(f. 27^{r}); *Ott the Jew*: wrestling (f. 67^{r}); *Mertein Hünczfeld*: armoured combat (f. 100^{v}); and finally *Andre Lignizer*: sword and buckler (f. 105^{r}).

Manuscript JMR is somewhat peculiar since it offers two versions of Liechtenauer's long sword verses with two versions of glosses; one in line with Master Sigmund, the other according to Master Lew. The manuscript is an autograph by fencing master Joachim Meyer who mentions himself two times (ff. 4^{v}, 123^{r}). The other names that occur are *Johannes Liechtennaur*: long sword verses, mounted and armoured combat (ff. 6^{r}, 58^{r}, 98^{r}); *Sigmundt Einring*: glosses for the long sword (f. 6^{r}); *Merten Siber*: long sword (f. 39^{v}); *Master Leb*: armoured combat (f. 58^{r}); *Merteinn Huntzvelt*: armoured combat (f. 67^{r}).

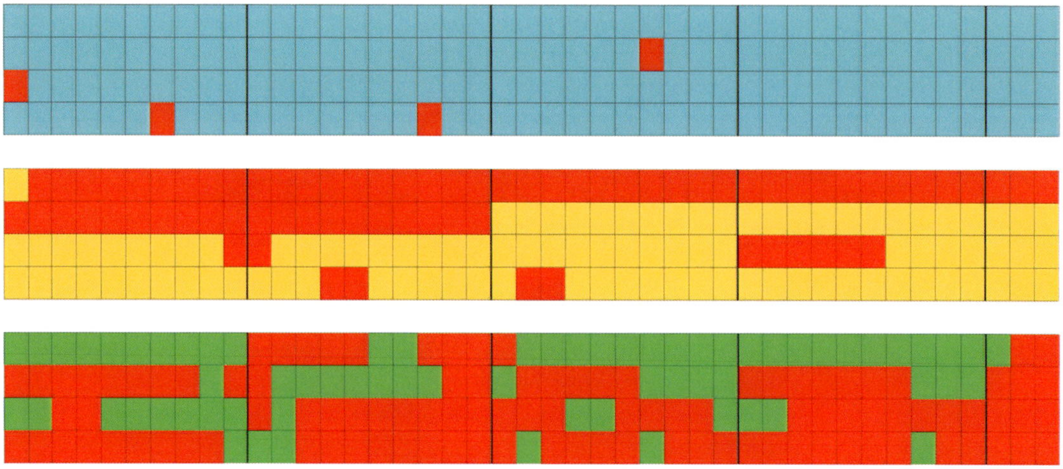

Figure 5: Overview of the long sword sections of SR (top), GF (middle), and JMR (bottom). For a complete concordance, see the Appendix.

8. Is it any good, though?

Manuscript SR truly shines when it comes to matters of scope. While it has severe deficiencies as far as accuracy is concerned, this is more than compensated for by the number of techniques featured in the manuscript—particularly in contrast to its closest relatives GF and JMR.

Figure 5 shows a consolidated number of 172 individual techniques or paragraphs, taken from the long sword sections of SR, GF, and JMR. A red rectangle in one manuscript signifies a vacancy in comparison to the other two witnesses.

It becomes clear that in regard to scope, SR is the superior source, with only four blank spots. GF shows many more red markings; but since the first quires of the manuscript are missing, the initial red section becomes understandable—with the remarkable exception of the very first paragraph that in SR and JMR serves as the introduction to the text; in GF, however, it is placed at the end. It is also noteworthy that some of the vacancies appear in pairs, hinting at a loss of one or more pages since in this particular manuscript each technique usually occupies an entire page.

Depending on the respective source in question, the amount of missing material ranges between considerable and neglectable.

In a broader perspective, when we compare SR with the aforementioned manuscripts PD and L from other traditional lines, there are several more extensive or even unique passages. The first example explains in a general lesson how to combine a strike of the sword with proper footwork.

SR: 11^{v}–12^{r}. The emphasis in bold shows an instruction that is more detailed than in the following two examples.

> *Remember, this is the first lesson of the 'long' sword: that you should learn to cut the cuts correctly from both sides, if you want to otherwise fence strongly and correctly. Learn it like this:* **when you want to cut from the right side, see that your left foot is forward, and when you want to cut from the left side, see that your right foot is forward.** *If you then cut a cut from over your right side, accompany the cut with your right foot. If you do not do this, then the cut is false and incorrect; since your right side remains behind, the cut is too short and cannot have its correct path downward toward the correct other side in front of the left foot.*[17]

PD: 10^{r}. The shorter text keeps the essence but is less informative.

> *Note: this is the first skill of the long sword: that, before all things, you should learn how to strike correctly if you should wish to fight strongly. And understand this thus:* **when you stand with your left leg forward** *and strike from your right side, if you do not follow the stroke with a step forward of your right foot, then the strike is false and incorrect. When your right side stays behind, thus is the strike too short and cannot have its correct path to the other side before and above the left foot.*[18]

L: 1^{v}–2^{r}. This witness has the shortest passage as there is an omission of several words which leaves the meaning mostly intact but with a certain ambiguity.

> *The first lesson of the longsword is that you shall first learn to strike correctly, and secondly that you want to fence strongly. So hear*

[17] *Merck dz ist die erst lere des langes schwercz dz du die hew von baÿden sÿtten recht solt lernen hawen Ist dz du annders starck vnd gerecht fechten wilt Dz ver nÿm allso* ***Wenn du wilt howen von der rechten sÿtten So sich dz dein lincker fuß vor stee Und wenn du wilt howen von der lincken sÿtten so sich dz dein rechter fuß vor stee*** *Häustu dann den ober haw von der rechten sÿtten so folg dem haw nach mitt dem rechten fuß tust du dz nicht / so ist der how falsch vnd vngerecht wann dein rechte sÿten pleibpt dahinden Darumb ist der haw zu kurcz vnd mag sein rechten gang vndersich zu der rechten anderen sÿtten vor dem lincken fuß nicht gehaben.* See Chapter 5, section 3.

[18] *Merck das ist die erst kunst des langenn swertz Das du vor allen dingen die häw solt lernen gerecht hawen ist das du anders starck vechten wilt vnd das ver nÿm also* ***wenn dw mit dem lincken fues vor stest*** *vnd haust vor deiner rechten seitten volgstu dann dem haw nicht nach mit deinem zu trit deins rechten fuess So ist der haw valsch vnd vngerecht wenn dein rechte seitt die pleibt da hinden dar vmb wirt der haw zu kurcz vnd mag seinen rechten gangk zu der anderen seitten vor dem lincken fuess vndersich nicht gehaben.* Translation by CHRISTIAN HENRY TOBLER from HAGEDORN & TOBLER 2021: 77.

> *this:* ***If you remain with your left foot forward*** *and strike from your right side, then the strike is wrong and unjust because the right side remains behind. Therefore the strike becomes too short and cannot have its correct way to the right side.*[19]

Another example that teaches the counter to a technique called *Abnehmen* ("taking off") shows that in some instances, SR is in fact the most verbose and detailed version.

SR: 20^{v}

> *Remember, when you bind with someone strongly on the sword, and he then pulls his sword upward, off from your sword above, and hews in again at you toward the other side of your head, onto the sword, then bind strongly with the long edge in above toward his head.*[20]

PD: 13^{v}

> *When he takes off above, then bind strongly on his sword above in to his head with the long edge.*[21]

(Manuscript L is completely void of this technique.)

The last example is a technique from the *Zwerhaw* ("crosswise cut") that exists exclusively in SR (and not even in the related and similar manuscripts GF or JMR) and does not appear in other traditional lines of gloss.

SR: 28^{v}

> *Likewise, when you bind on his sword with the crosswise, if he is Soft on the sword then press his sword down with the crosswise and lay the short edge forward, behind his arms on his neck.*[22]

[19] *Die ersten lere des langenn swerts ist Das du vor allen dingen die hew recht solt lernen hawen Ist es anders das du starck vechten wilt Unde das vernÿme also* ***wenn du mit dem lincken fuß vor steest*** *vnd hewest von der rechten seittenn So ist der hawe falsch vnd vngerecht wann die recht seitten pleibt da hinden Darumb wirt der hawe zu kurtz vnd mag seinen rechten ganck zu der rechten seitten nit gehaben etc.* Translation by FALKO FRITZ and THOMAS REHM, taken from HAGEDORN 2017: 93, 95.

[20] *Mörck wenn du mitt ainem starck am schwert bindest Ruckt er dan sein schwert übersich oben abe von dinem schwert vnd hawt dir zu der anderen sÿtten am schwert wider eÿn zuo dem kopffe So bind starck mitt der langen schnÿden Im oben eÿn zu dem kopffe.* See Chapter 5, section 14.

[21] *wenn er oben ab nÿmpt so pind an seinem swert starck oben ein ze seinem kopff mit der langen schneid.* Translation by CHRISTIAN HENRY TOBLER from HAGEDORN & TOBLER 2021: 85.

[22] *Item wen dü Im mit der zwer an sin schwert bindest Ist er dan waich am schwert So truck mit der zwer sin schwert nÿder vnd leg Im die kurtzen schnÿden hinder sinen armen vornen an den halß.* See Chapter 5, section 32.

What makes SR truly stand out, apart from these examples, are brief passages that appear in no other manuscript that is known today. These include a separate chapter about the long sword containing condensed and somewhat garbled techniques primarily from the *Zornhaw*, *Krumphaw*, and *Absetzen* chapters but with different wording, thus making this section an interesting alternative reading.[23]

9. Conclusion

With the fight book SR, we are in possession of a peculiar little manuscript that refuses to mention the individual names of its authors; that refrains from providing us with images that at some time almost certainly had belonged to its development history; and that introduces us to a fencing master with a name that in all likelihood had never existed in this form. It is written in an inattentive script with many corrections; deletions and omissions are frequent. Its cursory overall character is only slightly redeemed by the use of red ink in order to give it structure. Nevertheless, it has its undeniable merits. If nothing else, the contents of the long sword section is exhaustive and offers an interesting alternative to other traditional lines in the name of Master Liechtenauer.

One singular and distinguishing feature, however, is that it contains small sections of fencing and wrestling techniques that are absolutely unique in the entire corpus of German fight book sources, and for this reason alone manuscript SR deserves to be cherished by a wider audience. These sections are hidden gems that broaden the scope of the fencing lore and offer interesting additional insight into otherwise well-known explanations.

Although it will never become my favourite manuscript, I will not loathe it anymore. We shall be friends.

[23] SR: 55^{v}–59^{v}.

Figure 1: An image of a seated man holding a sword and gesturing at the beginning of the *Record* on the facing manuscript page (and quoted on the facing page in this book), assumed to represent Johannes Liechtenauer (PD: 2v).

1. Historical background

Michael Chidester

This chapter will give some background information on a few of the people connected to Ainring's gloss—the fencing masters Johannes Liechtenauer, Sigmund Ainring, and Hans Medel von Salzburg, and the prince[1] Albrecht of Bayern—to help contextualize these teachings.

1.1. Johannes Liechtenauer, the grand master

Ainring belonged to the tradition of Johannes (or Hans) Liechtenauer, who is widely credited as the founder of the most well-documented fencing tradition of the 15th and 16th centuries.[2] However, the records of his life are scant, and most of what's known about him comes from three sources.

The anonymous *Peter von Danzig fight book* (PD) states:

> *The* Record *of the chivalric art of fighting was composed and created by Johannes Liechtenauer (may God grant him grace), who was grand master of the art, beginning first with the extended sword, then with the lance and sword on horseback and with the retracted sword in the duel. Since the art belongs to princes, lords, knights, and soldiers, and they should learn and know it, he allowed this art to be written down. But because of frivolous fencing masters who would trivialize the art, it's written in obscure and cryptic words so that not just anyone will grasp or understand it..., and that way those masters can't make his art common or open among people who won't treat it with proper respect.*[3]

[1] 'Prince' is the usual translation for *Fürst*, a member of the high nobility. Any specific association with the children of kings is quite recent.

[2] This doesn't mean that his tradition was the most widespread or popular tradition in this era, but rather that Liechtenauer and his students kept more extensive written records than any others.

[3] *Alhÿe hebt dye zedel der Ritterlichen kunst des fechtens dye do getícht vnd gemacht hat Johans Liechtenawer der ain hocher maister In den künsten gewesen ist dem got genadig seÿ /Des ersten mit dem langen swert /Dar nach mit der glefen /vnd mit dem swert zw roß /Dar nach mit dem kürczen swert zw champf /vnd dar umb dÿ kunnst furrsten /vnd herren /Ritter vnd knechten zw gehört das sy dy wissen /vnd lernen sullen /So hat er die selbig kunst igleich besunder lassen schreiben mit verporgen /vnd verdackten worten /Dar umb das sÿ nit yderman vernemen noch versteen sol... /vnd hat das getan durch der leichtfertigen schirmaister willen /Dÿe Irr kunst gering wegen /das sein kunst von den selbi-gen maisteren nicht gemain noch geoffenwart sol werden solichen lewten die /dye kunst In wirden nicht behalten als ~~an~~ dann der kunst zw gehört.* PD: 3^{r}. All transcriptions are by Dierk Hagedorn and all translations are my own.

The equally anonymous *Nicolas Pol Hausbuch* (N) adds the following:

> *Above all else, you should notice and remember that there's only one correct art of the sword, and it was discovered and developed hundreds of years ago, and it's the foundation and core of all fighting arts. Master Liechtenauer understood and practiced this art completely and correctly; as I just wrote, he didn't discover or invent it himself, but rather traveled through many lands and searched for the true and correct art for the sake of experiencing and knowing it.*[4]

One other biographical note comes from the treatises of Paul Kal:

> *Liechtenauer (may God grant him grace) has created the art with his Fellowship and applied it with all knightly weapons: Master Johannes Liechtenauer; Master Peter Wildigans von Glatz; Master Peter von Danzig; Master Hans Spindler von Znaim; Master Lamprecht von Prag; Master Hans Seydenfaden von Erfurt; Master Andres Lignitzer; Master Jacob Lignitzer (his brother); Master Sigmund Ainring; Master Hartmann von Nürnberg; Master Martin Huntsfeld; Master Hans Pegnitzer; Master Philipp Perger; Master Virgil von Kraków; Master Dietrich Degenfechter von Braunschweig; Master Ott, a Jew, who was wrestler to the Lords of Austria; and the noble and steadfast Master Stettner, who is most of all the master of all students—and I, Master Paul Kal, was also his student. May God grant him grace before all others.*[5]

From these passages, a picture develops of a well-traveled, educated fencer and teacher who recorded a sophisticated fencing system; one who taught his art to people who would appreciate it (especially members of

[4] *...vor allen dingen vnd sachen, saltu merken vnd wissen, /das nür eyne kunst ist des swertes, /vnd dy mag vor manchen hvndert Jaren seyn fvnden vnd irdocht, /vnd dy ist eyn grunt vnd kern aller künsten des fechtens, /Vnd dy hat meister lichtnawer gancz vertik vnd gerecht gehabt vnd gekunst • /Nicht das her sy selber haben fvnden vnd irdocht, als vor ist geschreben //Sonder/ her hat manche lant, durchfaren vnd gesucht, durch der selben rechtvertigen vnd warhaftigen kunst wille, /das her dy io irvaren vnd wissen wolde...* N: 13ᵛ.

[5] *...dye kunst dye liechtenawer mit siner geselschaft gemacht vnd gepraucht hat, In aller Ritterlichen weer das Im got genedig sey Maister hanns liechtenawer Maister peter von dantzk Maister Peter Wildigans von glatz Maister hanns Spindler von tznaim Maister lamprecht von prag Maister Hanns Seydenfadm von Erdfurrt Maister Andre lignitzer Maister Jacob lignitzer geprueder Maister Sigmund åmring Maister hartman von Nuernbergk Maister Mertan huntzfeld Maister Hanns pägnitzer Maister Philipps perger Maister virgili von kraca Maister dietrich degenfechter von prawnschweyg Maister Ott Jud der der herren von Österrich Ringer gewesen ist Der Edel vnd fest Stettner der am maisten der maister aller Schüeler gewesen ist Vnd ich Maister pauls kall pin sein Schüeler gewesen das Im got genedig sey vor In allen.* PKB: 2ᵛ–4ʳ; compare PKM: 2ʳ and PKW: 9ʳ.

the military class), but also wanted it kept secret from those he deemed unworthy.

The assertion that he didn't invent the art he taught leaves one to wonder whether he pieced together the fencing method in the *Record* from many smaller bits of teaching that he learned in his travels or, alternatively, he were the inheritor of an oral tradition containing the entire *Record* and was the first master in his tradition to allow it to be written down.

Liechtenauer was also associated in some way with sixteen fencers and wrestlers that Kal labels the 'Society' or 'Fellowship of Liechtenauer' (*Gesellschaft Liechtenauers*); these men apparently played some role in developing or disseminating Liechtenauer's tradition.[6]

Unfortunately, these statements don't give much detail about his life.

The narrative can be fleshed out slightly further by analyzing some of the names presented so far. 'Liechtenauer' likely indicates a family that originated in a place called 'Liechtenau', and further implies a non-noble status—nobles would typically instead be styled with the word '*von*', i.e., *von Liechtenau*. (This rule isn't universally followed throughout German history, but it's a reliable baseline expectation.) There are several places in central Europe called 'Liechtenau' and there's no way to know which is being referenced from the name alone.

Paul Kal, the only writer quoted above who isn't anonymous, provides an important clue in solving this mystery, because he can be found in the service of the Free Imperial City of Nürnberg,[7] a city that was famous in this time for its fencing culture.[8] His list of the Fellowship includes not only the obvious connection of Hartmann von Nürnberg, but also Hans Pegnitzer, named for the river that runs through the city,[9] and Stettner, who fought in the First Margrave War, a conflict between Nürnberg and Albrecht III 'Achilles', Elector of Brandenburg.[10] Peter von Danzig records in his treatise that he resided in Ingolstadt, 90 km south of the city.[11] (Hans Talhoffer and Hans Leckùchner, other contemporary masters of the Liechtenauer tradition, were also residents of the city.)[12]

Based on these connections, it seems likely that Liechtenauer's family came from the Lichtenau in Ansbach, a castle and village about 40 km outside Nürnberg which was purchased by the city in 1406 to serve as part of

[6] The Fellowship of Liechtenauer is the subject of a forthcoming publication by BRIAN PUCKETT.

[7] TOBLER 2006: 4; ACUTT 2019: 41; BECKER 2020: 56–64. A longer biography of Paul Kal will be included in PAUL BECKER's forthcoming dissertation.

[8] BECKER 2020: 54–56.

[9] See also ACUTT 2019: 34 and BURGER 2021: 126.

[10] ACUTT 2019: 40.

[11] PD: 108^r.

[12] See BECKER 2020 and BURGER 2021 (respectively) for biographies of these men.

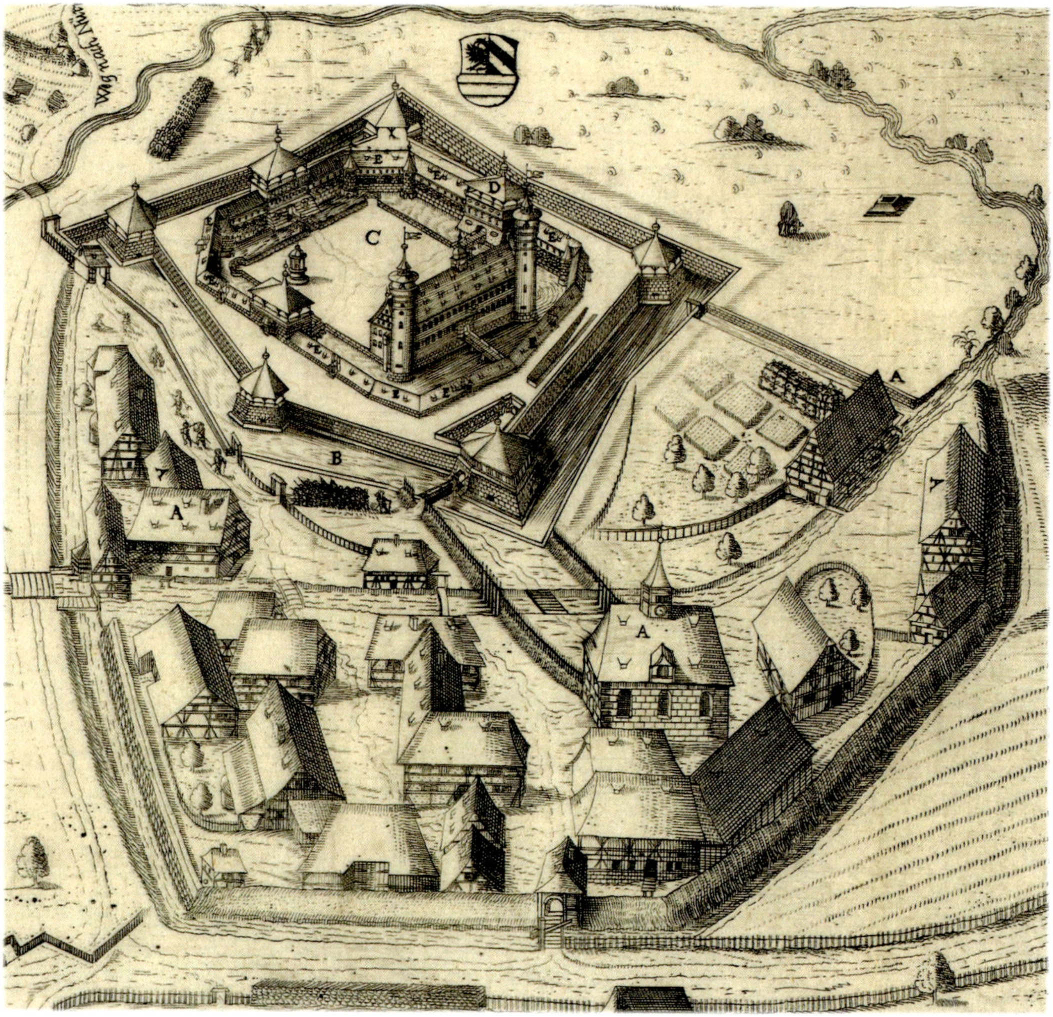

Figure 2: The village and fortress of Lichtenau, as depicted by Matthäus Merian in 1633.

its defensive perimeter.[13] This is only speculation until records are located which definitively place the fencing master there.[14]

Determining when Liechtenauer lived is also difficult. The most basic piece of evidence for determining this is the statement "may God grant him grace" (*Gott sei ihm gnädig*), a formulaic blessing which was commonly placed after the names of deceased persons in this time period—the absence of this statement doesn't necessarily mean that the subject was alive

[13] For a survey of other places called Liechtenau, see ACUTT 2019: 17–18.

[14] JOELI TAKALA has suggested Hans Rummel von Lichtenau († 1434), the Nürnberg aristocrat who occupied the fortress in the early 15th cen., as a candidate. This will be the subject of a forthcoming publication.

at the time of writing, but its presence is a good indicator that the author believed them to be dead.

Thus, in the first quotation above, Hans Rott (believed to be the scribe of PD)[15] indicates that in 1452, he believed Liechtenauer was deceased. This source similarly notes the deaths of Martin Huntsfeld, Andre Lignitzer, and Ott Jud (all the authors included in PD except for Peter von Danzig). The presence of the blessing at the beginning and end of the roll of the Fellowship of Liechtenauer (quoted above) may be intended to convey that all the men on the list were deceased (rather than just Liechtenauer and Stettner) by the time it was recorded in the late 1460s.[16]

The earliest dated witness of Liechtenauer's teachings is in a manuscript owned by Hans Talhoffer (HTG), which assigns it to the year 1448 and lacks the blessing on the dead,[17] so it's possible—though not certain—that Liechtenauer still lived at that time. N too lacks the blessing after Liechtenauer's name (as seen in the quote above), and it's sometimes argued to be much older than that, dating perhaps to 1390 or even earlier;[18] others date the manuscript to the early 15th century.[19]

Regardless of when N was created, the three masters who were deceased in 1452 would likely have been born around the turn of the 15th century (if not earlier). If the masters in the Fellowship were his students (which might explain why Kal gives Liechtenauer's name to the group), then Liechtenauer's career likely spanned the end of the 14th century and the beginning of the 15th.

1.2. Sigmund Ainring, the Teacher

Sigmund Ainring is barely less of a mystery than Johannes Liechtenauer himself. The single biographical statement about him comes from the beginning of his gloss in the *Hans Medel fight book* (HM), *Joachim Meyer-Rostock* (JMR), and the *Sigmund ain Ringeck fight book* (SR):

[15] Signed on PD: 3^r; also see SCHNEIDER 1978: 25.

[16] TOBLER 2006: 201. There has been much hand-wringing in the literature about why contemporary masters like Hans Lecküchner and especially Hans Talhoffer were left off the list, but this may simply be because they were still alive at the time of writing; Kal also doesn't include himself.

[17] HTG: 18^r.

[18] This was asserted by catalogers from WIERSCHIN to LENG based on a list of the years from 1390 to 1495 on f. 83^v, but has largely been rejected by Liechtenauer researchers in more recent years. MACIEJ TALAGA has reinvigorated the 'early date hypothesis' and presents novel arguments for it in multiple forthcoming publications, including his dissertation.

[19] VODIČKA 2019: 88; see also TOBLER 2010: 5–10.

> *Then Master Sigmund Ainring, currently fencing master to the highborn prince and noble Lord Albrecht, Count Palatine of the Rhein and Duke in Bayern, had those same obscure and cryptic words of the* Record *clarified, explained, and commented on (as you'll find written and illustrated here in this booklet), so that anyone who can already fence could learn and understand them well.*[20]

Ainring is described as a *Schirmaister* here, which is a slightly tricky term. A *Schir__m__meister* is a fencing teacher (using *schirmen*, the late medieval term for fencing, rather than *fechten*),[21] whereas a *Schir__r__meister* is a lord's stablemaster or a logistics officer in a military setting in charge of animals and anything pulled by animals (wagons, cannon, etc.).[22] '*Schirmeister*' could be a spelling of either one, but in context, it seems more likely to be establishing Ainring's credentials as a fencer or teacher of fencing than his skill tending animals. The descriptor "currently" seems to indicate that this wasn't a long-term position he made his career in.

Furthermore, this states that where Liechtenauer's art is recorded in obscure and cryptic words to conceal it from the unworthy, Ainring created a key to unlock those words in the form of a gloss (see Chapter 2).

While this might seem to defeat Liechtenauer's purpose, Ainring wasn't exactly shouting it to the world. Every manuscript is a unique work of art, and the only person who knows what a manuscript says is its owner and the people that they choose to show it to. A manuscript can only be copied if the owner allows scribes to have access to it. Perhaps Ainring considered giving a written record of his teachings to his students to be no different than giving them physical instruction in the first place.

To return briefly to the name game, DIERK has already pointed out the many spellings of Sigmund's second name—*amring*, *emring*, *einring*, and *ain ringeck*. Which spelling you consider most likely to be correct depends largely on how much authority you invest in Paul Kal's scribes and how much in the copyists of Sigmund's gloss. DIERK proposed an evolutionary process whereby *Amring* could become *Einring*, but the opposite path is also possible: *Ainring* could lose the dot over its first *i*, causing *Aın*- to be misread as *Am*-. A scribe might also expand the terminal *-g* to *-gck* in order to strengthen the consonant, and *Ainringck* could easily be misread later as *Ainringeck* or even *ain ringeck*.[23]

[20] See paragraph 1 in Chapters 3 and 5; GF has a similar but shorter passage at the end.
[21] For a discussion of the terms *schirmen* and *Schirmmeister*, see KELLET 2012: 34–46.
[22] See GRIMM for definitions of these and other terms raised in this and the next chapter.
[23] The idea that it should be read as 'a dude from the Rhineland' (*ein Rhein Geck*) started as a joke by JENS-PETER KLEINAU which unfortunately took on a life of its own and has even been repeated as fact in academic literature.

Since this book is about Sigmund and his glosses and not Paul Kal and his Fellowship, I'll treat the gloss as authoritative and refer to him as 'Ainring' (while acknowledging that any archival hunt for Sigmund should search for *Amring* as well).

Ainring is a town in present-day Germany just outside of Salzburg (on the Austrian border); it's about 120 km from München and about 260 km from Nürnberg. 'Amring' isn't a place-name as far as I can tell, but is an attested Germanic surname going back to the early Middle Ages.

1.3. HANS MEDEL, THE REVISER

Hans Medel von Salzburg is as much an enigma as his two predecessors in the tradition. He is introduced in this statement that the manuscript HM appends to the introduction to Ainring's gloss (quoted above):

> *[This gloss was] also marked and refined thereafter by other masters, and particularly by Hans Medel von Salzburg...*[24]

Even the name 'Medel' isn't definite, since the script is unclear and it may instead be 'Niedel'. The appellation *von Salzburg* indicates his family originated in the city in modern Austria and hints at a noble status (though as mentioned before, the *von* isn't quite that definitive).

Hans Medel is often considered to be the author of a gloss of Liechtenauer's teachings and a short treatise on fencing from seven stances. This gloss is sometimes identical with Ainring's, sometimes identical with an anonymous gloss nicknamed 'Nicolaus', frequently completely original, and occasionally offers a unique teaching alongside the text of one of these two older glosses.[25] As DIERK mentioned, the gloss refers to Ainring as "Sigmund Schining" (or "Schninig"); based on the proximity of Ainring and Salzburg, JAY ACUTT has suggested that this is evidence that the two masters were acquainted and thus Hans Medel was able to supply Sigmund's correct surname.[26] (This is can't be demonstrated with current evidence.)

The manuscript was dated 1539, possibly by Paul Hektor Mayr, but the script is more reminiscent of the late 1400s. Since the manuscript likely wasn't written by Medel personally, this tells us little about when he was alive. One clue in the gloss is that it includes three references to teachings of Hans Seydenfaden von Erfurt,[27] mentioned above in the roll of the Fellowship of Liechtenauer. (It also includes many teachings attributed to a

[24] See paragraph 1 in Chapter 5.

[25] These gloss variations are described in Chapter 2.

[26] ACUTT 2019: 44.

[27] HM: 29^{v}.4, 30^{v}.2, & 31^{r}.2.

"Master Hans" without specifying a surname, which could also be from Seydenfaden.) If Medel were a student of Seydenfaden, then he likely lived in the mid- to late-15th century.

1.4. Albrecht, the student

As quoted above, Ainring's gloss begins with a mention of a princely client identified as Albrecht, Count Palatine of the Rhein and Duke in Bayern. These titles indicate that he was a member of the Wittelsbach dynasty, a powerful noble house that has been a major player in central Europe for most of the past thousand years.

1.4.1. The many Wittelsbach Albrechts

The Wittelsbach were in an unusual situation in this time period in that they practiced a system of *multigeniture*. Unlike the familiar system of *primogeniture* in which one male heir (usually the eldest) inherits most or all the titles and wealth possessed by his father, Wittelsbach princes divided their titles and lands among all their male heirs.

Ludwig IV 'the Bavarian' (1282–1347), Holy Roman Emperor, was both Count Palatine of the Rhein and Duke of Bayern, and with the Treaty of Pavia of 1329, he assigned the Palatinate to the joint rule of his two nephews and Bayern to the joint rule of his six sons. Furthermore, the treaty stipulated that these fiefs couldn't be inherited by any but a Wittelsbach unless the entire family went extinct. Thereafter, it was customary to refer to all princes of the dynasty by both titles in token of this rule of succession—though none would actually claim both titles again until Karl Theodor in 1777. The heraldry of the Wittelsbach family likewise had the blue and white diamonds of Bayern quartered with the lion of the Palatinate.

As DIERK mentioned, there were four Albrechts who ruled Bayern in the time period that interests us.

Albrecht I (1336–1404) was Ludwig's fifth son; disputes among the brothers lead to the end of joint rule of a unified Bayern, dividing it into Upper and Lower regions in 1349 (each co-ruled by three of them), and further subdivisions and recombination happened over the subsequent decades. Albrecht was left ruling a partition called Bayern-Straubing with his brother Wilhelm (1330–1389) and later with his son Albrecht II (1368–1397), who turned 21 the same year Wilhelm died and was made co-ruler with his father.

Albrecht III 'the Pious' (1401–1460) was a cousin of the first two. He descended from Stephen II (1319–1375), another of Ludwig's sons, who co-

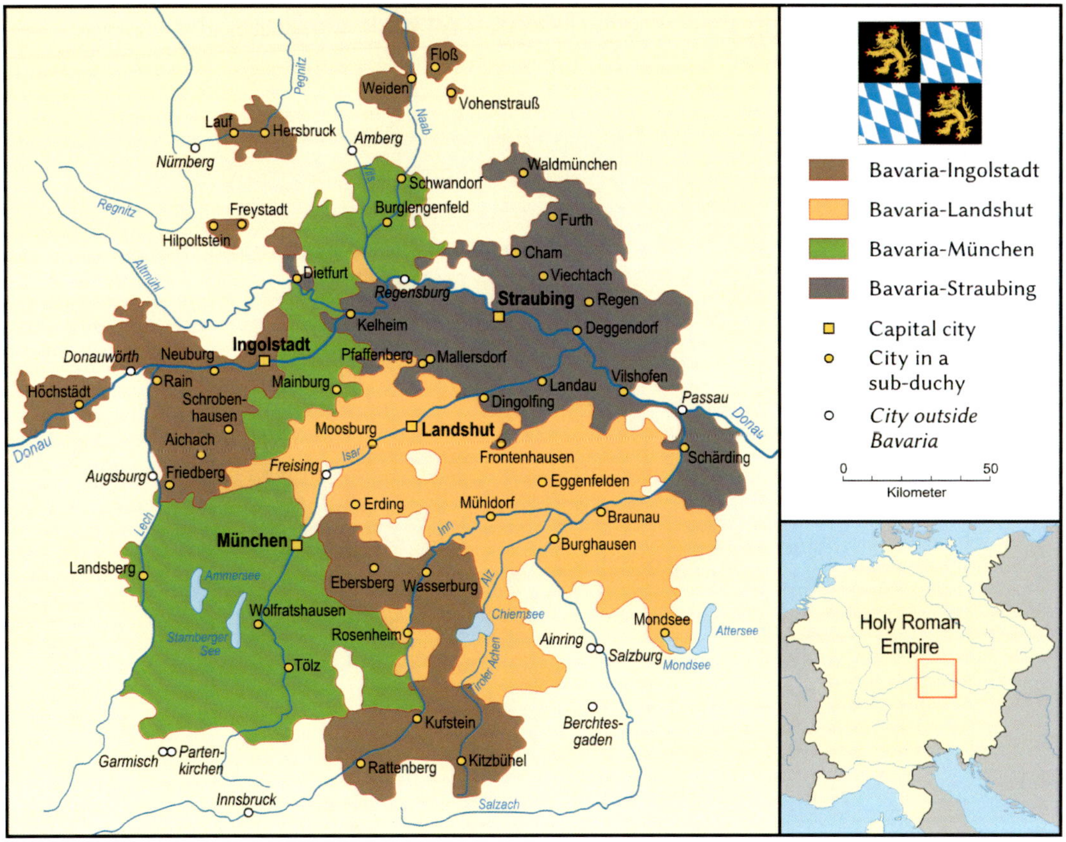

Figure 3: Bayern after the partitions of 1392 (modified from a map by Lencer, CC BY-SA 3.0).

ruled Lower Bayern with Albrecht and Wilhelm before they partitioned it in 1392 and he received the other half, Bayern-Landshut.

By Albrecht III's day, there were four Bavarian partitions, since Bayern-Landshut had been further subdivided to add Bayern-Ingolstadt and Bayern-München—Albrecht III's duchy—to the mix. As you may have started to see, this dynastic history is messy and complicated, and that's without going into the wars and feuds fought between these cousins along the way.

Albrecht III's fifth son, Albrecht IV 'the Wise' (1447–1508), spent his life re-unifying Bayern through political and military conquest (fighting not only his brothers and his Bavarian cousins but also his cousins on the Palatinate side), ultimately succeeding in 1505 after the War of Landshut Succession. As the sole Duke of Bayern for the first time in over 150 years, he decreed an end to multigeniture in 1506, and the Bavarian Wittelsbach family practiced primogeniture ever after until the end of their line.

These are the four potential Albrechts, but there's no simple way of determining which one is mentioned in Ainring's text.

Since Ainring had some kind of connection to Albrecht, he can help narrow it down further. If Ainring lived in the same period as other masters of the Fellowship, then he might have been *Schirmaister* any time from the end of the 14th century to the middle of the 15th. What has apparently led many researchers astray is the assumption that Ainring must have held a position like 'court fencing master' to a sitting Duke of Bayern, which would suggest he were either among the oldest of the Fellowship and held a position during Albrecht I's reign from 1347–1404, or much younger than the others and held a position during Albrecht III's reign from 1438–1460.[28]

I think this is mistaken, since the titles assigned to Albrecht were also used by non-reigning heirs of the Wittelsbach family, and the times when a nobleman would typically hire a fencing master were either to train him for a planned duel, or more commonly, to educate his teenage sons—the 'young knights' that Liechtenauer addressed his *Record* to.

If Ainring were not an employee but rather his teacher, then his Albrecht would have been a teenager in the 'Fellowship period', and that leaves Albrecht III, born 27 March 1401, as the best candidate and also means that Ainring would have been hired to teach him in perhaps the 1410s rather than the 1440s or '50s.[29]

1.4.2. Albrecht III 'the Pious'

I can only offer the briefest biography of Albrecht the Pious here.

He was born in 1401, the eldest child of Ernst (1373–1438), Duke of Bayern-München, and Elisabetta Visconti (1374–1432). His father and his uncle Wilhelm III (1375–1435) co-ruled their duchy and were in frequent conflict with their own uncle, Stephen III 'the Fop' (1337–1413), Duke of Bayern-Ingolstadt. Indeed, Albrecht was born during the München Unrest, a coup sponsored by Stephen upon the death of Ernst and Wilhelm's father which forced them to flee the city from 1397 to 1403 while riots and conspiracies played out.[30]

As a child, Albrecht was sent to Prague to be educated under the supervision of his aunt Sophia (1376–1428), Queen of Bohemia, returning before 1417. While she was sympathetic to the Hussite movement, he was apparently insulated from them because in 1419–20, he joined his uncle Wilhelm in a campaign in Bohemia at the beginning of the Hussite Crusade.[31]

[28] See, e.g., WIERSCHIN 1965: 79–84, HILS 1985: 55, and TOBLER 2010: 8–10.

[29] WILKENS 2016 agrees that it's Albrect III, but declares it as fact without offering any supporting evidence.

[30] ALTMANN 2006: 223–224.

[31] MITTERMÜLLER 1867: 3–4.

Figure 4: *Albertus III Pius...* by Joseph Anton Zimmermann (1773), after a painting by TS of ST.

If Ainring were hired to educate him in fencing as a youth, it would likely have been either in his final years in Prague or when he returned to München.

1420 brought another conflict between the Wittelsbach princes as Stephen's son Ludwig VII 'the Bearded' (1368–1447) attacked Heinrich XVI 'the Rich' (1386–1450), Duke of Bayern-Landshut. This would come to be known as the Bavarian War of 1420, and Albrecht's family entered in 1421 when they declared war on Ludwig and began conducting raids against Ingolstadt's territories.

The Bavarian War culminated in the Battle of Alling on 19 September 1422. Ludwig, in an attempt to turn the tide of the war, ordered a direct assault on München and its surrounding villages. His army was met at Alling by forces lead by Ernst, Wilhelm, and Albrecht and defeated decisively; over 400 prisoners were taken at the conclusion of the battle.

The account of Albrecht's horse being slain in the fighting and Ernst fighting his way to him and standing over him with a mace, shouting "What, did you think to capture a prince?" at the enemy knights is a great story, but probably apocryphal.[32]

Albrecht was never made full co-ruler with his father and uncle, but during the 1420s he was given increasing responsibilities, and in January 1433, he was allowed to set up his own court and residence in Straubing and administer the territories of the former duchy of Bayern-Straubing that had been absorbed by Bayern-München in 1425–29.

He was an avid hunter and tournament fighter, in between the various feuds[33] and other conflicts that his family became involved in during this

[32] ALTMANN 2006: 225–226.
[33] On the 15th cen. German practice of feuding, see ZMORA 2011 & SHARP 2024.

Figure 5: A 15th cen. clay statue of Agnes Bernauer.

time. It was at a tournament in Augsburg in 1428 that he may have first met Agnes Bernauer.

The daughter of a craftsman, Agnes was far below Albrecht's station, but his family took little notice when she moved into Blutenberg Castle (Albrecht's private residence) on the assumption she was merely a passing lover. However, when Albrecht moved to Straubing, she was introduced to his court as his duchess. Their secret marriage probably occurred in 1432–33, and Ernst was outraged when he discovered this since it put the family's dynastic ambitions in jeopardy—no child of such a marriage could inherit his noble titles or lands, and Wilhelm's son Adolf wasn't expected to survive to adulthood. Other noble families also took notice, and in 1434, Albrecht was barred from fighting in a tournament in Regensburg for conduct unbecoming a knight.[34]

Wilhelm had apparently been a moderating influence on Ernst, since after his brother's death on 12 September 1435, Ernst decided to take decisive, deadly action.

On 12 October, while Albrecht was away on a hunting trip with Heinrich, Ernst ordered the local authorities in Straubing to try Agnes for treason and execute her, which they did and she died by drowning in the Danube. (This is commonly described as a witchcraft trial, but there's no evidence for that charge.)

Albrecht was enraged and immediately traveled to Ingolstadt to join Ludwig in plotting a new war against his father. Intercession by the emperor and the church eventually reconciled father and son, and the following year, Ernst ordered a chapel built in Agnes' memory and Albrecht created an endowment for a perpetual mass and annual memorial service (all at the Carmelite monastery in Straubing).[35]

A year later, on 6 November 1436, Albrecht married Anna (1414–1474), daughter of Erich, Duke of Braunschweig-Grubenhagen-Einbeck, and Elisabeth of Braunschweig-Göttingen. They ultimately had ten children between 1437 and 1454, three daughters and seven sons.

[34] MITTERMÜLLER 1867: 13–15.

[35] MITTERMÜLLER 1867: 15–18. The annual memorial service continues to the present, now funded by the Bavarian government.

Figure 6: *Duke Albrecht III 'the Pious' of Bayern refuses the Bohemian Crown in 1440* by Hans Werl (1603); Bayerische Staatsgemäldesammlungen – Staatsgalerie in der Burg zu Burghausen.

Ernst died in 1438, leaving Albrecht as co-ruler with Wilhelm's four-year-old son Adolf (1434–1441). In 1440, Albrecht was offered the Kingdom of Bohemia by the parliament in Prague, but he ultimately declined the crown out of friendship with Elisabeth, the widowed Queen of Bohemia who was trying to win the crown for her infant son Ladislaus 'the Posthumous'.

Albrecht's reign was characterized by frequent low-level conflict, including feuds with various nobles (and his cousins, the other Bavarian dukes) and periodic efforts to eradicate robber-knights from his lands. He was also very religious (more than earning his nickname) and invested a lot of resources and attention in reforming and expanding Bavarian monasteries, including making visits to monasteries to personally appeal to brothers who were resistant to his reforms. He also hired Johann Prunner, Dean of Indersdorf, to write homilies and sermons that were read to him each day at lunch and dinner.[36]

The dark side of his religious fervor was manifest in 1442, however, when he expelled all Jewish people from his duchy, which was ultimately both costly for him personally and brought condemnation from some of the other princes (to say nothing of the suffering of his Jewish subjects).

[36] Mittermüller 1867: 25–26.

Figure 7: Albrecht III 'the Pious', as depicted in the 16th cen. (München, Bayerische Staatsbibliothek, Cgm 2822, f. 14r).

Jewish people would remain banned from Bayern until restrictions began to be relaxed in the 18th century.

Albrecht III died in München on 29 February 1460, apparently of gout.

He had decreed that his two eldest sons, Johann IV (1437–1463) and Ernst II (1438–1460), should rule together, but the latter died in Straubing on the same day as his father. Instead, Albrecht's third son Sigmund (1439–1501) reigned with Johann, and when Johann died in 1463, Albrecht IV took his place.

Sigmund abdicated in 1467 and wasn't replaced by the next son, beginning Albrecht IV's long (and often bloody) project to reunify all of Bayern under his sole rule.

1.4.3. Connections to Liechtenauer

Sigmund Ainring's teachings aren't the only thing connecting Albrecht III to the Liechtenauer tradition.

Johannes Hartlieb (ca. 1400–1468) served first as secretary and later personal physician to Albrecht. Hartlieb wasn't only a doctor of medicine (from the University of Padua), but also deeply interested in the occult.

Hartlieb authored a number of treatises on magical topics, but among the first, created some time before 1437, was *Über die Erhaltung des Sieges* ("On the Preservation of Victory"), a treatise on using *onomancy* (name-based divination) to determine what day and time to schedule a fight in order to give advantage to one of the fighters. This text was included in Hans Talhoffer's earliest manuscript (HTG),[37] and Talhoffer was also personally acquainted with Hartlieb—Hartlieb intervened on Talhoffer's behalf when he was on trial for his life in 1434.[38]

[37] HTG: 7r–17r.
[38] Becker 2020: 49–50.

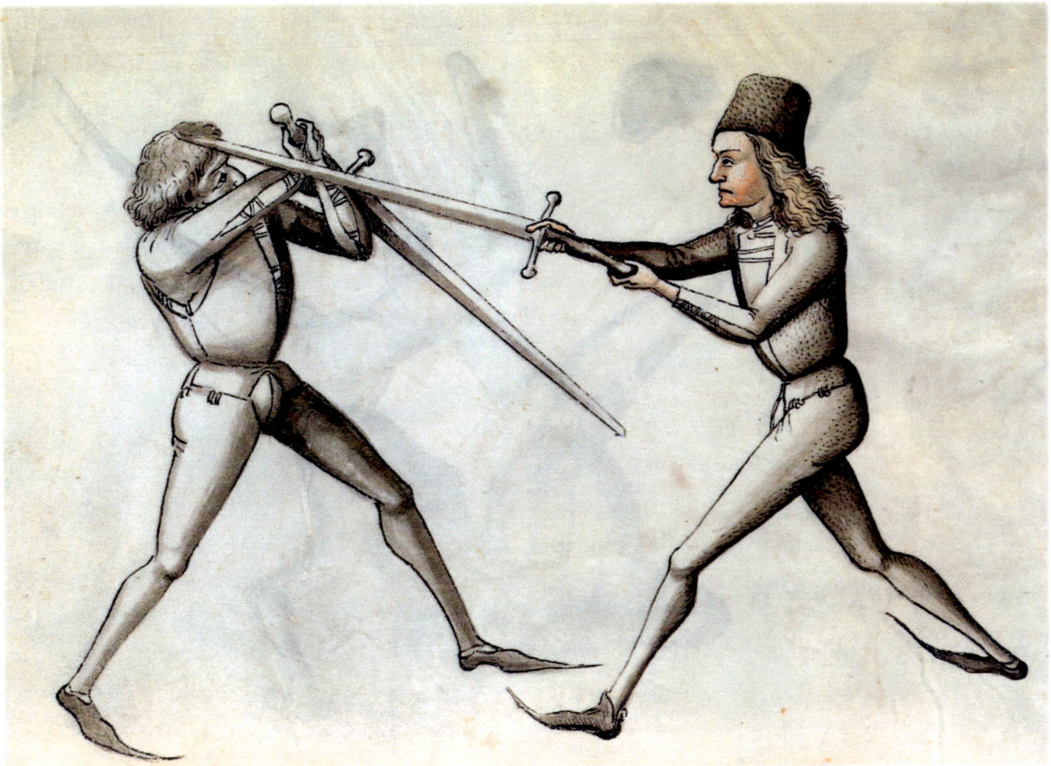

Figure 8: An illustration from one of Talhoffer's treatises that resembles the piece Berthold used against him (HTM: 11^{v}).

A more curious connection appears in PD, which states the following:

> *Master Berthold sliced Master Hans Talhoffer on the hand and struck him on the head with this piece in front of my lord, his grace Duke Albrecht, in München:*
> *Remember, when a fencer has bound you on their sword, if they strike crosswise around from your sword to the other side, then fall onto their hand or arm with your long edge and press their arm away from you with the edge of your sword. From the slice on their arm, strike them on their head with your sword.*[39]

[39] *Mit dem stuck hat maister pertolt maister hansen den talhofer vor meiner herren genaden Hertzog Albrecht zw minchen In die hant geschniten vnd auf den kopff geschlagen /Merck /wenn dw einem vechter an sein swert gepunden hast /Slecht er denn vom swert vmb mit der twer dir zw der anderen seitten /So val ÿm mit der langen schneid in sein hand oder auff die arem /vnd druck mit dem schnÿt sein arem mit swert mit tall von dir /vnd slach in auß dem schnÿt von seinem arem mit dem swert auf den kopff*. PD: 19^{v}. The red text was generally considered unreadable until DIERK HAGEDORN recovered it and announced it in HAGEDORN & TOBLER 2021: 13–15.

This passage is unique in the Liechtenauer corpus for apparently describing a historical incident in which the specific technique being taught was used. Furthermore, it invokes the name of a fencing master who is well-known today—Hans Talhoffer. And it happened in front of a Bavarian Duke named Albrecht.

While the date of this encounter is unknown, Talhoffer's activities are documented beginning in 1431[40] and **PD** is internally dated 1452,[41] so it must have been relatively early in Talhoffer's career—he would produce at least two and potentially up to four treatises after 1452: **HTK** in the 1440s or '50s, **HTB** in the 1450s, **HTKo** in 1459, and **HTM** in 1467. Given the known birth and death dates of our four Albrechts, this timeframe also means it must have been referring to Albrecht III.

This tantalizing passage raises endless questions. What was the nature of the fight: was it an exhibition, a *Fechtschule* bout, or even a judicial duel? Why was Albrecht in attendance (if not in a supervisory or sponsorship role)? Who is the speaker in this passage who apparently served Albrecht? And why was the red text rubbed off the page and then crossed out in black ink (see Figure 3 in Chapter 2), making it virtually unreadable?

The manuscript Hans Rott was copying is currently unknown so there's no way to know whether this text was part of that source, was added by Hans Rott based on his own personal knowledge, or was dictated by whoever the client of the manuscript was. It's equally unclear if it was effaced by Rott or by an early owner of the manuscript.

[40] BECKER 2020: 48.
[41] PD: 113^{v}.

2. Structure of the Gloss

Michael Chidester

This chapter will discuss the tradition of writing glosses of Liechtenauer's *Record*, outline the structure of his teachings, and offer some advice on interpreting the glosses of Ainring and his fellow masters.

2.1. Text and Gloss

Sigmund Ainring's teachings are recorded in the form of a *gloss*, but what does that mean?

In simplest terms, a 'gloss' is an explanation of a text; the process of writing interpretations is known as *glossing*.[1] It can range from a simple translation of a word in a foreign language to an extensive commentary on an entire book. In the medieval period, glosses would often be written in between the lines of text (if short) or in the margins (if longer).

Glossing authoritative texts was a common practice in the scholastic movement of the Middle Ages, and scholars would record their thoughts in the margins of their books: expanding upon a subject, arguing with the author, or adding commentary from other authorities. By the 12th century, comprehensive glosses for the entire Bible and many Roman texts had been assembled from the writings of experts of centuries past. These were called *glossae ordinariae* ("standard glosses") and included both marginal and interlinear commentaries, often dwarfing the actual text being discussed.[2] It became common for books to be made that had these glosses pre-installed, though owners would still expand them further with their own glosses.

By the 14th century, the tradition of creating formal glosses had faded away in scholastic circles and scholars were more likely to author entirely new treatises which merely referenced older authorities. However, for centuries thereafter, famous glosses continued to be commonly reproduced and sold as reference books—the *glossa ordinaria* of the Bible was even among the first books to be printed after the invention of moveable type.

(The practice of recording personal thoughts in the margins of books, of course, continues until the present day. Please fill this book with your glosses—we've tried to leave plenty of space for them.)

[1] This is also where the term 'glossary' comes from, and the glossary at the back of this book can be considered a collection of short glosses on keywords in it.

[2] The Jewish practice of recording *Midrashim* in the margins of the *Tanakh*, which continues to the present, is a parallel tradition that arose in this period from the same cultural influences.

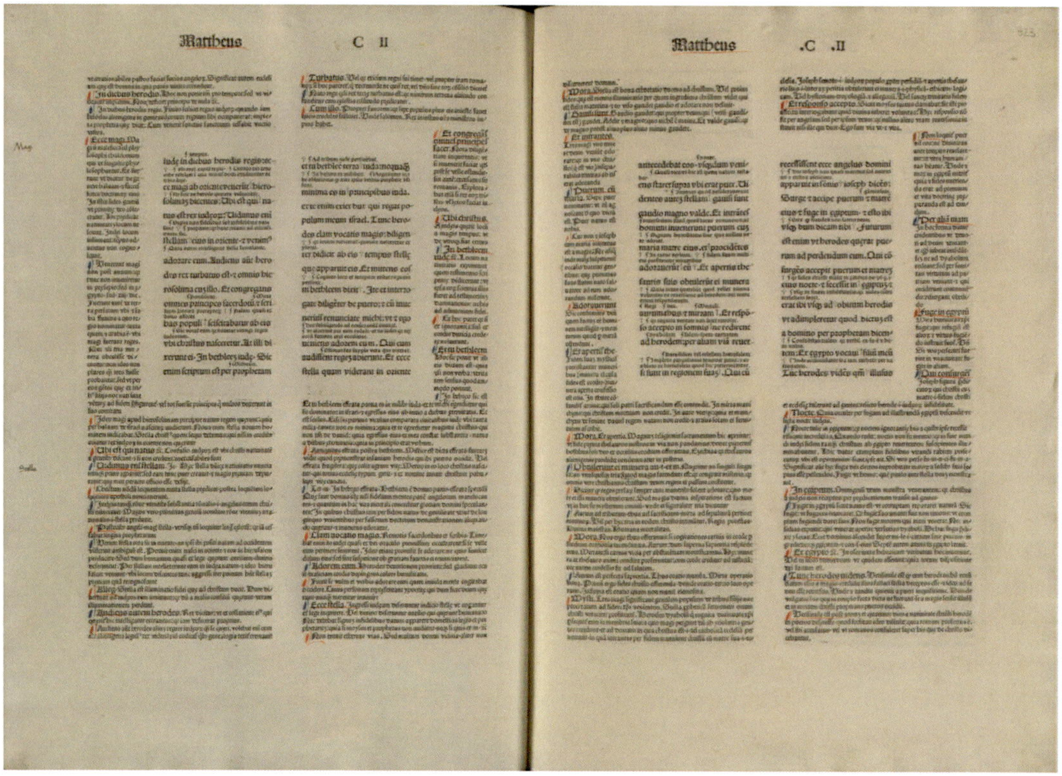

Figure 1: A Bible printed in Strasbourg in 1481 including the *glossa ordinaria*. The large print in the middle is the text of the bible, and the gloss is written around it and between the lines.

2.1.1. Hans Liechtenauer

Liechtenauer recorded his teachings in a poem titled the *Record* (*Zettel*)[3] with 178 couplets, including 109 on unarmored long sword fencing and 69 on armored dueling (40 mounted on horseback and 29 dismounted on foot); a set of 26 short phrases called 'Figures' are also associated with the *Record*.[4] This poem was deliberately cryptic to prevent the uninitiated from unlocking its teachings without the help of a teacher.

When the Liechtenauer tradition arose in the 15th century, many of its early authors chose to treat his *Record* as an authoritative text and create

[3] *Zettel* is a tricky word to translate. The closest English cognate is "schedule" (both come from the Latin *schedula*), but only in the more obscure legal sense of a formal list, not the familiar sense of a timetable. It's commonly used in modern German to denote a short list or a scrap of paper that could hold a list (like a receipt). *Zettel* is translated as "record" in this book, but other common translations include "didactic poem/mark-verses", "epitome", "notes", and "recital".

[4] N: 86^{r}–89^{r} also includes a treatise on wrestling that it attributes to Liechtenauer, but no other source corroborates this.

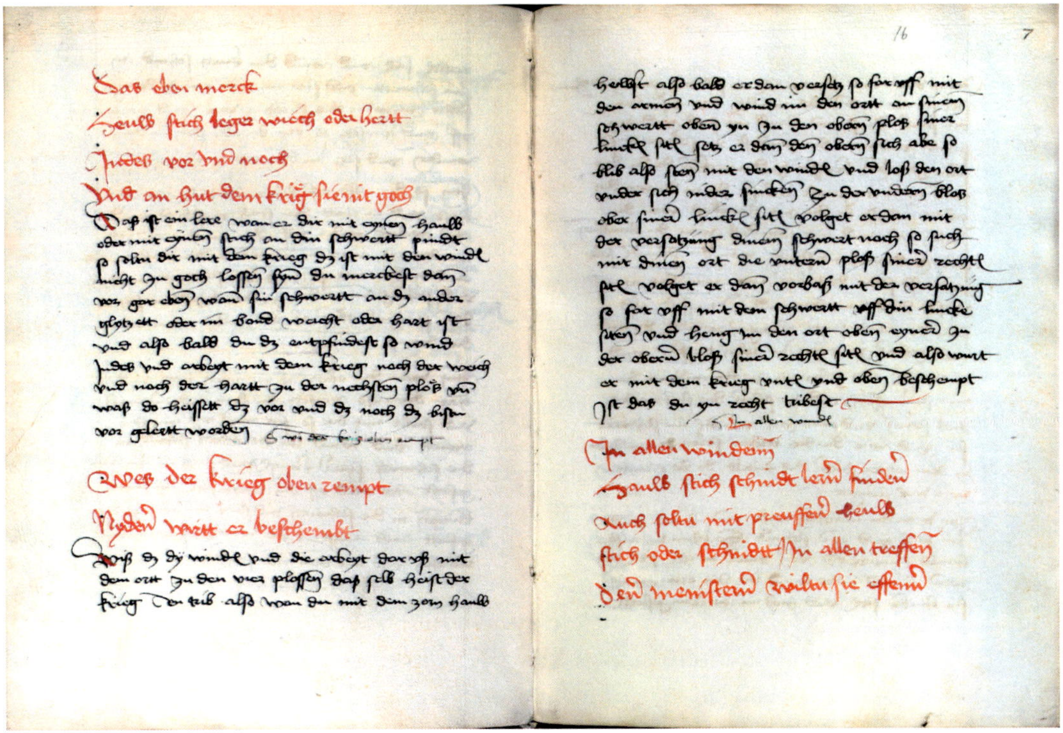

Figure 2: The *Record* (red text) and Lew gloss (black text), from HS: 6^{v}–7^{r} (1491).

glosses to explain their interpretations of it in the fashion of earlier generations of medieval scholars.

There are several different surviving glosses of Liechtenauer's *Record*, and they can be divided into *major glosses* and *minor glosses*. These texts all seem to have been created around the middle half of the 15th century, but the age of surviving witnesses varies quite a lot.

Major Glosses

Like the *glossa ordinaria* mentioned above, the major glosses include commentary on the entirety of one or more sections of the *Record*. They are associated with two masters of the Fellowship of Liechtenauer: Peter von Danzig and Sigmund Ainring.

Peter von Danzig's gloss is the shorter one, limited to a commentary on Liechtenauer's armored dueling on foot. The only known witness is PD.[5]

The other major gloss is fragmented into multiple variations, including one attributed to Sigmund Ainring[6] and three others that are anonymous, though their authors are assigned the nicknames 'Pseudo-Peter

[5] A 'witness' is a later copy or version of a medieval text; the original is called the 'archetype'.
[6] See Chapter 4 for a discussion of the witnesses to Ainring.

von Danzig',[7] 'Lew',[8] and 'Nicolaüs'[9] to make them easier to discuss. The Ainring and Pseudo-Danzig variations cover all three sections of the *Record*, whereas the Lew only covers the long sword and mounted fencing, and the Nicolaüs just covers the long sword.

This group is the only gloss of Liechtenauer's *Record* that's known to exist in multiple witnesses, and the Lew variant is the only gloss to have been translated into a second language (Latin).[10]

Textually, the Ainring and Lew variants are the most distinct in wording and content, whereas the Pseudo-Peter von Danzig variant is mostly aligned with Lew but has intermittent inclusions from Ainring. The Nicolaüs variant reads as sort of a mashup of Pseudo-Danzig and Lew, but with a few unique passages of its own. Ainring is the only variant that seems to have been illustrated in its original form,[11] though a mid-16th century illustrated witness of Pseudo-Danzig also exists (G).

Though the four variations diverge in their teachings in some places, they share phrases and even whole paragraphs freely (especially in the dueling section), and the technical differences in their teachings are small enough that it often makes sense to refer to them as a single unified work (usually labeled **RDL** or **RDLN**).[12] Hans Medel treats them similarly, referring to them as the *gemainer gloss* ("common gloss").[13]

Indeed, the **RDLN** variations are similar enough textually that they almost certainly had a common origin, though it's impossible to say whether this was a lost 'gloss Q' that they were all modified from or whether they were all the works of a single author, either adapting his teachings to different audiences or writing at different periods in his life as his teachings evolved. It's easy to imagine that Sigmund Ainring, for example, also authored the 'Lew' variation for a different client, and combined elements of both teachings together to create the 'Pseudo-Danzig' variation for a third,[14] based on their individual needs and interests.[15]

[7] Witnesses to Pseudo-Danzig are found in G, GF, LSA, O, PD, PFS, PKW, & SR.
[8] Witnesses to Lew are found in AD, HC, HS, JMR, L, LSM, PMD, PMM, & PMW.
[9] Witnesses to Nicolaüs are found in AR, GE, JWM1, LSA, LSM, LSWO, O, & PKW.
[10] Found in PMM & PMW.
[11] This is discussed further in Chapter 4.
[12] I coined the term 'RDL' (an abbreviation of Ringeck-Danzig-Lew) around 2014 when I began to understand that Pseudo-Danzig and Lew had enough interesting divergences to make them worth translating separately (it replaced 'Ringeck-Danzig' as a catchall). I added the -N when it became clear that the Nicolaüs gloss did too, but that hasn't yet caught on to the same extent.
[13] HM uses the term "common gloss" three times: once referring to Ainring (HM: 33^r.4), once to Nicolaüs (HM: 35^v.4), and once without being clear about which gloss (HM: 35^r.3).
[14] The effaced statement in PD naming Albrecht as the writer's master (quoted in Chapter 1 and shown in Fig. 3) may even be a small piece of evidence in favor of this.
[15] 'Nicolaüs' was probably created by combining Lew and Pseudo-Danzig.

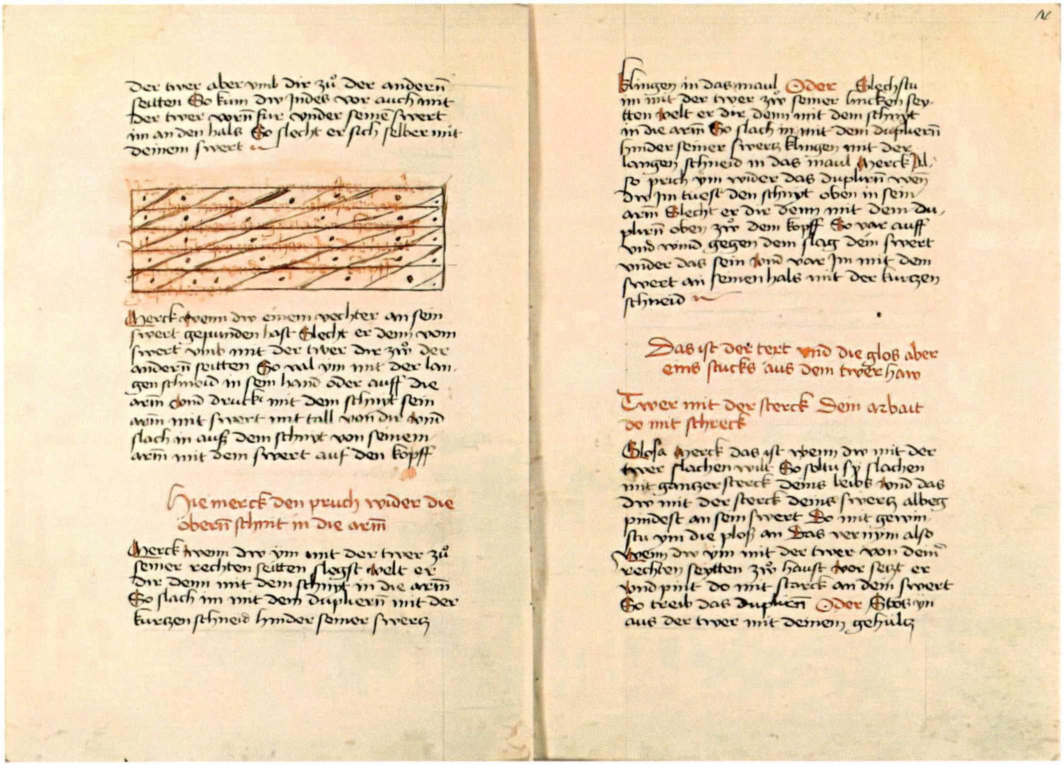

Figure 3: The *Record* and titles (red text) and Pseudo-Peter von Danzig gloss (black text), including the effaced reference to Talhoffer and Berthold, from PD: 19ᵛ–20ʳ (1452).

The 'single-author hypothesis'[16] is compelling because it doesn't require the existence of any hypothetical lost manuscripts beyond the archetypes of each variant, whereas the 'single-text hypothesis' requires far more manuscripts to have existed and then disappeared without a trace—generations of aggregate scribal errors and multiple interventions by unknown fencing experts who added, subtracted, and reworked text and techniques.

As a final note on the major glosses, it's worth remembering that only the Ainring long sword fencing and Danzig dismounted dueling glosses include any attribution (though the Lew mounted fencing also concludes with a statement sometimes interpreted as an attribution). The other three long sword gloss variations, two dismounted variations, and two mounted variations are anonymous.

For convenience, the mounted and dismounted dueling glosses that typically accompany the Ainring long sword tend to be attributed to him, the

16 We dubbed this the 'Sigmund the Lion' hypothesis on the *Fencing by the Book* podcast, after the idea that '*Loew*=lion' might be a nickname or title.

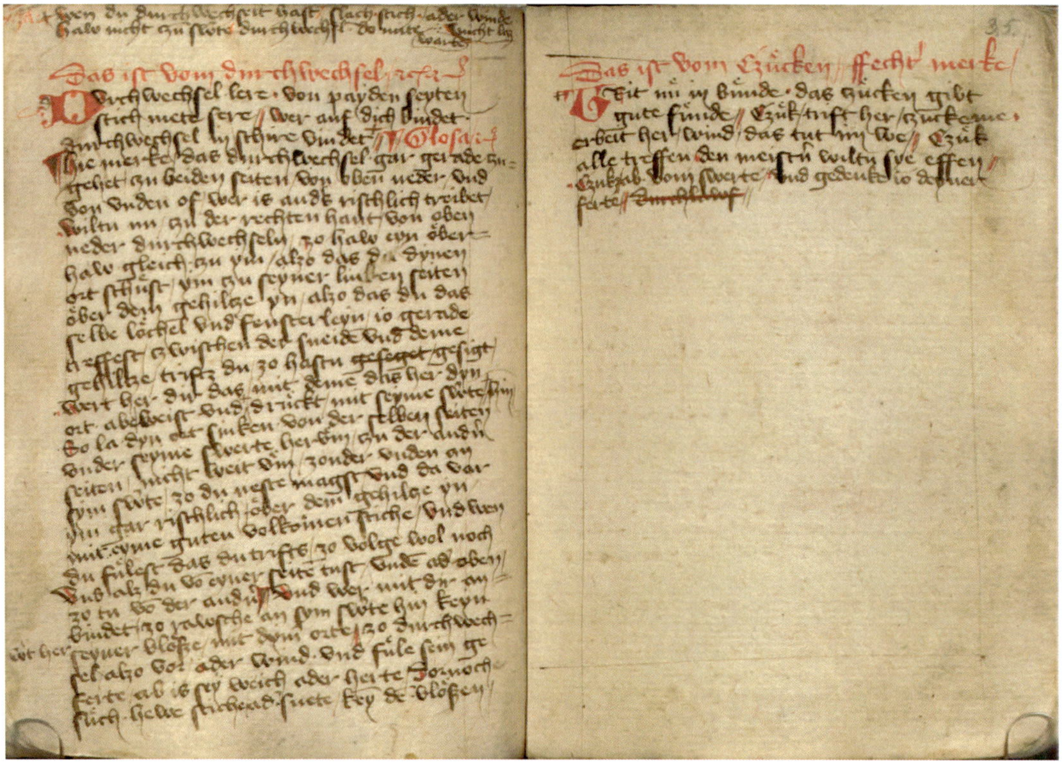

Figure 4: The incomplete Pseudo-Döbringer gloss; including text and gloss of the changing through (left) and the unglossed text of pulling (right), from N: 34^{v}–35^{r} (ca. 1400).

long sword gloss that usually accompanies the Lew mounted gloss tends to also be attributed to him, and the remaining three glosses (which also often appear together) are classified as Pseudo-Danzig.[17] These groupings by association are useful for easy reference, but it should be remembered that there's little evidence supporting them.

Minor Glosses

The minor glosses only offer commentary on shorter segments of the *Record*. Except for Meyer, they're all anonymous. Despite their incomplete nature, they're valuable for study because they often present different interpretations from the major glosses and can help develop a more nuanced understanding of Liechtenauer's teachings.

[17] I labeled this group 'Pseudo-Peter von Danzig' around 2010 because even though they're not true pseudepigrapha, they were often wrongly attributed to Peter von Danzig for decades by researchers who assumed the entirety of manuscript PD was created by Danzig personally (see, e.g., HILS 1985: 112), a view that's no longer commonly held (see, e.g., WELLE 2017). This term made it apparent what text was being referenced without agreeing with the attribution.

The longest of the minor glosses is the one found in **N**, whose author is nicknamed Pseudo-Hans Döbringer.[18] This one seems to have been an aborted attempt at writing a major gloss; all the verses were written into the manuscript with blank pages left between them to hold commentary, and the author described his work as a *glossa generalis*. However, despite writing a long introduction to Liechtenauer's teachings, the author only glossed about half of the sections of the *Record* on the long sword (and nothing on the mounted or dismounted dueling) before apparently abandoning the project.

Joachim Meyer also composed a minor gloss for *Joachim Meyer-München* (**JMM**) covering verses 17–20, 36–54, 58–61, 66, 71, 75–76, 78, & 93 of the long sword. This treats them out of order and appears to be based on the version of the *Record* in the first printing of Christian Egenolff's 1530s edition of Andre Paurnfeyndt's 1516 treatise.[19]

Three other minor glosses are quite short:

- The Dresden Gloss, found in **SR**, covers three couplets of the *Record* on the long sword (27–28 & 43).
- The Glasgow Gloss, found in **GF**, covers the first seven couplets of the *Record* on dismounted dueling.
- The Speyer Fragments are a series of nineteen paragraphs added into the witness of the Lew long sword gloss in **HS**.

Meta-Gloss

Finally, the gloss of Hans Medel is worthy of special attention.[20] It seems to have been a major gloss, but only the only known witness (**HM**) is incomplete. It discusses the first 87 couplets of the long sword and then stops abruptly at the end of a folio with the title *Von dem Zuckenn*; the remaining text is lost.

However, what's most interesting about the Medel gloss is that the author frequently quotes the glosses of Ainring and Nicolaüs verbatim, and sometimes even criticizes them and offers his own interpretations as counterpoint. He also offers three teachings that he attributes to Hans Seydenfaden von Erfurt, a master of the Fellowship of Liechtenauer from whom no writings are known to survive. This acknowledgement of and engagement with the prior glosses makes Medel's work unique in the gloss tradition.

[18] I devised the label 'Pseudo-Hans Döbringer' for this author around the same time and for similar reasons to 'Pseudo-Peter von Danzig'.

[19] VanSlambrouck 2023: xv–xvi.

[20] Note that the text exclusively refers to Medel in the third person, suggesting that it was compiled by one of his students rather than being authored by him personally.

Figure 5: Lecküchner's verse, gloss, and illustrations, from LM: 7v–8r (1482).

2.1.2. Hans Lecküchner

Inspired by the Lew gloss,[21] the priest Hans Lecküchner apparently created his own *Record* (expanded from Liechtenauer's) and a gloss to explain it in the 1470s. This text is far longer than any of the Liechtenauer glosses and, indeed, is the longest single fencing treatise of the 15th century.

Unlike the others mentioned already, Lecküchner continued expanding his poem and gloss over time. Two presentation copies exist (both dedicated to yet another Wittelsbach prince, Philipp 'the Upright', Elector Palatine), preserved in Heidelberg (**LH**) and München (**LM**). **LM** was probably produced a few years after **LH**, and is substantially longer and also extensively illustrated.[22] However, several other witnesses exist which are shorter than either and were probably based on an earlier draft of the text.[23] This includes a Latin translation in the works of Paul Hektor Mayr.

[21] Demonstrated by Grzegorz Żabiński, Russell A. Mitchell, and Falko Fritz in a translation circulated online in 2012; echoed in Forgeng 2015: xi.

[22] For more information about LH and LM, see Forgeng 2015 and Chidester, et al. 2021.

[23] These shorter witnesses to Lecküchner are found in AD, GE, HS, LSM, PKD, PKM, & PKW.

2.1.3. Joachim Meyer

The final prominent author in the Liechtenauer tradition, Joachim Meyer not only authored his own minor gloss for JMM, but also an original teaching poem for the long sword based on Liechtenauer's which he titled the *Free Fencer Record* (*Frei Fechter Zedel*) and a gloss of this poem. This is included in both *Joachim Meyer-Lund* (JML) and his 1570 book. Meyer also owned *Joachim Meyer-Rostock* (JMR), which includes both the Ainring and Lew glosses, but at least some of this writing was done before he acquired it.[24]

2.2. The structure of Liechtenauer's *Record*

A complete interpretation of Ainring's gloss is beyond the scope of this book,[25] but this section will briefly review the sections of the *Record* and offer some high-level concepts as a guide for interpretation.

One of the curious things about Liechtenauer's teachings is their essentially bifurcated nature. The mounted and dismounted dueling form one teaching and the long sword fencing another, and there's very little overlap between them apart from a little shared terminology that was likely just common, everyday German words.

Likewise, the long sword is itself bifurcated: the common lessons and the five strikes form a coherent sequence of teachings, beginning with basic principles and then building each section on the one before it, adding new complexities and permutations. After that, the main pieces start over from basic principles and present another coherent sequence of teachings. But these sequences are largely self-contained, and references between them are few (for example, apart from four of the five strikes being name-dropped in the section on parrying, only the crosswise is actually used in the main pieces). Likewise, the main pieces are all mentioned at least once during the permutations of the five cuts, but usually with little or no explanation (and just as often, the actions described without assigning them a name).

It's almost like these two halves of the teaching were invented separately and then married together later. There's a certain unity between these two sets of teachings, but that only really becomes apparent in the final two sections of the *Record*.

24 For the history of JMR, see Chapter 4.

25 Several partial or complete interpretations of Ainring's gloss have previously been published, including Lindholm & Svärd 2003, Lindholm & Svärd 2006, Smith 2021, and Tobler 2001.

The intended context of the mounted and dismounted armored teachings seems to be the martial duel that became popular in the late 14th century among members of the military class, but the context of the unarmored sword is less clear. Some have suggested that it was intended primarily for application in tournaments like the *Fechtschule*. However, the actual techniques are a poor fit for that context due to the predominance of thrusting.[26] Others have suggested a self-defense focus, but the circumstances of the techniques—two fencers holding equal weapons at a wide distance, without scabbards or the possibility of outside interference—make it a poor simulation of that. There's also no evidence of a widespread culture of formal dueling with the two-handed sword without armor.

It's possible that Liechtenauer's fencing with the two-handed sword wasn't intended as training for any specific context, but rather to develop basic skills that would be useful in a wide range of situations (from sportive tournaments to fencing for one's life against unarmored or partially-armored opponents).

2.2.1. The foreword

The **foreword** is usually attached to the long sword fencing section, and sometimes also repeated for the dueling section.[27] It seems aspirational in nature, addressing itself to the sons of noblemen that Ainring taught (and presumably Liechtenauer did as well), stating that skill in fencing would bring them glory, laying out the weapons that Liechtenauer taught (lance, spear, sword, and knife) in addition to wrestling, and ending with the proverb 'all arts need measure and moderation'.

Though it isn't discussed by Ainring or the other major glosses, N connects this proverb with the fact that much of Liechtenauer's teachings are organized around contrasting pairs of opposites.[28] Thus, in studying Liechtenauer's teachings, it's useful to look for these pairs; some examples include before & after, strong & weak, hard & soft, long & short (or extension & retraction), high & low, straight & crooked, etc.

2.2.2. Fencing with the long sword

The *Record* on fencing with the long sword grip consists of common lessons, a sort of table of contents, and then seventeen main pieces (the last of which is described as the conclusion).

[26] There hasn't been a lot of published research on 15th cen. *Fechtschulen or Schirmschulen*, but see DUPUIS 2013 for an example of a ruleset.

[27] See, e.g., N: 53^r & PD: 53^r.

[28] N: 22^r.3–22^v.4.

The common lessons

The **common lessons** (*gemainer lere*) present a small, tight fencing paradigm that creates a basic 'model fight' which much of Liechtenauer's teachings elaborate on. It breaks down into four sub-lessons:

1. Coordinate footwork and blade movement by stepping left and right with steep diagonal downward cuts (from above to the ground in front of your back foot).
2. Don't wait to parry your opponent's strikes; instead, preempt them by cutting to threaten with the point as you approach, provoke a parry, and then skirmish.
3. Coordinate the sides of your body with blade movement (either based on handedness or the side your sword is on—unclear).
4. The tactics in this section can be described using five words: Before, After, Strong, Weak, and Within.

This paradigm might be characterized as the 'direct' or 'straight fight', focused on approaching from wide distance with downward cuts, straight thrusts, and the extended point.

The contents of the *Record*

The next segment introduces the five strikes and then explains that Liechtenauer's art breaks down into seventeen **main pieces** (*Haupstucke*) in Ainring (or five strikes and twelve main pieces in DLN).

The five strikes

The five strikes are specific attacks, but each one also includes a family of techniques called "pieces" (*Stücke*) which fit together into a specific paradigm or approach to fencing. While the word *haw* is often translated as "cut" or "hew" (including in this book), it's a generic 'striking verb' that also includes attacks intended to hit with the point.

The **wrathful cut** (*Zornhaw*) is the "straight" (*schlecht*) cut and continues the 'straight fight' by explaining how to counter the common lessons: instead of responding with a parry, do the same action in opposition. From this equal engagement, various responses are described and 'the war' is presented as an alternative to the skirmish—instead of disengaging from the bind to strike a new exposure, remain in the bind and work from there.

The **four exposures** (*vier Blossen*) isn't one of the main pieces, but is listed separately in some witnesses of the *Record* and teaches universal principles of the system (like a fifth common lesson). The main lesson here is that the proper way to fence is by threatening your opponent's exposures in only one of three ways: shooting the point (as in the common lessons and the wrathful cut) or pursuit (which is explained later) as you approach, and turning (which was just introduced) in the bind.

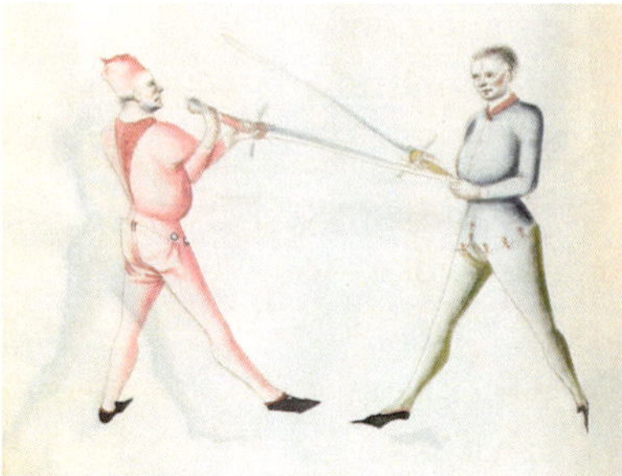

The **curved cut** (*Krumphaw*) introduces the 'crooked fight', and each of its pieces breaks one or more of the 'rules' in the common lessons—even the basic cut itself can be described as a diagonal cut on the 'wrong' side. The curved cut is a clean-up tool that can handle whatever messy fencing your opponent throws at you, and it has pieces for almost any distance from the opponent. At the end of the section, it returns to the 'straight fight' and the war.

The **crosswise cut** (*Twerhaw*) is a shallow diagonal rising or descending cut (or sometimes even horizontal), and the 'crosswise fight' is all about rapid changes from side to side and high to low—more of the skirmishing mentioned in the common lessons—at very close distance with the arms remaining high. The 'crosswise fight' also includes feinting from side to side to provoke parries and stealing away the sword before contact. The 'crosswise fight' and the 'straight fight' are paired opposites with contrary strategies.

The **cockeyed cut** (*Schilhaw*) is described as Liechtenauer's secret technique and introduces us to the 'cockeyed fight', which is useful against an opponent closing distance against you, either as a "buffalo" (a fencer who relies on strength instead of feeling) or with the extended point. The cockeye also includes teachings about the extension and retraction (which expand upon the common lessons again). The 'cockeyed fight' and the 'crooked fight' are mirrors of each other.

The **part cut** (*Schaitelhaw*) doesn't introduce a new paradigm, but instead completes the circle and brings us back to the 'straight fight'. It applies the common lessons against an opponent standing in a low guard. It's a very short section, and unsurprisingly, the plays are similar to the wrathful cut but with a new variation (slicing).

The (other) twelve main pieces

As mentioned before, the teaching sort of starts over again at this point. These twelve main pieces can be divided into three groups of four, each with its own focus, and together they arguably form a complete fencing curriculum (distinct from the five strikes). Where the first half of the *Record* begins by teaching you how to approach a fight, this half begins the way many HEMA schools are inclined to structure beginner courses: telling you how to stand, how to parry, how to attack, etc.

The four **guards** (*vier Hutten*) or **lairs** (*Leger*) are high and low guards with the point directed toward the opponent or away from them. (There's a lot of variation between different Liechtenauer texts in the details of how the guards are formed, but the basic features remain the same.)

The next main piece is **parrying** (*versetzen*), and even though readers often focus their attention on why using the five strikes to flush an opponent out of their guard is referred to as 'parrying', it's mostly about the proper way to parry and how to counter an opponent who likes to parry.

The third main piece is **pursuit** (*nachreisen*), which was referred to back in the four exposures as one of the three ways to correctly attack. The gloss focuses on two opportunities to attack safely: either before the opponent attacks (just as they enter a guard to attack from), or after they attack (just as their attack misses you). This section also explains how to work from the bind that results when you pursue their attack and they parry (called 'outer taking'), and (re-)introduces the sharp word "Within" (*Indes*).

The fourth main piece is **overrunning** (*uberlauffen*), which is about position and leverage when you attack. The piece in Ainring focuses on aiming at a higher target than your opponent, but the (arguably more important) teaching in Lew is about controlling the bind with your sword on top of your opponent's.

The next group of four pieces begins with **setting aside** (*absetzen*), which consists of luring your opponent's attack so you can parry it and then attack with a thrust.

Next comes **changing through** (*durchwechseln*), in which you avoid the opponent's sword by dropping underneath it and then attacking the other side; this can be done from an attack or from the extended point. Lew also uses this section to refer back to the common lessons, advising that if your opponent keeps their point aimed at you, you shouldn't change through against them but instead rely on turning in the bind.

The seventh main piece is **pulling** (*zucken*), in which you extract your sword from of the bind and either cut or thrust on the other side; note that Ainring has both a cut and a thrust, whereas other glosses only thrust in this piece. (The earlier piece called 'taking away' is very similar to pulling and cutting and could be the same play with two different names.)

The eighth main piece and last in this group is **running through** (*durchlauffen*). This is where most of the wrestling techniques in Liechtenauer's long sword are taught. It includes both 'body wrestling', in which you throw an opponent who rushes in against you with arms high, and 'arm wrestling', in which you lock up an opponent who closes with arms low.

The third group of pieces begins with **slicing off** (*abschneiden*) and then **hand-pressing** (*Hende drucken*). These two main pieces are the most poorly-differentiated in the RDLN gloss; combining the two sections, there are three sets of pieces: pulling away from a low bind to attack high, slices to the four exposures, and transitioning from a low slice to a high slice. In Lew, it can be argued that the first set is slicing off and the other two sets are hand-pressing. But Ainring assigns the first set to the section on parrying, so either the four slices are his slicing off or he doesn't have one. Ainring's hand-pressing is especially interesting for one reason: other glosses frame it in terms of an opponent rushing in against you, but Ainring has you rush in and then launch the technique. This opens the door to the idea that *any* time the gloss tells you to do something as a response to your opponent's actions, you can also simply create the situation yourself.

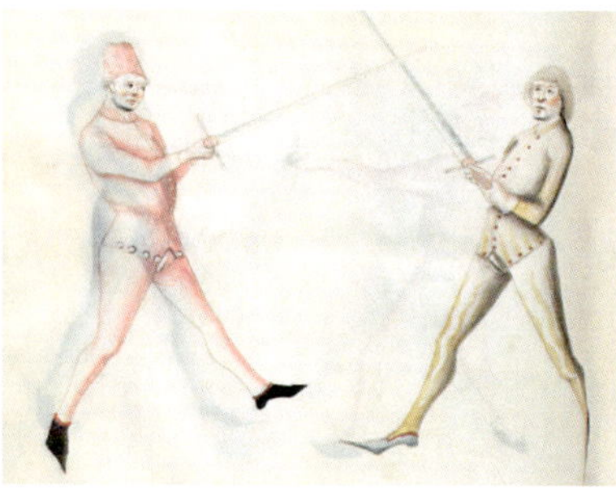

The eleventh main piece is about **angling** (*hengen*), but this gets quickly side-tracked by a discussion of the **spreading window** and a fifth guard called the **extended point**. This brings us back to the common lessons and 'straight fight' a final time—the extended point is the position assumed in the common lessons (with your point near your opponent's face or chest) and the spreading window is the resulting bind; it describes how to respond to an opponent who *binds onto* your sword (i.e., strikes a wrathful cut) rather than parrying badly, and most of the pieces in this section are responses to the pieces of the wrathful cut that countered the common lessons (and to other applications of the extended point).

The final main piece is described as the summary or resolution of the Record and is about **turning** (*winden*), bringing us back to the final teachings in the wrathful cut, but this time in more detail and sketching out twenty-four possible ways to turn. It also has more information about the angles which was missing from the previous section (they're the point-forward guards on both sides).

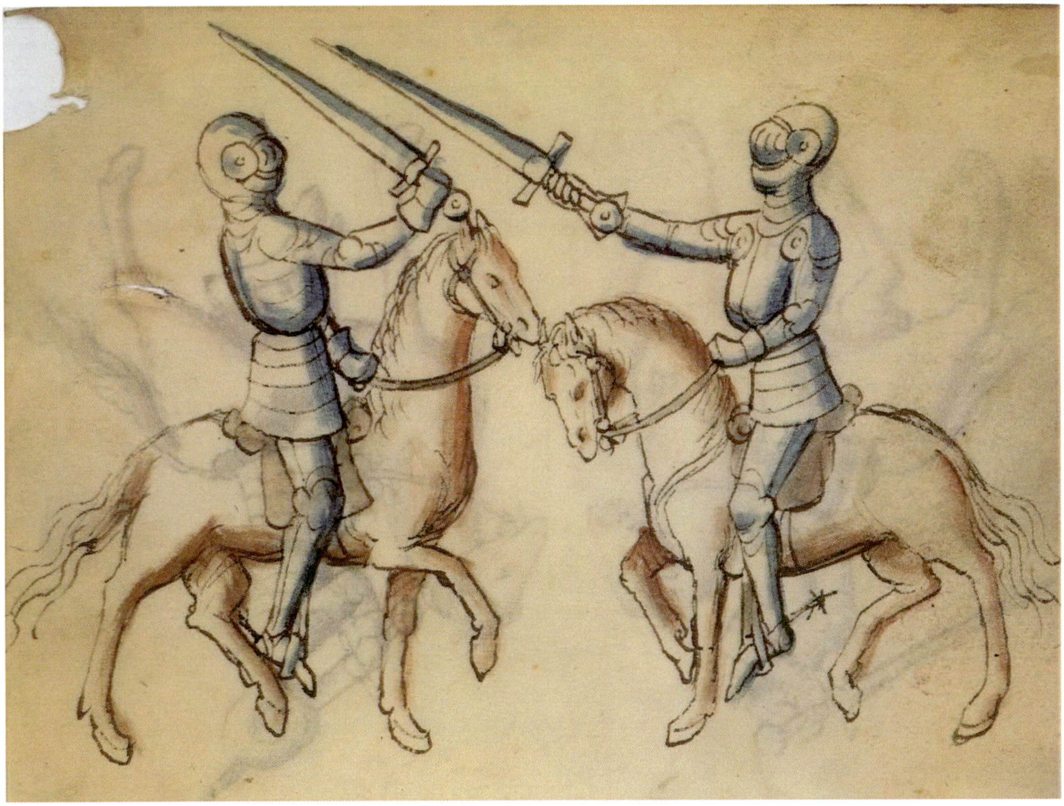

Figure 6: Armored combat on horseback, from W: 45v (2nd quarter, 15th cen.).

2.2.3. ARMORED DUELING ON HORSE AND ON FOOT

Where the unarmored teachings outline a theory of fencing with sophisticated organization, the armored teachings (mounted and dismounted) are completely different. The organization is by weapon, and the pieces are mostly arranged according to which guard they're easiest to initiate from. Instead of strategies that constrain the opponent and control the flow of the fight, mounted fencing is organized around short exchanges as your horse moves past your opponent's, while armored fencing revolves around hardware exploits—finding and targeting the vulnerabilities in the other's armor.

Structurally, these teachings follow the progression of a martial duel or deed of arms:[29] beginning with the lance on horseback, proceeding to swords

[29] I describe this in CHIDESTER 2021: 11–25, but detailed discussion of dueling under the *jus belli* (Law of War) has been sadly neglected in the literature; it's also frequently conflated with the tradition of trial by combat, muddying the waters for both. The principal work on the subject is MUHLBERGER 2005 (expanded further in MUHLBERGER 2019), but this work is concerned →

Figure 7: The 26 Figures of mounted fencing arranged in a tree diagram, from PD: 7^{v}–8^{r} (1452).

(including spear against sword) and mounted grappling; then the dismounted section begins with an unhorsed fighter against a mounted one (Ainring skips this step), after which they proceed with combat on foot with spears, spear against sword, sword against sword (primarily using the short sword grip), wrestling, and dagger (also omitted by Ainring).

Most treatises on armored fighting from the 15th century follow this progression, including the Gladiatoria group (foot combat only),[30] the treatises of Paul Kal,[31] the German manuscripts nicknamed '*die Blume des Kampfes*',[32] and two witnesses of the work of Fiore de'i Liberi.[33]

Martial duels were divided into two basic categories: duels to satisfaction (*Schimpf* in German, *ad piacenza* in Italian) and duels to the extremity (*Ernst* in German, *ad oltranza* in Italian). A duel to satisfaction was fought

primarily with the Anglo-French sphere. For Central Europe, it's discussed briefly in CAVINA 2016 and ELEMA 2019, and at only slightly more length in HUGHES 2007. I'm working toward an eventual book on this subject based largely on the ca. 1472 dueling treatise of Paride del Pozzo.

30 Witnesses of *Gladiatoria* are found in BF, GK, GN, GW, GWo, & P.

31 Witnesses of Paul Kal are found in PKB, PKG, PKG1, PKM, PKP, PKS, & PKW.

32 Witnesses of '*Blume des Kampfes*' are found in BKW1, BKW2, & LEE.

33 FLM & FLP.

Figure 8: The 26 Figures of mounted fencing arranged in a tree diagram, from HTG: 22^{v}–23^{r} (1448).

for glory and followed agreed-upon rules with some arbitrary end condition(s)—number of passes or blows, first fighter unhorsed or knocked to the ground, first blood, etc. A duel to the extremity, however, was based on a grave accusation and a denial that accusation, and could only be resolved by one of the fighters dying or recanting that accusation/denial.

Ainring's gloss includes pieces suited for both kinds of duels—some which are merely uncomfortable or inconvenient and some which are lethal. The gloss also shows awareness of different contexts, including describing victory by forcing your opponent to the edge of the ring and wrestling techniques described as forbidden for friendly encounters; it also includes advice for fighting opponents with different physical attributes.

One unique thing about the mounted fencing is the presence of 26 Figures, short pieces of text that are often listed as part of the *Record* and, in early sources like HTG and PD, arranged in a tree format where they have a visual relationship with each other. The Figures are invoked in the gloss at the end of certain pieces, and when these pieces are grouped according to the tree, they create an alternative sequence of ideas from the one in the verses of the *Record*.

Figure 9: Armored combat on foot with spear, sword, and dagger ready, from GN: 4r (1430s).

The structure of the trees only exists in the earliest witnesses, though, and seems to have been lost on most copyists. Both G and the Lew gloss merely list the Figures, while GF retains the circles but places them in a grid pattern. The numbering isn't consistent either, with Figures sometimes moving around in the list; Lew even splits one of the Figures in half (resulting in 27). This is particularly odd considering the sequence of verses in the *Record* remains consistent throughout the tradition.

Liechtenauer's armored teachings make some motions toward integrating concepts from the long sword teachings, such as Before and After and the pursuit, but by and large they have little in common. It's been suggested that the art of the 'long sword' could be combined with the armored teachings in order to effectively fight partially-armored opponents (who would far outnumber heavily-armored opponents in most combat settings), but there's no advice on this in the text[34] and the two methods of fencing aren't easily joined together.

[34] The unglossed witness of the *Record* on mounted fencing in N adds an admonition to remember the long sword lessons (N: 53v), but this seems to be the only such advice in the corpus.

3. Combined translation

Michael Chidester

This chapter combines all nine witnesses into a single eclectic translation.[1] The intent is to produce a text that is credibly similar to what Ainring originally wrote. It begins with the gloss of the unarmored fencing with the extended sword, and this is followed by the gloss of the armored dueling on horseback and on foot.[2]

When a 15th cen. reader looked at a gloss like Ainring's, what they saw was a poem (crude in some places, elegant in others) accompanied by explanations of its meaning—sometimes illustrated, sometimes not, but written entirely in their native language using fairly simple, straightforward words and grammar. I've tried to capture some of that reading experience in this chapter.

If you compare this translation to a transcription of any of the witnesses, you'll notice that here, little effort was made to preserve German word order or speech patterns, clauses are moved around or collapsed, and sometimes whole sentences are out of order. (Furthermore, words are often translated differently in different contexts.)

All text that rhymes in the original is also rhymed here. Rather than try to replicate the inconsistent meter of the original, I chose to cast all of the poetry in iambic tetrameter (a common English-language meter).

This is done with the goal of producing an approachable and understandable text, similar to DIERK's and my book on the *Nicolas Pol Hausbuch* (N),[3] For this same reason, I've limited the footnotes to only those necessary for understanding the text—if you have questions about a decision in the translation or a detail of the transcription, you should consult the Glossary or the footnotes in Chapter 5.

Ainring's gloss of the 'long' sword was intended to be extensively illustrated, though only the *Glasgow fight book* (GF) includes any actual illustrations along with the many references to them. Unfortunately, the library has declined to scan GF and the only available photos are too low-resolution to publish in a useful size; even if they weren't, it doesn't have illustrations for every passage that calls for them since the manuscript itself is incomplete (see Chapter 4).

[1] For a discussion of the different manuscripts containing Ainring's work, see Chapter 4.
[2] In SR, the dismounted dueling precedes the mounted, but this isn't how Liechtenauer's *Record* is typically presented and obscures the fact that these form a continuous teaching. Other manuscripts, including GF, place the mounted fencing first.
[3] CHIDESTER & HAGEDORN 2021.

In order to fulfill Ainring's intentions for this gloss, I've identified illustrations in other available 15th cen. manuscripts from the Liechtenauer tradition that either match those in **GF** or match Ainring's description (when **GF** doesn't have one). In doing this, I prioritized the *Paris fight book* (**P**), which was clearly influenced by Ainring[4] (and in one case used a 16th cen. illustration from *Hans Medel* (**HM**) for the same reason), but otherwise these are an assortment of illustrations from the treatises of Peter Falkner, Paul Kal, Hans Talhoffer, and other anonymous texts. I also took the liberty of adding a few illustrations for pieces that are missing from **GF** and **JMR**, since there's no way to know if they would have been illustrated or not.[5]

In the same vein, Liechtenauer's 26 Figures of mounted fencing are presented differently than in the single witness we have. **GF** places each figure in a circle, but merely arranges them in two 3×4 grids (with the last two at the top of the next folio). The fact that they're drawn in circles suggests that they were copied from a diagram like the one in the *Peter von Danzig fight book* (**PD**), ff. 7v–8r, which arranges them in two "trees" of thirteen representing the two modes of Liechtenauer's mounted fencing (presumably the scribe didn't understand the significance of the arrangement).[6] This arrangement has been replicated in simple form here to reflect what I presume was the author's original intent.

Finally, it is a common custom in our community to leave certain 'special' terms untranslated. I despise this practice and believe that the translator's job is to translate, and that while a reader is free to choose whether to use German or English terms in conversation, a translator who refuses to translate a term deprives the reader of that choice. Furthermore, these magic words often don't appear in the text as nominative nouns or infinitive verbs, but rather in grammatical forms that aren't easily untangled. Turning *setz im an* ("set upon him") into "perform an *Ansetzen* against him" distorts the language to an extreme degree. For these reasons—and to give readers a taste of the experience of the original readers, as I mentioned above—I haven't left any words untranslated in this chapter.

However, to make this book more interoperable with those of other translators and to help new readers navigate the book, I've decided to place a few German terms in parentheses at the beginning of their sections—Liechtenauer's five words, the names of the five strikes, the verbs associated with the twelve key pieces, and a few others of particular importance.

[4] See Chapter 4 for a discussion of P's role in the tradition.

[5] When no equivalent to an illustration from GF was available, I did my best with digital upscaling and manipulation to make the low-resolution photograph useable. To examine the illustrations from GF in the best resolution currently available, see Chapter 5.

[6] The diagram from PD is on p. 48. The 26 Figures as a medieval tree diagram is the subject of a forthcoming paper by Jessica Finley. She previously presented her research in Finley 2019.

1 The explanation of the *Record* (*Zedel*) begins here:

In it is written the chivalric art of the extended sword[7] that was composed and created by Johannes Liechtenauer (may God grant him grace), grand master of the art. He had the *Record* written in obscure and cryptic words so that the art wouldn't become common knowledge.

Then Master Sigmund Ainring, currently fencing master to the highborn prince and noble Lord Albrecht, Count Palatine of the Rhein and Duke in Bayern, had those same obscure and cryptic words of the *Record* clarified, explained, and commented on (as you'll find written and illustrated here in this booklet), so that anyone who can already fence could learn and understand them well.

He did this so that princes, lords, knights, and soldiers—to whom the art belongs—would love it all the more.

2 The foreword of the *Record*:

Text

1 Young knight, learn first: have love for God,
And honor women—that I laud.
2 So that your honor great may grow,
Practice chivalry and know
3 Arts that adorn you well in play
And bring you fame in war some day.
4 Gain wrestling's artful holds with might;
With lance, spear, sword, and knife do fight,
5 And wield them all with gallant hand,
So that against you none may stand.
6 Cut here, guard there, and close with this;
Charge in whether you hit or miss.
7 The wise ones hate and will disdain
All those who praise themselves in vain;
8 So grasp when all is said and done:
All art needs measure, moderation.

[7] The literal translation here is "long sword", but since it isn't the sword that's long and instead it's holding the sword with both hands on the grip that 'lengthens' it, "extended sword" makes the meaning clearer (reading *langen* as a shortened form of *verlängerten*). Compare "retracted sword" in the armored lessons, which refers to placing one hand on the blade and thus 'shortening' it.

3 This is the text of many good common lessons (*gemainer lere*) of the extended sword:

Text

9 If you would bring the art to light,
See left advance and strike with right,
10 For you will find that left with right
Is how you may most strongly fight.

Gloss

Remember: the first lesson of the extended sword is to cut the cuts correctly from both sides if you want to otherwise fence strongly and correctly.

Learn it like this: make sure your left foot is forward when you want to cut from the right side, and make sure your right foot is forward when you want to cut from the left side. Then, if you cut from your over right side, accompany the cut with your right foot. If you don't do this, the cut is false and incorrect—since your right side stays behind, the cut is shortened and can't follow its correct path downward toward the other side (in front of your left foot). Likewise, the cut is also false when you cut from the left side but don't accompany the cut with your left foot.

So remember: whichever side you cut from, accompany the cut with the same foot so you can drive all your pieces strongly. Also cut all the other cuts like this.

4 Again, a lesson:

Text

11 Who waits and follows after blows,
Will little joy in this art know.
12 Cut close to them; your will revealed,
No changer comes to breach your shield;
13 Toward head and body, forth you race,
The skirmish you should then embrace.
14 And always fight with all your strength,
Your body tight at any length.

Gloss

Remember this when you approach to fence with them: don't watch for their cuts nor wait for how they'll drive them against you, because all fencers who watch and wait for someone else's cuts and only want to parry, they allow themselves little joy in the art and they are often struck by it.

Likewise, always remember to fence with the strength of your whole body, and cut in close to them (toward their head and body) so they can't change through in front of your point. When this cut leads to a bind of swords, make harrying strikes toward the nearest exposure (which will be explained later on in the five cuts and in other pieces).

5 Again, a lesson:

Text

15 A simple rule you should not slight:
Fence not from left when you are right.
16 If on your left is how you fight,
You'll be quite clumsy on the right.

Gloss

Remember: this lesson is about two people, one right and one left.[8]

Learn it like this: when you approach to fence with them, if you're right and mean to strike them, don't cut the first cut from the left side because that's weak and you won't be able to withstand someone who binds on you strongly. Instead, cut from the right side so you can artfully work whatever you want on the sword, strongly [and] with art.

Similarly, don't cut from the right side if you're left, since the art is very awkward for someone who is left to drive from the right side (just like it is for right from the left side).

6 This is a lesson on Before (*Vor*) and After (*Nach*):

Text

17 Before and after, these two things:
The font from which all true art springs.
18 With strong and weak, you'll rule the bind;
The word "within" bear well in mind.
19 Learn onward in this art until
You can defend and work with skill.
20 But if you easily take fright,
You'll find it hard to learn to fight.

Gloss

Remember above all else: you should understand the Before and the After, since these two things are the source from which all the art of fencing flows.

Learn it like this: the Before means that you should always come forth with a cut or thrust toward their exposure, and do it in such a way that they must parry you (before they do this to you). In this parry, work nimbly in front of yourself with your sword (from one exposure to another) so they won't have a chance to make their own pieces before your work.

But if they run in toward you, then come forth with the wrestling.

[8] This segment is typically interpreted as referring to right- and left-handed fencers and translated accordingly, but the text just says "right" and "left" and it's not clear whether it's referring to handedness, which side of the body the sword is held on (regardless of handedness), or which foot is forward (thus echoing the first lesson, just as the fourth lesson echoes the second).

7 On what is called 'the After':

Remember: if you can't come Before then wait until After, which includes the counters for all pieces that they drive against you.

Learn it like this: when they advance in such a way that you must parry them, then Within your parry, work nimbly toward the nearest exposure with your sword. In this way, they won't have a chance to make their own piece before you meet them.

Once you've won the Before and they remain After (as above), then in the Before and After, also remember how to work with the word "Within" (*Indes*) according to the Weakness and Strength of their sword.

Learn it like this: the Strength (*Stercke*) of the sword is from the sword's hilt to the middle of the blade; with it, you can withstand someone that binds you on it. The Weakness (*Schwache*) of the sword is further on, from the middle to the point; you can't withstand anything there.

When you correctly understand these things, you can artfully work and protect yourself, and also teach princes and lords so that they may succeed with this same art (in play and in earnest). But you shouldn't learn the art of fencing if you startle easily, because it'll do you no good if your heart becomes fragile or despondent when struck by any art.

8 The five cuts:

Text

21 Five strikes you must now learn to heed,
Perform them with your right hand's speed;
22 Those fencers skilled whose art we vow
To reward well will show us how:

Gloss

Remember: the *Record* lays out five obscure cuts, and many masters of the sword have nothing to say about them. Don't learn to cut any other cuts from the right side against an opponent who stands against you in defense, since if you select one of the five cuts, you can meet them with the first strike.

Whoever can counter you without injury will be avowed by the masters of the *Record*, because their art will be more successful for them than another fencer who can't fence against the five cuts.

How to cut with the five cuts will be described later on (in the five cuts).

9 The pieces of the *Record*:

Text

23 The wrathful one hates crook and cross;
So cock your eye and parting cause.
24 A fool will parry all strikes, so
Pursue, run down, and set the blow.
25 Change it through, pull your point back,
Run through, slice off, or hands attack.
26 Then hang and wind, expose them more,
Strike, catch, sweep, and push your point fore.

Gloss

Remember: the main pieces (*Haupstucke*) of the art of the extended sword are correctly named here; each is called by its individual name so that you can understand them better. They are seventeen in number, and they begin with the five cuts.

Now remember the 1st cut, called 'the wrathful cut' (*Zornhaw*)

2. The curved cut (*Krumphaw*)
3. The crosswise cut (*Zwerhaw*)
4. The cockeyed cut (*Schilhaw*)
5. The part cut (*Schaitelhaw*)
6. The four guards (*Hutten*)
7. The four parries (*verseczen*)
8. Pursuit (*nachraisen*)
9. Overrunning (*uberlauffen*)
10. Setting aside (*absezen*)
11. Changing through (*durchwechseln*)
12. Pulling (*zucken*)
13. Running through (*durchlauffen*)
14. Slicing off (*abschnÿden*)
15. Pressing hands (*Hend trucken*)
16. Angling (*hengen*)
17. Turning (*winden*)

How to expose someone with the angles and turns and how to drive all these pieces will be described later on.

10 This is the wrathful cut (*Zornhaw*) with its pieces:

Text 27 When from above they cut in near,
The point of wrath makes danger clear.

Gloss Learn it like this: when someone cuts in over you from their right side, cut wrathfully in along with them, ſtrongly with the long edge, also from your right shoulder. Then, if they're Soft on the sword, shoot the point in ahead of you, extended toward their face, and threaten to ſtab them (as illuſtrated below).

Picture *PKB: 19r.*

11 Another piece of the wrathful cut:

Text 28 If they should sense the threat and shove,
Fear not, just take it off above.

Gloss This is when you shoot the point in at them with the wrathful cut (as illustrated previously).

If they become aware of your point and parry the thruſt ſtrongly, then pull your sword upward, off from their sword above, and cut back in above on their sword toward the other side, in toward their head (as illuſtrated here).

Picture *PKB: 19ᵛ.*

14[9] A counter againſt the take-away (*abnemen*):

Remember: when you ſtrongly bind with someone on their sword and they pull their sword up and off from your sword above, and then they cut in at you again toward the other side of your head and onto your sword, then turn ſtrongly with the long edge in above toward their head.

[9] Moved here for clarity, but you can see it in its proper sequence in Chapter 5.

12 Another piece of the wrathful cut:

Text

29 Or if they're strong, then more strength show
And thrust—if seen, take it below.

Gloss

Remember: this is when you cut the wrathful cut in (as illustrated previously).

If they parry you and stay on your sword strongly, then be strong against them on the sword and go up into the Weakness of their sword with the Strength of your sword, and on the sword, turn your sword with your hilt in front of your head and thrust in above toward their face (as illustrated here).

Picture *P: 13r (© RMN-Grand Palais/Art Resource, NY).*

13 Again, a piece of the wrathful cut:

When you thruſt in at them above with turning (as illuſtrated previously), if they raise their hands up high and parry your high thruſt with their hilt, then ſtay like this in the turning and set your point down on their cheſt in between their arms (as illuſtrated below).

Picture *PFW: 8ᵛ.*

15 Remember a good lesson here:

Text

30 Bear this in mind once in the fray:
Cut, thrust, then hard or soft you lay,
31 "Within", and then drive after more,
But calmly—do not rush to war;
32 If at the war above they aim,
Go down below and bring them shame.

Gloss

This means that when you bind with someone on your sword (with a cut, thrust, or otherwise). Clearly notice whether they're Hard or Soft on your sword, and when you've sensed that, you should know Within it which is best for you: whether to work on them with the Before or the After.

But don't rush to run in with the war (which is nothing more than turning while on the sword).

Drive the war like this: when you cut the wrathful cut in, as soon as they parry, go up high with your arms and turn your point in toward their upper exposure (while on their sword). If they parry this thrust, then stay in the turn and seek their lower exposure with your point. If they further follow after your sword with this parry, then go through under their sword with your point and angle it in above toward the other exposure on their right side.

In this way, they'll be shamed above and below by the war (if you can drive the threats correctly).

16 How to correctly find cuts and thrusts in all turning:

Text

33 Down every path that you may wind,
Cut, thrust, and slice you'll learn to find;
34 You also must learn to assess
Which one of them would serve you best,
35 So that, whenever steel meets steel,
The masters you'll confound with zeal.

Gloss

This means that you should correctly find cuts, thrusts, and slices in all turning. When you turn, immediately determine which of the three is most appropriate so you don't cut when you should thrust, nor slice when you should cut, nor thrust when you should slice.

When they parry one, remember to meet them with another like this: if someone parries your thrust, then drive a cut; if someone runs in, then drive a slice under their arms. Remember this in all meetings and binds of the sword and you'll confound the masters who go against you.

17 On the four exposures:

Text

36 Know there are but four exposures;
Clear your path and aim for closure,
37 And however they appear,
You may attack without a fear.

Gloss

Remember: an opponent has four possible exposures and you should always fence toward them. The first two exposures are the right and left sides above the navel. The other two exposures are also the right and left sides, below the navel.

In the approach, notice clearly which exposures they expose to you and safely and artfully clear and target them by shooting in your extended point, by pursuit, and by turning on the sword (and otherwise with all threats), and don't worry about how they threaten you or act against you.

In this way, you fence wisely and strike simply and hit with this, and they won't have a chance to make their own pieces.

18 On the redoubling and the transmutation; how they exploit the four exposures:

Text

38 Should they attack, revenge is sweet;
Exploit these four exposures neat:
39 To land above you should redouble;
Transmute below to cause them trouble.
40 Now let me make this plain and clear:
No one defends without a fear.
41 And if these truths you'll learn and know,
Then scarcely can they come to blows.

Gloss

This is when you want vengeance on someone and you want to quickly and artfully exploit their four exposures. Drive the redoubling toward the upper exposures against the Strength of their sword, and the transmutation to the other exposures. I say to you honestly that in this way, they can't guard themselves from this and won't have a chance to make strikes nor thrusts.

19 The redoubling (*duplieren*):

When you cut in above (with the wrathful cut or otherwise), if they parry you ſtrongly, then Within it push your pommel under your right arm with your left hand and ſtrike in on the sword with crossed hands, behind their blade (between your opponent and their sword), through their muzzle. Or ſtrike them on the head with this piece.

Picture *PKB: 21ʳ.*

20 Remember the transmutation (*mutieren*):

Drive the transmutation like this: when you bind on their sword (by cutting over them or otherwise), then turn your short edge onto their sword and go up high with your arms, and angle your blade over their sword (in from outside) and thrust in toward their lower exposure.

Drive this to both sides.

Picture *PKB: 21ᵛ.*

21 The curved cut (*Krumphaw*) with its pieces:

Text

42 Throw a curve with crooked grace;
Onto their hands your point will race.

Gloss

This is how to cut crookedly toward their hands.

Drive the piece like this: when they cut at you (toward your exposure) with a cut over or under your right side, leap away from the cut with your right foot (far to their left side against them), and strike in with crossed arms, with your point on their hands.

Also drive this piece when they stand against you in the ox guard.[10]

Picture *PKM: 66ᵛ.*

[10] This piece isn't described in Ainring; for a description, see the gloss of this verse in Lew.

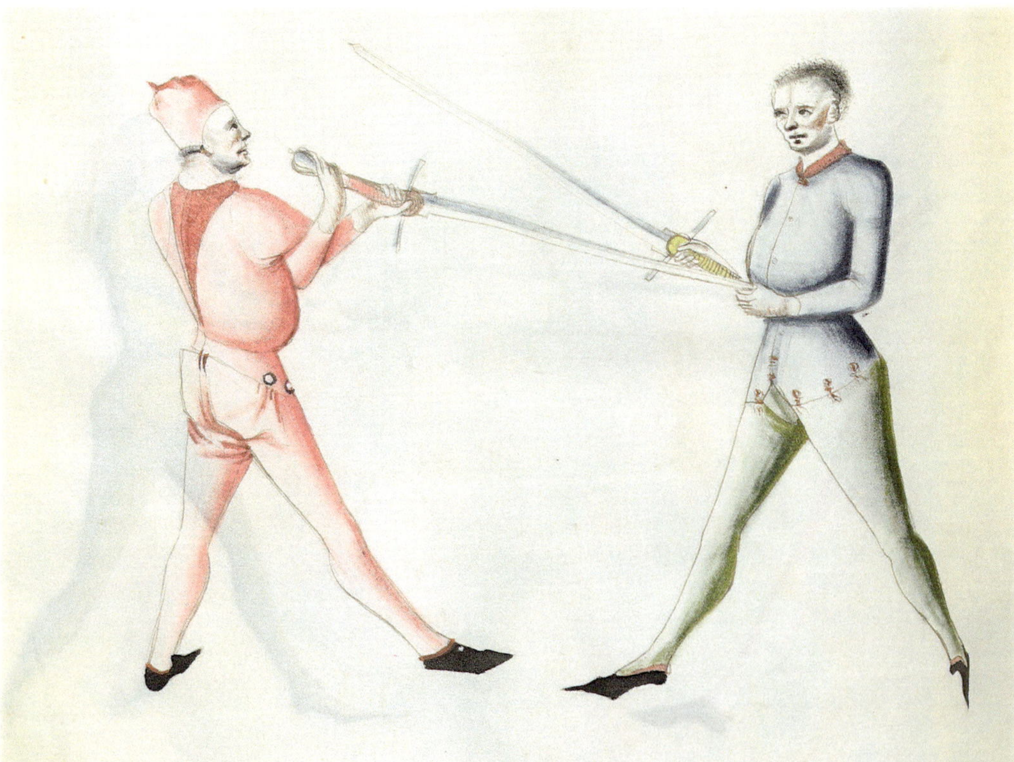

22 Again, a piece of the curved cut:

Text

43 Lure them out well and curve down low,
And step to hinder many blows.

Gloss

This is how to use the curved cut to set aside cuts from above.

Drive it like this: when they cut in toward your exposure from above their right side, step toward their left side with your right foot and [fall][11] over their sword (with your point on the ground in the lattice fence guard).

Drive this to both sides.

You can also strike them on the head from this setting aside.

Picture *P: 17^{r} (© RMN-Grand Palais/Art Resource, NY).*

[11] This missing verb is filled in based on Hans Medel.

23 Again, a piece of the curved cut:

Text 44 Cut crookedly up toward the flat,
Subvert the maſters' ſtrength with that.

Gloss This is when you want to weaken a maſter.

Drive the piece like this: when they cut in at you from above their right side, cut crookedly againſt their cut (onto their sword) with crossed hands.

Picture PFW: 4[v].

24 Again, a piece of the curved cut:

Text

When steel on steel above should spark,
Abscond and I will praise your art.

Gloss This is when you cut in onto their sword with the curved cut: strike quickly up from the sword with your short edge, in toward their head above, or with your curved cut, turn the short edge on their sword and thrust in toward their chest.

Picture *HTM: 11^{v}.*

25 Again, a piece of the curved cut:

Text 46 Curve not: cut short, your plan concealed,
Then with it, changing through reveal.

Gloss This is when they want to cut at you from above their right side: act like you want to bind on their sword with the curved cut, but shorten and go through under their sword with your point, and turn your hilt over your head on your right side and thrust toward their face (as illustrated here).

This piece counters the... [ox guard, too.][12]

Picture *PKB: 20ʳ.*

[12] This piece isn't described in Ainring; for a description, see the gloss of this verse in Lew.

26 Remember how to counter the curved cut:

Text

47 Who down a crooked path will lead,
The noble war ensnares indeed:
48 They'll truly have no way to know
Where they'd be safe from any blow.

Gloss

Remember: this is when you cut over or under them from your right side: if they cut crookedly onto your sword from their right side (with crossed arms) and they divert your strike with it, then keep your sword strongly on theirs and shoot your point in under their sword and extended toward their chest.

27 Another counter for the curved cut:

Remember: this is when you shoot your point in under their sword toward their chest (as described and illustrated previously), or when you cut in from above your right side and they also come onto your sword from their right side with crossed arms.

If they press your sword downward against the ground, then turn your sword on your right side and go up high over your head with your arms, and set your point above onto their chest (as illustrated below).

If they parry you, then keep your hilt in front of your head like this and work nimbly with your point (from one exposure to another).

This is called 'the noble war'. With it, you confuse them so much that they truly don't know where they should guard themselves from you.

Picture *No illustration available.*

28 The crosswise cut (*Twerhaw*) with its pieces:

Text 49 The cross takes what from sky comes down,
Rewarding all with glory's crown.

Gloss Remember: the crosswise cut counters all cuts that are cut down from above.

Drive the cut like this: stand with your left foot forward and hold your sword on your right shoulder, and when they cut in at you above (toward your head), then leap against them on their left side (away from the cut) with your right foot, and in the leap, turn your sword with the hilt high in front of your head (so that your thumb comes underneath), and strike in toward their left side with your short edge so that you catch their cut in your hilt and you press in toward their head (as illustrated here).

Picture *P: 18r (© RMN-Grand Palais/Art Resource, NY).*

29 A piece of the crosswise cut:

Text 50 The cross in strength performed its deed,
Your work remains, take careful heed.

Gloss This is how to work from the crosswise with the Strength of your sword.

Do it like this: when you cut in with the crosswise, think to turn the Strength of your sword strongly on theirs. If they hold strongly against it, then strike in on their sword with crossed arms, behind their blade on their head (as illustrated here), or slice them through the muzzle with this piece.

Picture *PKB: 21ʳ.*

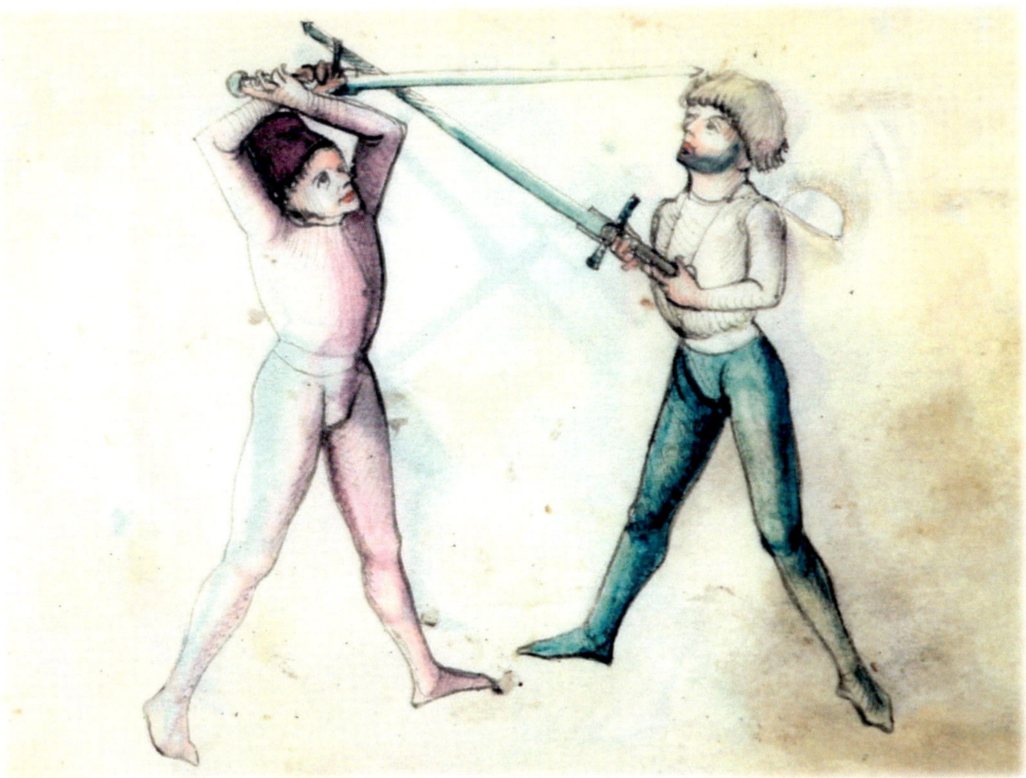

30 Again, a piece of the crosswise cut:

Remember: when you bind on their sword from the crosswise with the Strength of your sword, if they hold ﬆrongly againﬆ it, then push their sword from below with your hilt (on your right side, as illuﬆrated here) and swiftly ﬆrike the crosswise around again againﬆ the right side of their head.

Picture *HTM: 12^{v}.*

31 Again, a piece of the crosswise cut:

When you bind them on their sword with the crosswise, if they're Soft on your sword, then lay your short edge in on the right side of their neck and leap behind their left foot with your right, and pull them over it with your sword.

32 Another piece:

When you bind on their sword with the crosswise, if they're Soft on your sword, then press their sword down with the crosswise and lay your short edge forward, behind their arms on their neck.

33 [Counter to the lay-on:][13]

If they take you by the neck on your right side, then release your sword from your left hand and push their sword away from your neck with your right, and step in front of both of their feet with your left foot (against their right side), and go over both their arms close by the hilt with your left arm and make them dance, or thrust underneath toward their groin (between their legs).

13 Title from Hans Medel.

34 A counter against the upper crosswise cut:

Remember: when you bind on their sword from your right side (by cutting over them or otherwise), if they strike the crosswise around toward the other side, then Before it, come in under their sword onto their neck, also with the crosswise (as illustrated below), so that they strike themselves with your sword.

Picture *P: 18ᵛ (© RMN-Grand Palais/Art Resource, NY).*

35 A counter against the lower crosswise strike:

Remember: when you bind on their sword from your right side (with a cut over them), if they strike the crosswise around from your sword toward the other exposure on your right side, keep your hilt over your head and turn your blade downward against their cut, and thrust in toward the lower exposure (as illustrated below).

Picture *HTM: 8*^{v}.

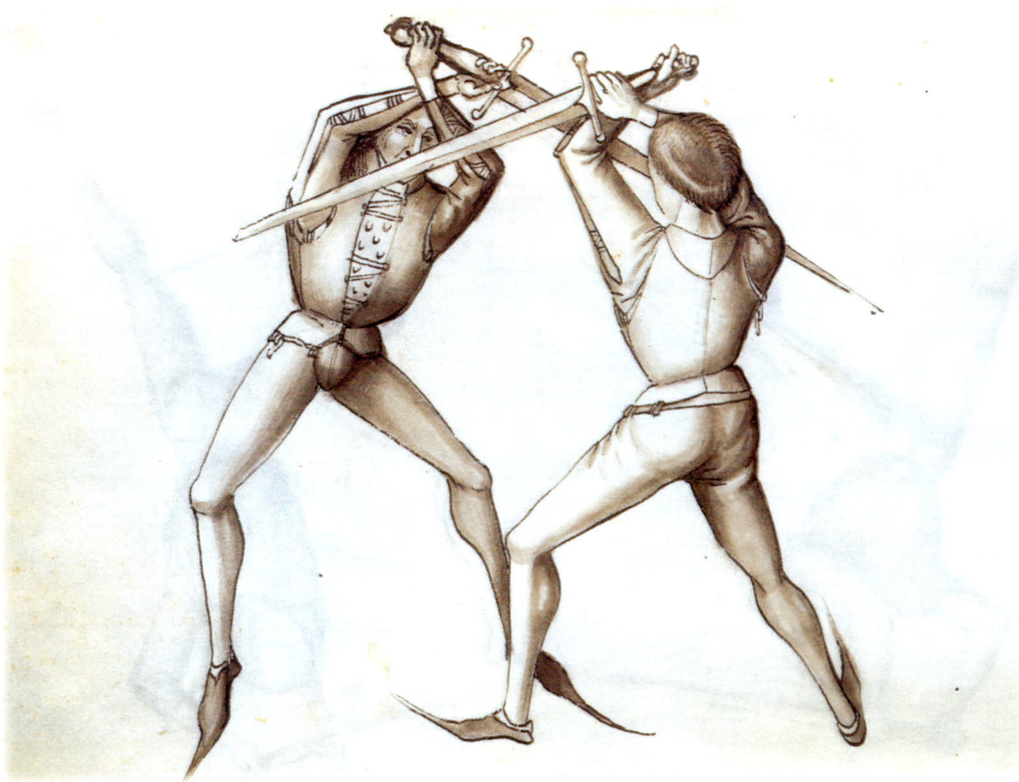

36 How to strike toward the four exposures with the crosswise:

Text

51 When to the plow you drive across,
Yoke it hard then to the ox.

Gloss

This is how to strike at all four exposures with the crosswise in one round.

Learn it like this: when you approach to fence with them, notice when it seems right to you and leap toward them and strike the crosswise in toward their lower left exposure (as illustrated below).

This is called 'striking toward the plow'.

Picture *P: 19ʳ (© RMN-Grand Palais/Art Resource, NY).*

37 Again, a piece of the crosswise cut:

Remember: when you've struck the crosswise in toward their lower exposure (as illustrated previously), quickly strike the crosswise up above toward the other side, in toward their head.

This is called 'striking toward the ox'.

Then nimbly strike a further crosswise strike toward the ox and another toward the plow, crosswise from one side to the other, and then cut in over them, above toward their head, and with that, pull yourself away.

Picture *PFW: 6ʳ.*

38 A lesson:

Text 52 Cross yourself and take a leap,
And threaten heads while yours you keep.

Gloss This means to leap out far on the side with each crosswise strike when you want to strike in toward him; thus, you may well hit their head. And see that you're well-protected above, with your hilt in front of your head.

39 Again, a piece of the crosswise, and it's called 'the misser' (*Feler*):

Text

53 Mislead by missing skillfully;
Go low and harry willfully.

Gloss

Remember: all fencers who like to parry will be misled and struck by the misser.

Drive the piece like this: when you approach to fence with them, act like you'll cut freely over them or strike a crosswise cut toward the left side of their head, but pull your sword away and strike a crosswise cut in toward their lower right exposure (as illustrated here).

Thus they are harried and struck underneath as you wish.

Picture *P: 20v (© RMN-Grand Palais/Art Resource, NY).*

40 Again, a piece of the crosswise; it's called 'the inverter' (*Verkerer*):

Text

54 Inversion forces and constrains;
Run through and wrestling holds you'll gain:
55 Their elbow take; be sure, then leap;
Done right, their balance you will reap.

Gloss

Remember to also drive this piece: when you bind on their sword with a cut over or under them, invert your sword (so that your thumb comes underneath) and thrust above toward their face in a way that constrains them so that they must parry you. In the parry, grab their right elbow with your left hand and leap in front of their right foot with your left and push them over it like this (as illustrated below).

Or run through them with the inverter and wrestle (as you'll find described later on in the running through).

Picture *P: 21*ᵛ *(© RMN-Grand Palais/Art Resource, NY).*

41 Again, a piece of the misser:

Text 56 Now miss them twice, and if you hit,
Then make a classic slice with it.

Gloss This is called 'the double misser', because you drive a double misleading in one approach.

Drive the first like this: when you approach to fence with them, leap against them with your right foot and act like you'll strike the crosswise strike in toward the left side of their head, but pull the strike away and strike in toward the right side of their head (as illustrated below).

Picture *P: 22^v^ (© RMN-Grand Palais/Art Resource, NY).*

42 Again, a piece of the misser:

Text 57 The second time you miss, I say
To step in left, and don't delay.

Gloss This is when you've struck in toward the right side of their head after the first misleading (as illustrated previously): strike in again quickly, around toward the left side of their head, and go over their sword with your short edge with crossed arms and leap in left (that is, on their left side), and slice in through their muzzle with the long edge (as illustrated).

Picture *PKB: 21ʳ.*

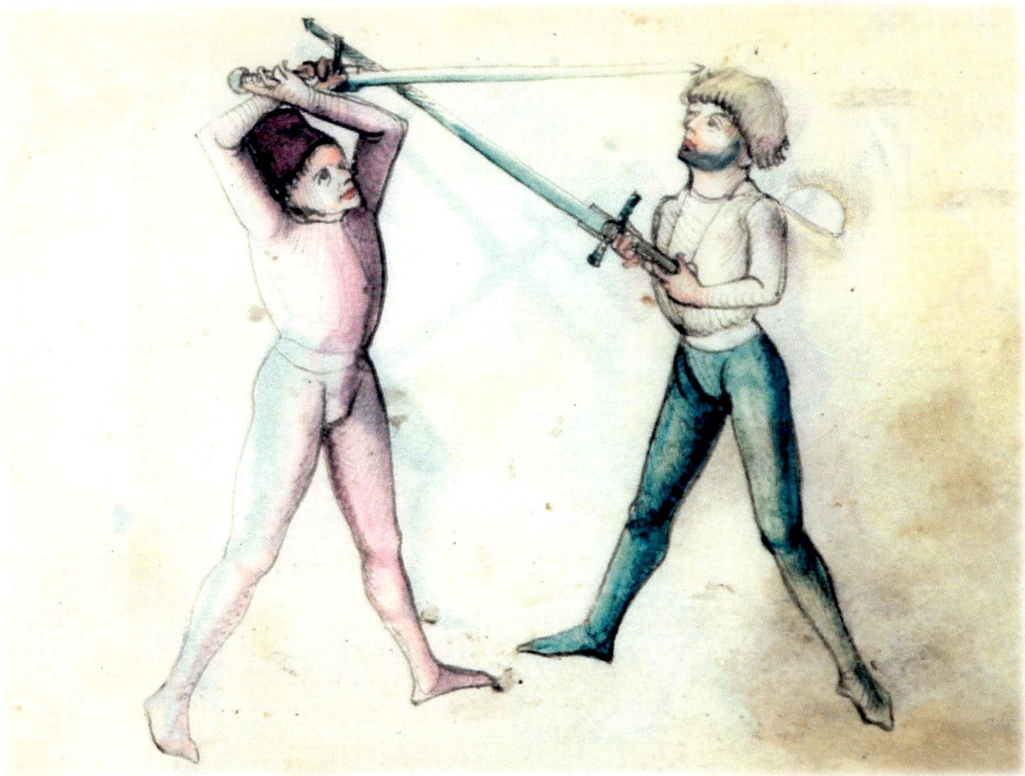

43 This is the text and the gloss on the cockeyed cut (*Schilhaw*) and on the pieces to drive from it:

Text

58 When buffalo will cut or thrust,
The cockeye breaks and enters thus.
59 If, with the changer, threat they lay,
The cockeye robs them anyway.

Gloss

Remember here: the cockeye is a cut that counters the cuts and thrusts of the buffalo (one who takes mastery with violence).

Drive the cut like this: when they cut in from above their right side, then cut from your right side too, with your short edge against their cut, into the Weakness of their sword with outstretched arms, and strike them on their right shoulder. Thus you strike and parry together, and hit them with the cut (as illustrated here).

If they change through, then with the cut, shoot [your point] in extended toward their chest.

Also cut like this when they stand against you in the plow guard, or when they want to thrust toward you underneath.

Picture *P: 24ᵛ (© RMN-Grand Palais/Art Resource, NY).*

44 Again, a piece of the cockeye (*Schiller*):

Text
60 If you see them shortchanging you,
Take your revenge by changing through.

Gloss
Remember: this is a lesson to cock an eye (with your face) and clearly see whether they fence shortened against you.

If you recognize that they don't outstretch and extend their arms when they cut toward you, then you should cut too, and in the cut, go in through under their sword with your point, turn your hilt over your head on your right side, and thrust in toward their face (as illustrated here).

Likewise, freely change through with your point extended (from cuts and thrusts) [against] all fencers who fence shortened: from the ox, from the plow, and with all turnings in front of you. With that, you retain them at the sword so that they must let you come to bind on them and be struck.

Picture *PKB: 20ʳ.*

45 Again, a piece of the cockeyed cut:

Text

61 Down to their point you cock your eye,
But, fearless, take their neck up high.

Gloss

Remember: the cockeye counters the extended point (by deceiving with your face).

Drive it like this: when they stand against you and hold their point against your face or chest with outstretched arms, stand with your left foot forward and cock an eye toward their point (with your face) and act like you want to cut in toward it, but cut strongly onto their sword with your short edge, and with that, shoot your point in, extended toward their neck, while stepping in with your right foot (as illustrated below).

Picture *GF: 6ᵛ.*

46 Again, a piece of the cockeyed cut:

Text 62 Or cock your eye up to their part,
If you would spoil their hands with art.

Gloss Remember: when they want to cut in above you, cock your eye (with your face) like you want to ﬅrike them on their head, but cut againﬅ their cut with your short edge, and ﬅrike in onto their sword blade with your point on their hands (as illuﬅrated below).

Picture *PKB: 21^v^.*

47 This is the text and the gloss on the parter (*Schaitler*) and on its pieces:[14]

Text 63 Cut from your part to seek your prize
And threaten them under the eyes.

Gloss Remember: the parter is dangerous to their face and chest.

Dive it like this: when they stand against you in the foolish guard, cut in toward their head from the part of your hair (down from above) with the long edge. If they parry, keep your arms high and angle your point in over their hilt toward their face (as illustrated below).

Picture *P: 13ʳ (© RMN-Grand Palais/Art Resource, NY).*

[14] The part cut is the section with the most disagreement between witnesses in Ainring's entire gloss, and putting together the pieces requires the most speculation. Even the way the *Record* is divided up varies. See the reading apparatus in Chapter 5 for the full range of variations.

48 A piece of the parter:

Text

64 Then turn and take it down below,
And threats against their heart bestow.

Gloss

Remember: this is when you angle your point in toward their face with the parter: if they push your point firmly upward with their hilt, then invert your sword with the hilt high in front of your head (so your thumb comes underneath) and set the point on their chest, under their hands (as illustrated here).

Picture *PFW: 8ᵛ.*

49 How the crown (*Kron*) counters the parter:

Text

Whatever from your part descends,
Their lofty crown can well defend.

Gloss

Remember: when you cut the parter in above: if they parry with the hilt high above their head, the parry is called 'the crown', and they run in at you with it.

Picture *HTM: 12ᵛ.*

50 How the slice counters the crown:

Text
66 Slice through their crown—refuse to kneel—
Its glory broken by your steel.
67 With sweeping cuts press your attack;
Slice through and then pull yourself back.

Gloss Remember: when they parry the parter (or other such cut) with the crown and run in with it: take the slice underneath their hands and arms, and press firmly upward (as illustrated here). In this way, the crown is countered well. Then turn your sword from underneath into the slice over, and pull yourself away with it.

Picture *GF: 9r.*

51 This is the text and the gloss on the four lairs (*Leger*):

Text

68 In four lairs only should you lie;
Hold there and vulgar guards decry.
69 The ox that plows, the foolish one,
And from the day you should not shun.

Gloss Remember: this means to not hold any lairs other than the four named here.

52 The first guard

The first guard is called 'the ox' (*Ochs*); arrange yourself in it like this: stand with your left foot forward and hold your sword next to your right side with the hilt in front of your head and angle the point against their face (as illustrated below).

53 The second guard

The second guard is called 'the plow' (*Pflug*); arrange yourself in it like this:

Picture PD: 1^{v}.

stand with your left foot forward and hold your sword next to your right side with crossed hands, over your knee, so that your point is against their face (as illustrated previously).

54 **The third guard**

The third guard is called 'foolish' (*Alber*); arrange yourself in it like this: stand with your right foot forward and hold your sword with arms outstretched away from you, with the point on the ground (as illustrated below).

55 **The fourth guard**

The fourth guard is called 'from the day' (*vom Tag*); arrange yourself in it like this: stand with your left foot forward and hold your sword on your right shoulder (as illustrated below), or hold it with arms outstretched over your head.

You'll find all about how to fence from the guards described in this book.

Picture *PD: 2ʳ.*

56 This is the text and the gloss on the four parries (*versetzen*) that injure or counter the four lairs:

Text

70 Now four displacements learn with care,
Which also flush them from their lair.

Gloss

Remember: you've heard previously that you should only fence from the four lairs, and you should also know that the four parries, they are four the cuts.

The first cut is the curved cut; it counters the ox guard.[15]

The second cut is the crosswise cut; it counters the guard from the day.

The third is the cockeye; it counters the plow guard.

The fourth is the parter; it counters the foolish guard.

Picture *P: 16^r (© RMN-Grand Palais/Art Resource, NY).*

[15] Ainring doesn't describe how the curved cut counters the ox or how the crosswise cut counters the day. Illustrations of countering the ox and day are included below as a starting point, but both of these pieces are described most clearly in Lew and you can read about them there.

57 A lesson:

Text /1 Be mindful of displacement's game,
It guards you well or brings you shame.

Gloss Guard yourself before parrying as other fencers parry: they allow the point to go out low or high to the side in front of their opponent, and whoever does that can't seek the exposure on their opponent.

So don't do this parry, but remember: when they cut, then you should cut too, and when they thrust, then you should thrust too. You'll find how to cut against their cuts described in the five cuts, and you'll find how to fence against their thrusts described in the setting aside.

Picture *HM: 27r.*

58 A piece against parrying:

Text

72 Should you become displaced at last,
However this has come to pass,
73 Then listen now to what I say:
Wrench off, cut in, and don't delay.

Gloss This is when you are parried (however it happens).

Remember: if you cut over someone and they parry and come in close, then in this parry, go against them with your pommel over their front hand and wrench downward with it, and with this wrench, strike him on the head with your sword (as illustrated below).

Picture *P: 30^{r} (© RMN-Grand Palais/Art Resource, NY).*

59 Again, a piece against parrying:

Remember: when you cut from underneath your right side, if they fall on your sword with theirs so you can't come up with it, then go nimbly over their sword with your pommel, wrench your blade from under theirs, and strike in by snapping the long edge toward their head (as illustrated below).

Or, if they fall on your sword against your left side, then strike in with your short edge.

Picture *PFW: 9r.*

60 Again, a piece against parrying:

Text 74 Set on to four extremities,
Learn to remain and end with ease.

Gloss This is when you cut in from above your right shoulder.

If you want to quickly end with them, then when they parry, remember to quickly strike around crosswise, and with this strike, grab your sword in the middle of the blade with your left hand and set the point in their face (as illustrated below).

Or set upon (*ansetzen*) them (toward whichever of the four exposures you can approach best), etc.[16]

Picture *P: 32ʳ (© RMN-Grand Palais/Art Resource, NY).*

[16] This piece isn't described Ainring, though Pseudo-Danzig, Lew, and Nicolaüs all cover it; it's described most clearly in the Lew gloss.

61 **Again, a piece againſt parrying:**

Remember, when you set the point in their face with the half sword (as illustrated previously): if they parry that, then bash your pommel in toward the other side of their head, or leap behind their left foot with your right and go around in front of their neck and in over their right shoulder with your pommel, and pull them over your right leg with it (as illuſtrated below).

Picture *GF: 12^{v}.*

Counter

Take the elbow. Step behind his foot and push.[17]

[17] Counter taken from Hans Medel and Paris.

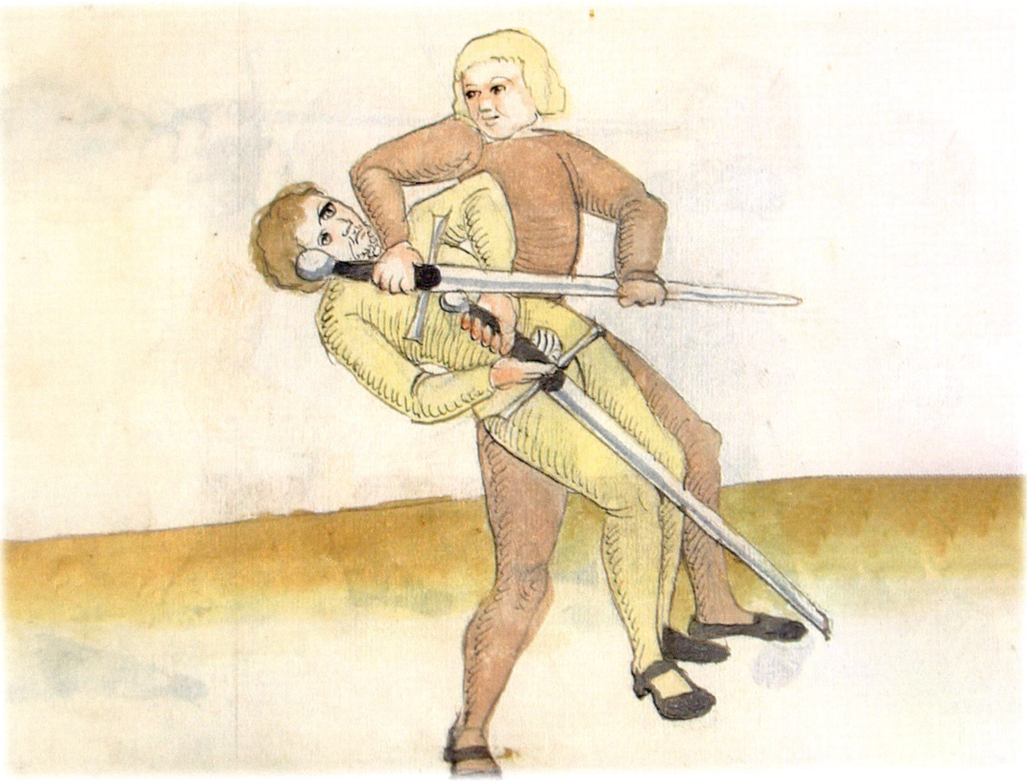

62 On the pursuit (*nachreisen*):

Text 75 Learn to pursue, then learn it twice;
Or into their defenses slice.

Gloss Remember: this means that there are two pursuits and you should learn them well.

When they want to cut in above you, drive the first like this: while they pull their sword upward to strike, pursue them with a cut or thrust and land in toward their upper exposures (before they come down with the cut). Or fall into their arm from above with your long edge, and with that, press them away from you.

Or, if they want to stab you, remember: while they pull the sword toward themselves for the thrust, pursue and thrust them (before they bring their thrust).[18]

Picture *PKB: 20^v^.*

[18] This paragraph is grouped with the second pursuit in the original text, but I suspect this is a scribal error so I'm moving it here.

63 Again, a pursuit:

When they cut toward you from above, let their sword go down toward the ground with the cut, and pursue them with a cut in above toward their head (before they come up with the sword). Thus they are ſtruck.

Picture *P: 34ʳ (© RMN-Grand Palais/Art Resource, NY).*

64 On the outer take (*Ausser nym*):

Text

76 When they're outside, there take them on
In two forms; start what work you want.
77 Then gauge each threat with hand and blade:
If pushing hard or softly laid;

Gloss

Remember: the two outer forms are two pursuits upon the sword.

Drive them like this: when they mis-cut in front of you and you pursue them, if they parry you, then keep your sword on theirs and gauge whether they're Hard or Soft in the threat.

If they rise up strongly with their sword, then extend your sword out over theirs and thrust in toward their lower exposure.

65 The second outer take:

When you fence cautiously against them by cutting underneath (or otherwise threatening underneath), if they lay over you and turn upon your sword before you come up with it, then keep your sword strong against theirs underneath.

If they work toward your upper exposure, follow after it with your sword and take the Weakness of their sword with your long edge, and press down and thrust toward their face.

66 On feeling (*fulen*) and the word "Within" (*Indes*):

Text

78 For this you must learn how to feel.
The word "within" cuts deep as steel.

Gloss

This means that you should learn and understand the feeling and the word "Within" well, since these two things belong together and are the greatest art in fencing.

Learn it like this: when one of you binds the other one on the sword, then meanwhile, as the swords flash together, you should immediately feel (that is, sense) whether they've bound you Soft or Hard. As soon as you've sensed that, think about the word "Within"; that is, that in this same sensing you should work nimbly toward the nearest exposure with your sword after the Hard and Soft so that they'll be struck before they know it.

Likewise, remember to think of the word "Within" in all binds of the sword, because:

A1 "Within" above redoubles blows,
And transmutes strikes to go below.
"Within" runs through, their space invades,
And changes through from off their blades.
A5 "Within" takes slices to suppress,
And wrestles equally no less.
"Within" takes swords when it requires,
And grants all that your heart desires.[19]

"Within" is a sharp word, and all fencers who don't know the word will be cut down with it. The word "Within" is also the key that will unlock all the art of fencing.

[19] These eight lines are rhyming couplets in German and appear at this point in every gloss of the extended sword, but they aren't part of Liechtenauer's *Record*, so they may be a remnant of another teaching or a separate mnemonic.

67 Another pursuit:

Text 79 Pursue them twice, and if you hit,
Then make the same old slice with it.

Gloss Remember: this is when they mis-cut in front of you, so pursue them with a cut toward the upper exposure. If they go up and bind under you on the sword, then remember: as soon as one sword flashes on the other, fall in over their arms with the long edge (from the sword) and press them away from you like this (as illustrated here). Or slice in through their muzzle from the sword.

Drive this to both sides.

Picture *PKM: 70ʳ.*

68 This is the text and the gloss on overrunning (*uberlauffen*):

Text

80 If down below your sword they aim,
Flow over them and bring them shame.
81 When steel on steel above should spark,
Stay strong and I will praise your art.
82 Work onward then with skill and ardor,
Or press them hard and press them harder.

Gloss Remember, this is when they begin to fence and aim at your lower exposure with a cut or thrust: don't parry it, but see that you run over them with a cut in above toward their head, or set upon them above with your point (as illustrated here). Since all cuts and settings upon from above overreach those from below, you'll shame them in this way.

Picture *P: 34ᵛ (© RMN-Grand Palais/Art Resource, NY).*

69 **This is the text and the gloss, how one should set aside (*absetzen*) cut and thrust:**

Text

83 Learn well to lure them out at will,
Then hinder cuts and thrusts with skill.
84 Whoever tries to stab at you,
Your point meets theirs and breaks on through.
85 From either side, both left and right,
Your swords will meet if forth you stride.

Gloss

This means that you should learn to set aside cuts and thrusts with art like this, so that your point hits them and theirs will be countered.

Learn it like this: when someone stands against you and holds their sword like they want to stab you underneath, stand against them in the plow guard on your right side and offer an exposure with your left. If they thrust toward that same exposure underneath, then turn against their thrust with the sword on your left side and step toward them with your right foot so that your point hits and theirs misses (as illustrated below).

Picture *PKB:* 21^{r}.

70 **Again, a piece of setting aside:**

When you stand against them in the plow guard on your left side, if they cut toward your upper left exposure, then go up with your sword, and turn against their cut on your left side with the hilt in front of your head, and step toward them with your right foot and thrust in toward their face (as illustrated below).

Picture *P: 13ʳ (© RMN-Grand Palais/Art Resource, NY).*

71 This is the text and the gloss on changing through (*durchwechseln*):

Text

86 Learn to change through, your sword untied,
Then thrust sharply from either side.
87 Whomever tries to bind on you
You'll quickly find by changing through.

Gloss

This means that you should learn the change through well.

Drive it like this: when you begin to fence and cut in or thrust toward them, if they want to bind on your sword (with a cut or parry), then let your point sneak through under their sword and thrust in sorely toward their other side with it. In this way, you find their exposure quickly (as illustrated).

If they become aware of this thrust and go after it with the parry, change through toward the other side again.

Another: as you approach them, set your left foot forward and hold your point extended against their face. If they cut toward your sword (above or below) and want to knock it away, let your point sink underneath and thrust toward the other exposure of their other side, and do this against all cuts.

Picture *P: 35^{v} (© RMN-Grand Palais/Art Resource, NY).*

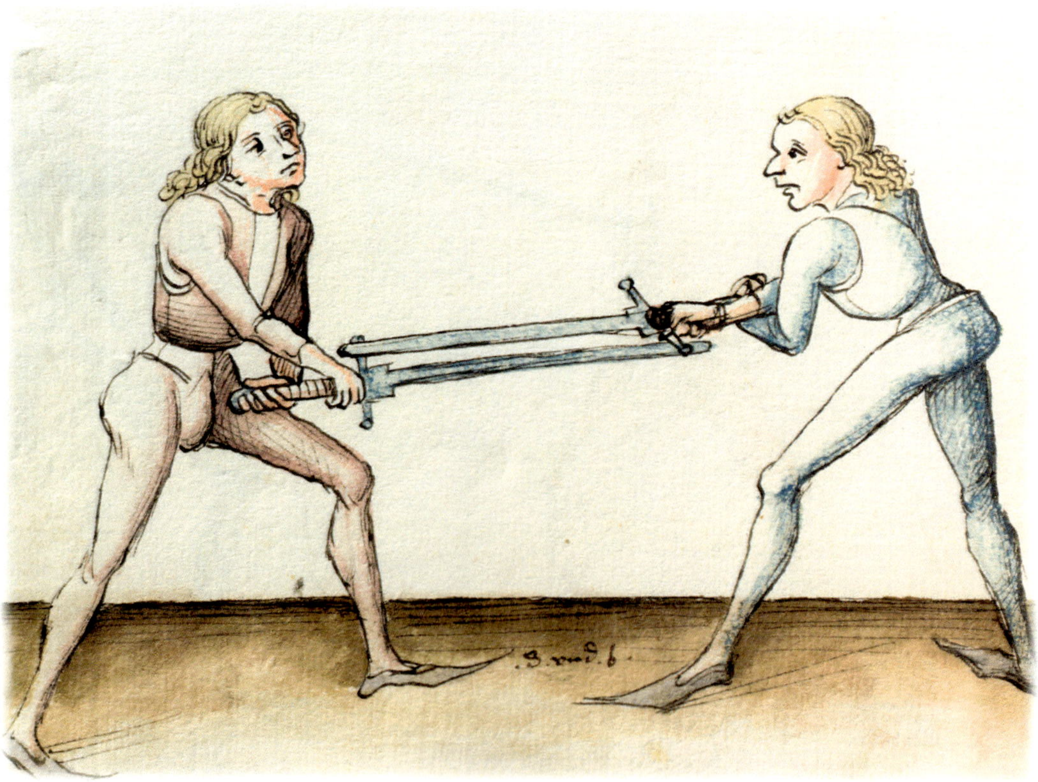

72 This is the text and the gloss on the pulling (*zucken*):

Text

88 Now step in close, engage the bind,
Then pull, and what you seek you'll find.
89 Pull back; if sword you meet, pull more;
Devise a work that hurts them sore.
90 Pull back whenever steel meets steel
And masters you'll confound with zeal.

Gloss

This is when you approach to fence with them: cut strongly from above your right shoulder in toward their head. If they bind on your sword (by parrying or otherwise), then step close toward them in the bind and pull your sword off from theirs above, and cut in above toward the other side of their head (as illustrated below). If they parry you on the other side, then strike in above toward the other side again and work nimbly after the upper exposures however you can (with the redoubling and with other pieces).

Or act like you want to pull, but stay on their sword and quickly thrust in again on their sword toward their face. If you don't correctly hit with the thrust, then work with the redoubling (or with other pieces).

Picture *PFW: 3*^{r}.

73 This is the text and the gloss on running through (*durchlauffen*):

Text

91 Hold pommel high, let point hang down,
Run through and wrestling abounds.
92 When strength would press and oppress you,
Remember this: just run on through.

Gloss

Remember, this is when one of you runs in on the other: if they go up with their arms and want to overwhelm you above with strength, then go up with your arms as well and hold your sword by the pommel with the left hand (over your head) and let your blade hang down behind you (over your back), and run through them with your head under their right arm. Leap behind their right foot with your right, and with this leap, go in with your right arm far around in front of their body, and in this way, hold them on your right hip and throw them in front of you (as illustrated here).

Picture *PFW: 17r.*

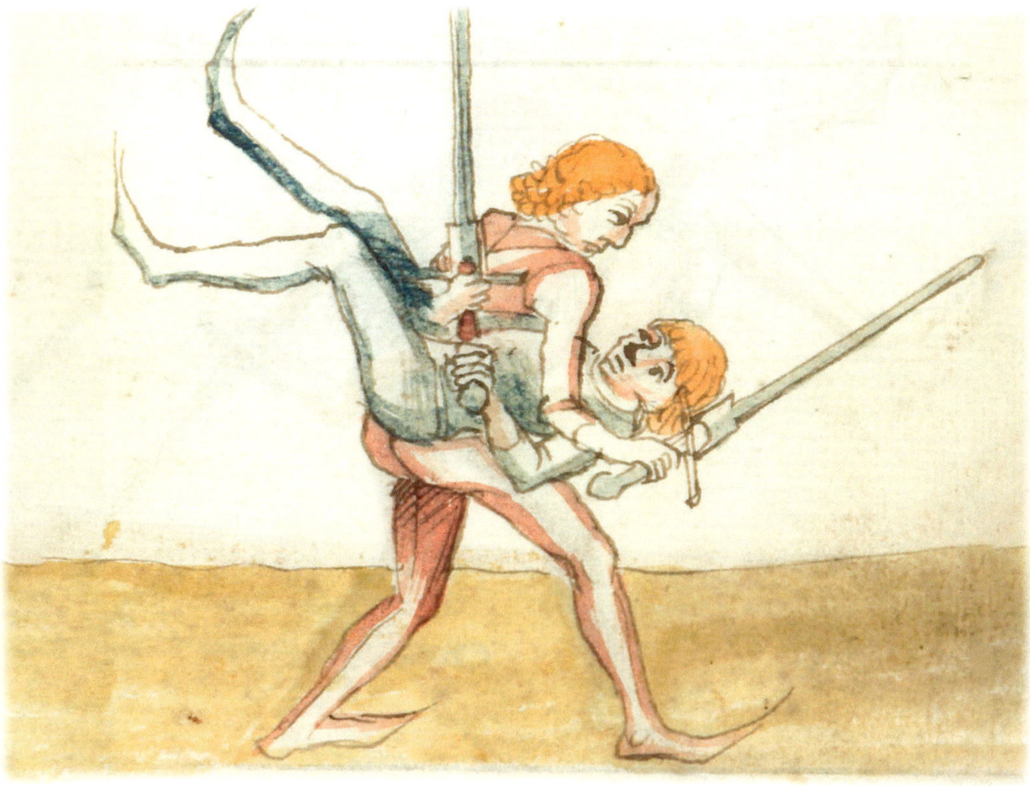

74 Again, a run through:

Remember, when they also go up with their arms in the running in with the sword, wanting to overwhelm you above with ſtrength: hold your sword by the pommel with your left hand and let the blade hang down behind you (over your back), and run through with your head under their right arm. Keep your right foot in front of their right, and go with your right arm far around behind their body and hold them on your right hip, and throw them behind you (as illuſtrated here).

Picture *GF: 17ʳ.*

75 **A wreſtle at the sword:**

When one of you runs in on the other: release your sword from your left hand and hold it in your right, and push their sword away from you on your right side with your hilt, and leap in front of their right foot with your left, and go in with your left arm far around behind their body and hold them on your left hip, and throw them in front of you (as illuſtrated here).

Picture *PFW: 6^{v}.*

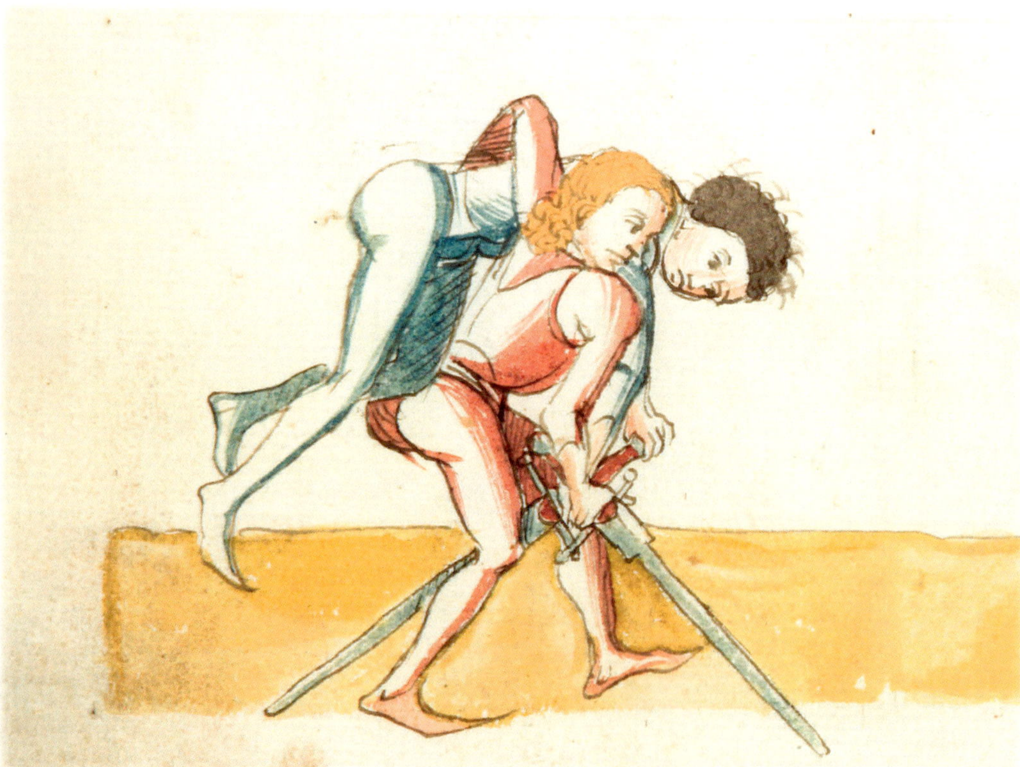

76 Again, a wreſtle at the sword:

When one of you runs in on the other: release your sword from your left hand and hold it in your right, and push their sword away from you on your right side with the hilt, and leap with the left foot behind their right, and go in with the left arm far around their body in front under their breaſt, and throw them backward over your foot.

Picture *P: 39ʳ (© RMN-Grand Palais/Art Resource, NY).*

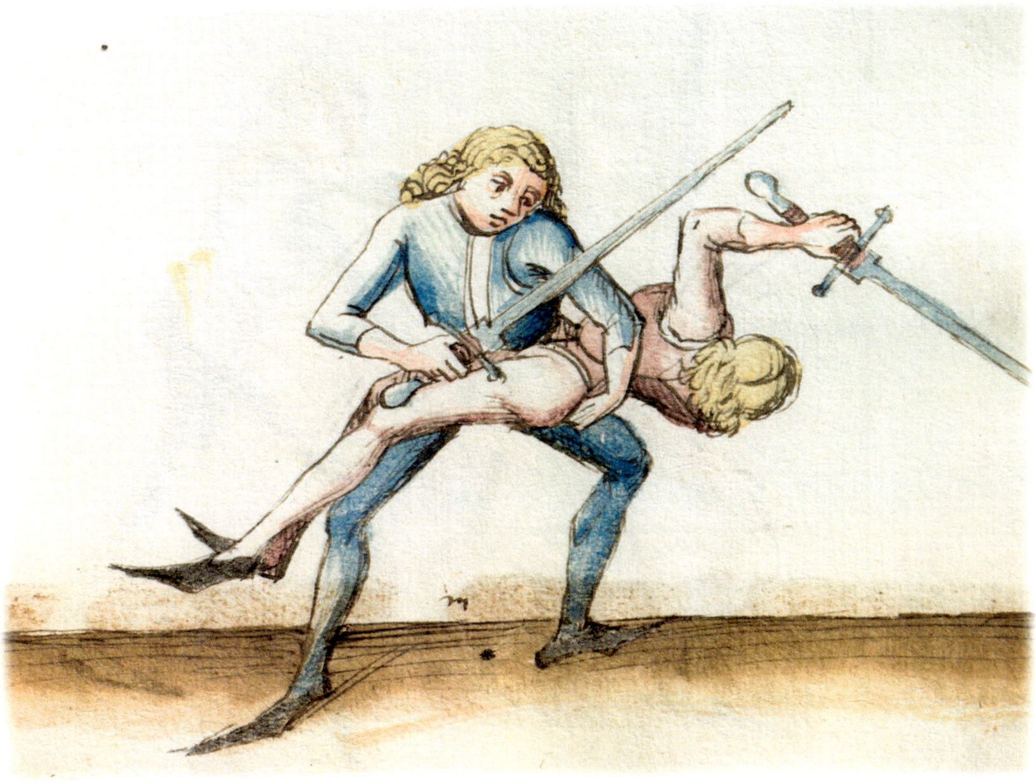

77 Again, a wrestle at the sword:

When you run in on someone: release your sword from your left hand and hold it in your right, and go in with your pommel outside over their right arm and pull downward with it, and grab their right elbow with your left hand, and leap in front of their right foot with your left, and in this way, pull them over your foot on your right side (as illustrated here).

Picture *PKM: 67^r.*

78 Again, a wrestle at the sword:

When one of you runs in on the other: go over their right arm with your inverted left hand and grab your right arm with it, and press their left arm over your left with your right arm, and leap with your right foot behind their right, and turn away from them on your left side, and throw them over your right hip (as illustrated here).[20]

Picture *PFW: 13r.*

79 Another wrestle:

When someone runs in on you at the sword, etc.: let your sword fall, invert your right hand and grab their right outside with it, and with your left, hold them by the right elbow. Leap in front of their right foot with your left and push their right arm over your left with your right hand, and rise upward with it. In this way, they're locked and you can break their arm like this or throw them in front of you over your leg.

[20] This piece shows your/their and left/right confusion between SR and GF, so check the reading apparatus in Chapter 5 for differences.

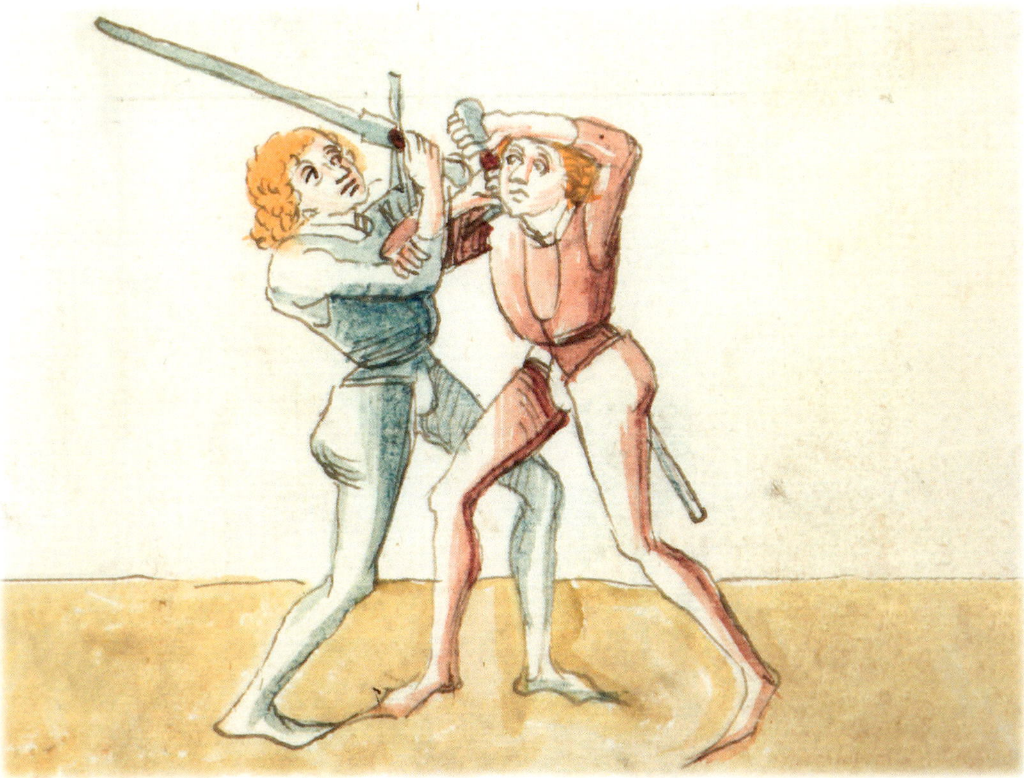

80 A disarm (*schwertnemen*):

Remember, when one of you runs in on the other: invert your left hand and go over their right arm with it, and grab their sword by the handle between both their hands and pull it on your left side. In this way, you take their sword (as illustrated here).

Picture *PFW: 15*^{v}.

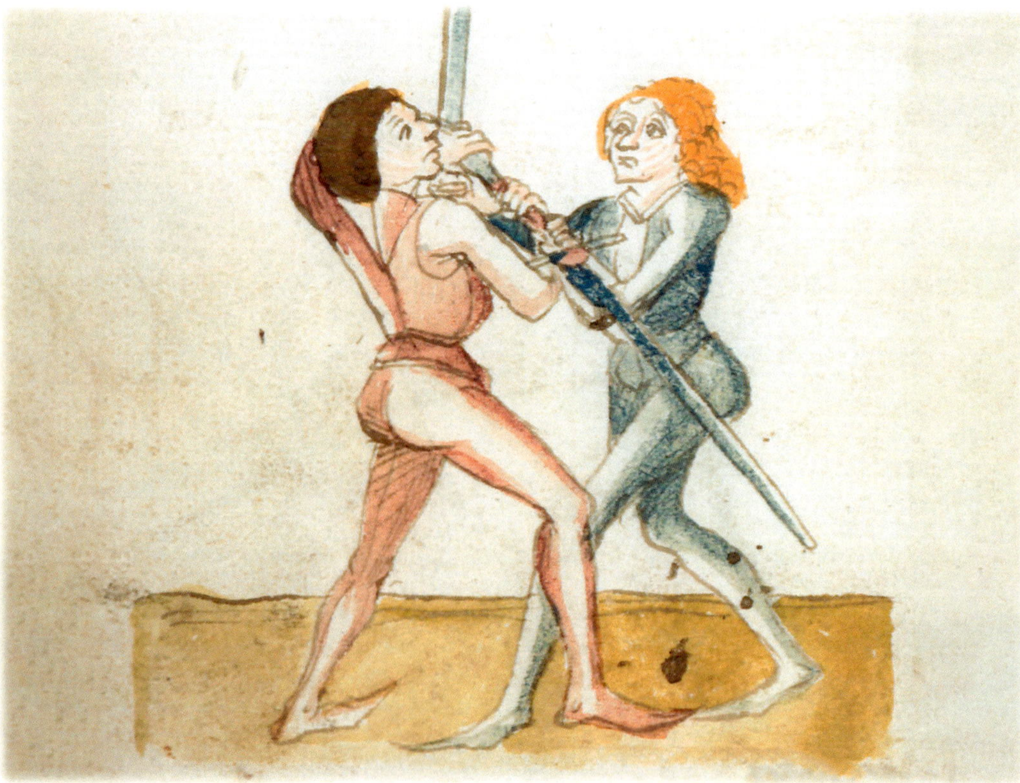

81 Again, a disarm:

Remember, when they bind onto your sword (by parrying or otherwise): grab both swords by the middle of the blades with your inverted left hand, and hold [the blades] fast together, and go through underneath them with your pommel, with the right hand against your left side, over both their hands, and pull it upward over you on your right side. In this way, you end up with both swords (as illustrated here).

Picture *P: 42ʳ (© RMN-Grand Palais/Art Resource, NY).*

82 This is the text and gloss on slicing off (*abschneiden*):

Text

93 Whenever hardness blocks your plays,
Slice off from underneath both ways.
94 There are four slices you must know:
Two falling high, two rising low.

Gloss

Remember: there are four slices.

Drive the first like this: when they run in on you and go up high with their arms and want to overwhelm you above with strength (against your left side), turn your sword and go in underneath their hilt with crossed hands and your long edge on their arms, and press upward with the edge (as illustrated here).

Or, if they run in against your right side, then go in onto their arms with your short edge and press upward as before.

Picture *P: 43^{r} (© RMN-Grand Palais/Art Resource, NY).*

83 **Again, a slice:**

When you bind strongly on their sword (with a cut or otherwise), if they let their sword snap away from yours and strike above at your head, turn your sword with the hilt in front of your head and slice through underneath their arm, and with this slice, set your point on their chest underneath.

84 **This is the upper slice:**

Drive the slice like this: when someone binds against the sword on your left side and they then strike around from the sword toward your right side (crosswise or otherwise), leap away from the cut on their right side with your left foot and fall in above over both arms with your long edge, and press them away from you (as illustrated here).

Drive this to both sides.

Picture *PKB: 20ᵛ.*

85 This is the text and gloss on the transformation of the slice:

Text

Now turn your slice with great finesse
And crush their hands as in a press.

Gloss

Remember, this is when you run in and, during it, you come under their arm with the edge (so your point goes out against their right side): press firmly upward with your edge, and with the pressing, leap on their right side with your left foot and turn your sword with the long edge above over their arms (so your point goes out against their left side), and press their arm from you with it. In this way, you've transformed the lower slice into the upper.

Drive this to both sides.

86 This is the text and the gloss on two angles (*hengen*):

Text

96 Two hangers from the ground now guide,
Out from each hand, both left and right:
97 In every threat once in the fray,
Cut, thrust, then hard or soft you lay.

Gloss

Remember: there are two angles from the ground, from each hand and from each side.

Drive them like this: when you bind on their sword on your left side with the lower setting aside, angle your pommel against the ground and thrust upward out of the angle (toward their face from below). If they push your point upward with a parry, then stay on their sword like this and go up along with him, and angle your point down from above toward their face.

You should nimbly drive all threats in the two angles like this, and cut, thrust, and slice afterward as you gauge (or sense) whether they are Hard or Soft in the bind of swords.

The angle from the ground is the plow on both sides.

87 On the spreading window (*Sprechfenster*):

Text

98 Spread windows wide that speech may flow,
And hear their case with joyful glow,
99 But snap the windows shut upon
Whomever tries to cut and run.
100 Now let me make this plain and clear:
No one defends without a fear.
101 And if these truths you'll learn and know,
Then scarcely can they come to blows.

Gloss Remember: this is called 'the spreading window'.

Make it like this: when they bind onto your sword (with a cut or a parry), keep your long edge strongly on their sword, arms outstretched, with your point in front of their face, and stand cheerfully and examine their intentions (whatever they want to drive against you).

If they strike around from the sword with a cut over your other side, then bind in against their cut with strength, in above toward their head with your long edge.

Or, if they strike around from the sword crosswise, then fall in over their arms with the edge.

Or, if they pull their sword back on themselves and want to thrust at you below, then pursue them on their sword with your point and set upon them above.

If they don't want to pull away from the sword or strike around, then work on their sword with the redoubling, and otherwise with other pieces afterward (as you sense the Hard and the Soft on the sword, etc.)

Picture *PKM: 70v (facing page).*

88 Hereafter, remember that which is called 'the extended point' (*langen Ort*):

Remember: when you approach to fence with them, set your left foot forward with outstretched arms before they bind on your sword, and hold your point extended against their face or chest.

If they cut down from above toward your head, then turn against their cut with your sword and thrust toward their face.

Or, if they cut toward your sword (down from above or up from below) and want to strike your point way, then change through and thrust in toward the exposure on the other side.

Or, if they hit your sword with strength in the cut, then let it snap around so that you hit toward their head.

Or, if they run in on you, then drive a slice or await the wrestling.

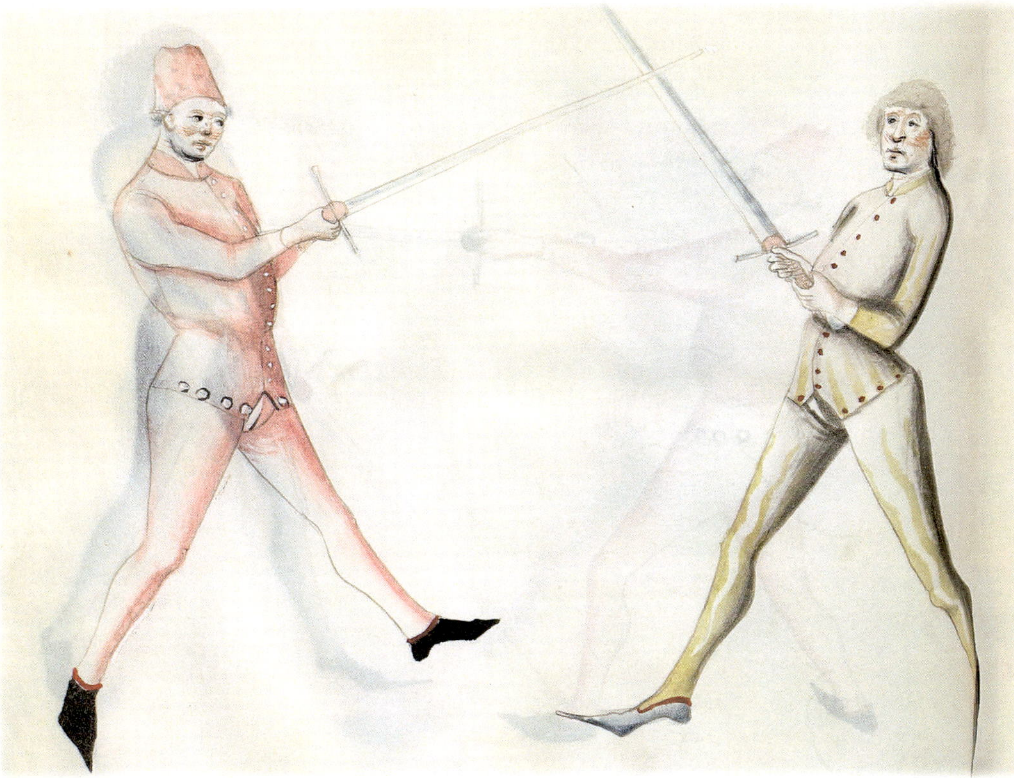

89 This is the text of the conclusion of the *Record*:

Text

102 If you lead well and break through right,
To this end you may guide the fight,
103 And breaking in with flashing steel,
Three wonders of the sword reveal.
104 So hang your point in straight and true,
And wind your sword to follow through.
105 Now eight winds note with thoughtful mind,
And weigh the paths that each may find:
106 In each and every wind of sword,
Three wonders wait to be explored.
107 They thus expand to twenty-four—
Count one by one, you won't need more.
108 From either side, both left and right,
Learn these eight winds when forth you'd stride.
109 Then gauge each threat with hand and blade:
But pushing hard or softly laid.

Gloss

This is a lesson in which the art of the *Record* is artfully grasped in short.

Learn it like this: practice and take account of the art very well, and in this way, nimbly lead all threats and pieces against whoever you fence with, so you know to correctly drive your counters against their pieces and you can work individually with three wonders from each counter.

Likewise, also angle on the sword well and bring eight turns from these angles, and consider and correctly weigh the turns like this so that you know which of the same three to drive.

90 Here remember how to drive the angles and the turns (*winden*):

Learn it like this: there are four binds of the sword, two over and two underneath. You should drive two particular turns well from each bind of the sword.

Do it like this: when you approach to fence with them, if they bind above against your left side, then turn your short edge on their sword and go up high with your arms, and angle your point in above and thrust toward their face. If they parry the thrust with strength, then angle your point in above on their sword and turn on your right side and thrust. These are two turns on one side of the sword.

Or, if they bind above against your right side, then turn against your right side as well (with your long edge on their sword) and go up high with your arms, and angle your point in above and thrust toward their face. If they parry your thrust with strength, then angle your point in above on their sword and turn on your left side and thrust. These are four turns from two binds (over the left and right sides).

91 You should now know to also drive four turns with all threats from the two binds underneath (as from the over); in this way, the turns from above and below will be eight. Recall that you should drive a cut, a thrust, a slice from each particular turn, and these are called 'the three wonders'.

92 From the eight turns, one can drive twenty-four things, and you should learn to drive the eight turns from both sides. Step with each particular turn in this way, and gauge their threat with it—nothing more than whether they're Hard or Soft on the sword—and when you've sensed these two things, drive the piece that correctly belongs to the turn. If you don't do this, you'll be struck by all turns.[21]

[21] GF ends with a paragraph (93) that's similar to the first paragraph in HM, JMR, and SR, but much shorter.

1. Hunt from your chest toward their right hand.
2. Turn around with your horse; pull their right hand with your left.
3. With scraping, take the saddle-strap or weapon.
4. Set upon high, swing, move through, or counter the sword.
5. The cover goes before all meetings, cutting, thrusting.
6. Grab onto the strong with both hands.
7. Here someone begins to seek the belt pouch cut.
8. Turn the right hand to them, set your point toward their face.
9. Defend against the thrust, turn, catch their right hand in your left.
10. Press firmly, push from the reins and seek their knife.
11. Seek the exposure: arms, leather, gloves, under the eyes.
12. Learn to sweep against all weapons with empty hands.
13. The bucket hold defends against all grips, wrestle under their nose.
14. Turn around again at length, thus the horse steps back.
15. Up close, catch the hand, reverse their face toward the nape.
16. Catch the weapon at length when riding against.
17. Hunt toward the right hand with your art.
18. If you hunt left, fall upon the pommel, push in under their nose.
19. Set the point against the face.
20. Cover against all meetings that ever happen.
21. Arrange yourself in the strength in the beginning.
22. Now this is the spear, run to the other meetings under their nose.
23. The nameless hold takes the weapon or fells them.
24. If they hunt you on both sides, turn left so they come on the right.
25. Think about taking and holding the knife.
26. Grab over whoever falls on you or do it riding against them.

1 **Here begins the chivalric art of fencing on horseback:**

Master Johannes Liechtenauer's fencing on horseback begins here, which he has allowed to be written with obscure and cryptic words, and it's interpreted and glossed here in this book so that any fencer can learn it well (if they otherwise can or know how to fence).

2 **[A common lesson:]**

Text

1 Direct your lance with honor plain;
Against you all will ride in vain.

Gloss

Remember, this is when you have a lance and someone else has one too, and they want to ride together with you: you should know to arrange yourself with your lance like this so that you divert theirs with it, and you land in with the thrust and they don't. And also know how to drive the pieces out of two guards with the lance which will be named later on, etc.

Figure

The 21st figure speaks about this: "Arrange yourself in the strength...", etc.

3 **The pieces out of the first guard:**

Text

2 If from the path you'd fall away,
The end will see great disarray.

Gloss

Remember, this is the piece out of the first guard: when you ride together with them, hold your lance under your arm at the thrust and, when they approach the meeting, act like it's too heavy for you and allow it to sink down in front with the point on your left side.

If they ride upon you with a thrust, then raise your lance upward onto theirs with strength so that you hit them and they don't hit you, since their lance goes off by the side of your own, etc.

Figure

The 1st figure speaks about this: "Hunt from your chest...", etc.

4 **This is a piece out of the second guard:**

Remember: when you ride together with them, hold your lance crosswise in front of you on the saddle bow with both hands in the middle.

If they ride upon you with a thrust, then strike their lance away from you on your right side with the front part of yours, and with that, turn your lance under your right arm, and in this way, you hit them and they don't hit you.

Figure The 17th figure speaks about this: "Hunt toward the right...", etc.

5 [A lesson on wrestling:]

Text
3 Strike here and there as you close in;
Draw not your blade from scabbard then,
4 But pull them left and seize them right—
You need not fence to win the fight.

Gloss This is if you both miss with the lance in the riding in: immediately let yours fall and don't draw sword or knife, and ride in on them and turn yourself with your left side on their right and drive the wrestling written later on.

When you approach with your left side on their right, if they grab you with their right hand in front and want to wrestle, then grab their right arm with your right (in front next to their hand), and drive the nameless hold or the obscure hold.

Or, if they've drawn their sword and go up with it and want to strike, then grab their right elbow with your left hand and push them from you with it, and lift their right foot with your left foot; in this way, they fall.

Or, once you've grabbed their right elbow with your left hand, grab their pommel with your right and pull it toward you; in this way, you take their sword.

6 Here remember another:

That is, if you can't approach with your left side on their right, then hold yourself with your right on their right. If they grab onto you with wrestling, then drive the bucket hold or the sun show.

When you approach with your right side on their right, if they have their sword drawn and have gone up with it and want to strike, then go down from above with your right arm, outside over their right arm, and press their arm into your right side and ride forward; in this way, you take their sword.

Or go up with your right arm from inside, over their right, and press their arm in front on your chest and ride forward; in this way, you take their sword again.

Grab their right hand with your left and pull it in front of your chest, and turn your horse from them; in this way, they fall.

Or grab their right elbow with your right hand and lift their right foot with your right foot; in this way, they must fall.

Figure The 2nd figure speaks about this: "Turn around with your horse...", etc.

7 [A lesson:]

Text

Learn this to joust or fence with lance:
Break calmly through in your advance.

Gloss When someone rides upon you with a lance, you should ride against them quite calmly without running with your horse. In this way, you can use the pieces written previously to undo all of their riding at you, whereas when you run, you can't come to the art or to your pieces and you are uncertain on your horse.

Figure The 22nd figure speaks about this: "Now this is the spear, run...", etc.

8 The belt pouch cut (*Taschenhaw*):

Text

6 If sudden shift the fight then shows
And swords begin to trade their blows,
7 Then tie the strong up well and good,
And mark them as a cutpurse would.

Gloss Remember, this is if you've both abandoned the lance and will fence with sword: lay your sword on your left arm in the [first] guard and ride straight at their right side (under their nose).

If they cut over you, then go up with your sword and parry their cut strongly with your long edge and thrust toward their face. If they parry your thrust and go up high, then cut in at their left hand or their reins with your long edge. And if their horse will press you, then cut at their right leg while running away.

Figure The 7th figure speaks about this: "Here someone begins...", etc.

9 [How to constrain an opponent:]

Text

8 Now learn to cover, strength displayed,
And then whenever blade meets blade,
9 Fear not, but set upon your foe:
Hang in at them from head to toe.

Gloss Remember: this means to bind boldly with your sword (be it with cutting or thrusting), don't pull yourself away from their sword, and use your point to constrain them into the pieces described later on.

Figure The 20th figure speaks about this: "Cover against all meetings...", etc.

See if you can set upon them with your sword. If they parry on their left side and ride upon you with it, then go through from underneath their sword with your pommel, over their left shoulder, around their neck, and come behind the pommel with your left hand to help, and pull them to you on the side.

Figure The 6th figure speaks about this: "Grab onto the strong...", etc.

When they parry your cut, go under their jaw with your hilt and grab them by the helmet with your left hand, and pull toward yourself with it, and push them from you with your hilt so they fall.

If they parry your cut or thrust and run toward you, then grab their right hand with your left and set your point in their face with your right.

Figure The 8th figure speaks about this: "Turn the right hand to them...", etc.

If you cut over them and they parry crosswise, then go out high with your right hand and angle your point in over their sword (toward their face or chest) and set upon them.

10 [A thrust and a turn:]

Text
10 Or if you'd harry and distress,
Then hunt them long and grant no rest.
11 Whoever to defense will rise,
Your wind of sword will hurt their eyes.
12 If they will yet defend at bay,
Catch well the reins and don't delay.

Gloss Hold your sword in the [third] guard near your right leg and ride in like this, and thrust extended toward their face, arms outstretched. If they parry your thrust, then go up with your right hand and turn on their sword, and keep your point in front of their face.

Figure The 4th figure speaks about this: "Set upon high...", etc.

If they parry with the sword and run in on you, then go inward with your hand (in over their right arm) and grab your reins under their arm with your left hand inverted. In this way, you engulf their hand with your reins.

Figure The 10th figure speaks about this: "Press firmly...", etc.

11 Here remember the setting aside with the sword on horseback:

Remember, when you ride on someone and you have your sword in a guard: then notice which side they cut toward you on. If they cut toward your left side from above, then turn your sword against their cut (also on your left side). Or, if they cut at you on your right side, then turn your sword on your right as well (such that your thumb always comes underneath). And with the parry, always set your point in their face, and also drive this against the lance.

Figure The 19th figure speaks about this: "Set the point against...", etc.

When they cut in at you or parry: if they stay near you in this way, then go out over their right hand with your pommel, and push your hand with the hilt in front of you (toward your saddlebow), and grab their pommel with your left hand and ride forward. In this way, you take their sword.

12 [The exposures:]

Text

13 Consider where they are exposed;
Your knife, not pommel, finds them most.

Gloss

This is when you want to fence with someone in armor: above all else, you should know which places you can best win in.

That is, seek all these exposures with thrusting (not cutting): in the face, or under the armpit, or outside on the hand in the glove, or inside in the palm of the hand, and in all joints of the armor at the arms and the legs.

If you can't work with the sword when you approach close to them then work with the dagger, and if you can't get to your dagger, then see if you can take theirs and work toward the exposure with it.

Figure

The 11th figure speaks about this: "Seek the exposure: arms, leather...", etc.

13 [A lesson with the empty hand:]

Text

14 Two sweeps now learn so you may stand
Against all arms with empty hands.

Gloss

Above all else, learn and know how to work with your free hand on horseback, and most importantly with wrestling. Therefore, you should set your reins in such a way that you can change from one hand to the other and seek your advantage with it in this way, and this is the greatest art on horseback.

Figure

The 12th figure speaks about this: "Learn to sweep against...", etc.

14 The bucket hold (*Schaff Griff*):

Text

15 And when to wrestling they would hasten,
Learn to hold them like a basin.
16 When underneath their very nose,
Correctly grab as you fly close.

Gloss

This is the best wrestling for anyone on horseback.

Drive it like this: when you want to wrestle, ride equally on your opponent's right side (under their nose), and grab onto them with wrestling. If they fall on you in turn, then grab their right arm with your inverted left hand (in front next to their hand) and pull it under your chest, and go over it with your right arm and grab the saddlebow with it, and ride forward so that they must fall.

Figure

The 13th figure speaks about this: "The bucket hold defends..."

15 The wrestling:

Text
17 Whoever falls on you with might
When riding 'gainst you: hold them tight,
18 And let yourself hang toward the earth
While gripping over—show your worth.

Gloss This is when you ride together with someone.

If they approach your right side with theirs and fall into your neck in front with the right arm, then go around theirs as well like this with yours, and go around behind them with your left hand and come in to help with your right, and pull them toward you on the side. Or strike your right arm above over their right and throw them with the bucket hold.

16 Another:

When you ride together with someone, if they approach your right side with their left and fall around your neck from behind with their left hand, then go up strongly behind you with your right arm over their left and come to help your right hand with your left, and press their left arm firmly behind the back of your neck. If they swerve with their arm, then grab their left hand with your left and drive the nameless hold or the obscure hold.

Or, if they grab around behind you with their left hand and want to wrestle, then strike strongly down from above with your right arm, outside, into their left arm joint, and ride forward.

Figure The 26th figure speaks about this: "Grab over...", etc.

17 A lesson:

Text
19 On either side, both left and right,
Against them learn to ride and fight.

Gloss This means that whichever side you approach someone on, you should hold yourself close to them and drive the art like this: if you approach them on your right side then nimbly drive the pieces that belong to the right side, and if you approach them on your left side then drive what belongs to the left side. In this way, they can't come to their pieces.

18 The left side:

Text

20 If you should choose to charge and ride
Your horse onto the other side,
21 First cover strongly and suppress,
Then set upon and cause distress.
22 When you defend, their sword catch wide
And carry near, their handle tied.

Gloss This is if you want to ride on someone's left side: you should also bind strongly on them with your sword (with cutting or thrusting), and always aim at their exposures with your point just as you've done on the right side. With this, you constrain them to the pieces that belong to the left side, since there are some pieces that you drive on the left side which you can't drive on the right.

Figure The 5th figure speaks about this: "The cover goes before all…", etc.

This is when you ride on their left side: cut a strong cut in over them. If they also cut in strongly and want to set upon you, then you should parry them with the sword while they are far from you, or grab their right hand with your left if they come close to you.

Figure The 16th figure speaks about this: "Catch the weapon…", etc.

19 A lesson:

Text

23 Or wheel about to hunt again,
Prepared to harry and restrain;
24 With every skill that hunters learn,
Seek your advantage as you turn.

Gloss This is if your horse carries you forward away from them so that you can't drive any piece on them. Turn yourself toward the side of them where you can drive the best advantage.

Figure The 14th figure speaks about this: "Turn around again at length…", etc.

20 [When you unintentionally ride to the left side:]

Text

25 Should you ride forth without relent
But end up left without intent,
26 Then worry not, but their sword find,
And wrestle well: push hard and bind.

Gloss

This is if your horse carries you away (or however it happens that you unintentionally ride to their left side): lay your sword on your left arm. If they cut toward your head, go up with your sword and parry with the long edge. If you come close to them with your parry, then go over their right hand with your left arm and press it firmly onto your left side and bash them in their face with the pommel.

Figure

The 18th figure speaks about this: "If you hunt left, fall upon...", etc.

21 Note:

If you press their right arm into [your] left side and ride away from them, you take their sword. You can also catch their hand by going over it with the reins if you want.

When they cut over you and you parry (as is written previously), then cut in freely over above toward their head.

When you have your sword on your left arm in the [first] guard, if someone rides at your left side with a lance, go well up with your pommel and let your blade angle on your left side and set aside their lance with it like this, and cut toward their head or set upon them.

Or, if they ride at your right side with the lance, then sweep straight up onto their lance with your sword and turn into the upper angle and set upon them.

22 [The setting upon behind:]

Text

27 If someone hunts you on the right,
Turn halfway 'round, prepare to fight.
28 With arms to catch and drive away,
No harm can reach you in the fray

Gloss

This is if you're pressed to flee and you're armored and have nothing but a sword, and someone sets upon you with the lance (behind your right side): turn yourself away from the thrust against them (on your left side), and turn yourself with the sword against their lance, and set upon them.

Or, if they set upon you behind your left side, then turn yourself against them on your right and turn with your sword as before, and set upon them.

Figure The 24th figure speaks about this: "If they hunt you on both sides...", etc.

In the flight, also know how to nimbly turn yourself around in the saddle from one side to the other, and stab behind you, and set aside sword and lance to both sides.

23 Remember a nimbleness with the lance:

When you are hunted and have a lance, if the one who hunts after you also has one, hold your lance on your right shoulder with your right hand, and when you are sure that they are close behind you, lift your lance over your head onto your left shoulder and keep your point behind you, and turn yourself against them on your left side, and with that, slam your lance under your arm so you come equally with them (under their nose).

24 [The sword taking:]

Text 29 Take sword away like it's a knife
And learn the holds that cause them strife:

Gloss Here remember how to take their sword or knife: ride toward their right side and seek the exposure however you can (with cutting or thrusting). If they parry and come close to you, grab their right arm behind their right hand with your inverted left hand, pull it in front of you and hold them firmly by it, and bend your left arm outward at the hilt of their sword in such a way that they must let their sword fall. And when you hold them by the arm like this, you can strike them with the sword or throw them with the bucket hold.

Figure The 25th figure speaks about this: "Think about taking...", etc.

25 The nameless hold (*vngenatten Griff*):

Text 30 Apply the hold that has no name
To move the strong and bring them shame.
31 Spoil strikes and thrusts with this defense;
Have your revenge—no need to fence.

Gloss Remember: this is the nameless hold.

When you approach them on your left side, if they've drawn their weapon and want to strike you, or grab your front with their right hand with wrestling, then grab their right arm with your right hand (in front next to their hand) and pull it under your chest, and lie on it with your body and ride forward. In this way, you break their arm.

If you don't want to break their arm when you've pulled it under your chest, then grab their right elbow with your left hand and push them away from you with it, and grab their sword by the pommel with your right hand and pull toward yourself with it. In this way, you take their sword.

Figure The 23rd figure speaks about this: "The nameless hold…", etc.

26 The sun show (*Sunnen zeichen*):

Text

32 If you would grab them as they ride,
Don't fail to ride up alongside.
33 To show the sun and make them bow:
Take first their left arm firmly now,
34 Then grab their head in front with skill,
And press it upward hard until
35 They sink down low and show respect
(And then might never stand erect).

Gloss Remember: this is the best wrestling on horseback.

When you ride together with them, if you approach their right side on your right, then hold them close to you, reach around behind them, grab them by their left arm with your left hand and pull them firmly around toward you, and grab underneath their jaw with your right hand and push it firmly away from you, upward, against their left side. In this way, you turn their face toward the sun. You win their momentum with this so they can't stop themselves.

If you approach their right side with your left, then grab them as before and throw them behind you on your left side.

This wrestling is called 'the sun show'.

Figure The 15th figure speaks about this: "Up close, catch the hand…", etc.

27 [How to counter the sun show:]

Text

36 If taking you down low's their aim,
Then grab them high and bring them shame;
37 Press arm to head, their grip defeat—
This often robs them of their seat.

Gloss Remember here: this is the counter against the sun show.

If someone approaches your right side with their right and wants to throw you by showing you the sun, then remember: when they grab onto your jaw with their right hand, strike your right arm over their right and pull it firmly onto your chest, and lay on it with your body and ride forward. Thow them like this, or throw them with the bucket hold.

If they approach your right side with their left and reach around behind you and grab your left arm with their left hand, then go up behind from underneath, over their left arm, and press firmly into the back of their neck. If they yield with their arm, then grab their left hand with your left and throw them with the nameless hold.

Figure The 9th figure speaks about this: "Defend against the thrust, turn...", etc.

28 Another:

Text

38 But should you seek a measured course
To catch and hold them on their horse,
39 Then with this wrestling they'll be found,
And without rope they will be bound.

Gloss Remember: this is called 'the obscure wrestling' (*verporgen ringen*).

If you're willing to make it common and let it be clearly seen, then drive it like this: ride with your left side on their right. If they grab you in front with wrestling, then grab their right arm with your right hand (in front next to their hand) and pull it forward, and grab their right elbow with your left hand and push it upward, and push their right arm up over their left arm with your right hand, and raise their right arm upward with your left arm like this. In this way, you've caught and bound them without any bonds.

Drive this to both sides.

29 [The Before and After:]

Text

40 Remember well the grab that leads,
To break through strength and work great deeds.

Gloss This is to always advance with the grappling and wrestling you've learned well before they do, and especially with the four main wrestlings (with which you let them come to no pieces)—these are the bucket hold, the nameless hold, the sun show, and the obscure hold—and don't forget the catches with the reins. And when you can do the wrestling well, no one can throw you from the horse without injury.

1 **The earnest duel (*ernstlich Kampff*) on foot begins here:**

Master Johannes Liechtenauer's fencing in dueling armor, which he allowed to be written in obscure and cryptic words, begins here. They are interpreted and glossed here in this book so that any fencer can learn the art (if they can otherwise fence).

2 **This is the foreword:**

Text

41 When you dismount or are unhorsed,
The fight on foot must take its course.
42 Take up your spear in steady hands
And face your foe in proper stance;

Gloss Remember, understand it like this: when two people want to fence with each other in armor on foot, each one should have three weapons—a spear, a sword, and a dagger—and the first occurrence should begin with the long spear. To defend correctly, arrange yourself with it in two stances in the beginning (as you'll hear below).

3 **The two stances:**

Text

43 Its sharpened point will serve your needs,
So boldly throw the thrust that leads,
44 Then leap and wind and set upon—
If they defend, pull and you've won.

Gloss This is the first stance with the spear: when you're both dismounted from your horses, stand with your left foot forward and hold your spear ready to throw, and step toward them like this (always keeping your left foot in front). See that you throw [your spear] before they do, and quickly follow after your throw with your sword toward them so they can't throw at you safely, and attack them with your sword.

4 **The other stance in the spear:**

Remember: if you don't want to throw your spear, then hold it by your right side in the lower guard and go toward them like this, and boldly thrust toward their face from underneath before they do it to you. If they thrust simultaneously with you or parry, then go up into the upper guard with your spear in such a way that their point stays on your left arm, and with that, angle your point in over their arm in toward their face. If they then go and parry with the left arm, then pull and set your point in underneath into the exposure of their left armpit.

5 **How to pull (*zucken*):**

Text

45 If you would pull and thrust with skill,
Then learn to break through guards at will.

Gloss Remember like this: learn to pull well so that no one can set upon you while you pull through.

Learn it like this: when you thrust toward them out of the lower guard and they parry with their spear (so their point goes out by your side), pull through and thrust toward their other side.

Or, if they keep their point in front of your face with another parry, then don't pull; keep your spear on theirs and work with the turning toward the nearest exposure that may be yours.

6 **[How to counter the pulling:]**

Text

46 Now if your foe fears injury
And backs away and wants to flee,
47 Then close with them and seek your chance
To catch and hold as you advance.

Gloss Remember, this is how to pursue with the spear: when you advance with a thrust and they parry and want to pull away from your spear, follow after them with your point. If you hit with it, then crowd (*dringen*) them in front of you. Then, if they want to flee backward from your point and turn one side to you, then see that you run in on them on that same side and surely grab them with wrestling and with arm breaks (as you'll find described later on).

7 **Wrestling in the duel:**

Text

48 And should you wrestle, learn to leap:
Your foot behind their front leg sweep,
49 Or deftly lock the leg you catch
Just like the closing of a latch.

Gloss Remember, this is when you approach to wrestle with them: know how to leap in front or behind the leg which they've set forward, and this should occur with no more than one step.

Remember, do it like this: when you grab onto them with wrestling and they to you, see whichever foot they set forward. If they've set the left foot forward then strike their left hand away with your right, and with this striking away, leap behind their left foot with your right and press into their left joint with your right knee, and pull them over the same knee with both hands.

Remember another one: when you leap behind their left foot with your right, step between both their legs with your left and hold their left knee between both your knees (and hold fast with it), and push the front of their headgear with your left hand and pull them backward to one side with your right.

8 [How to wrestle on both sides:]

Text

From either hand, both left and right,
Perform your art and end the fight.

Gloss

This means that you should know how to drive all wrestling from both sides so that you'll end artfully as soon as you approach them.

Learn it like this: when you leap with your right foot behind their left, if they step backward with their left foot during your leap, then quickly follow after them toward the other side with your left foot behind their right. Throw them over your knee, or lock their knee with your legs (as described above).

9 How to arrange yourself when the spear is against the sword:

Text
51 Should it come to pass that here
The sword is drawn against the spear,
52 Observe the way they thrust, then leap
And catch and wrestle as you sweep.

Gloss Remember: this is when you've thrown your lance and they hold onto theirs.

Perceive precisely how they hold it: whether they extend or retract the point in front of their forward hand, and whether they want to thrust toward you above or below with it.

If they hold it retracted and want to be in the upper guard, then arrange yourself against them with the sword like this: grab the middle of the blade with your left hand and lay the sword in front of you in the [third] guard with the middle on your left knee. (Or hold it by your right side in the lower [second] guard.)

If they thrust toward you underneath, then set aside the thrust in front of your left hand and leap behind their left foot with your right, and go in with your pommel in front over their right shoulder, around their neck, and throw them over your knee.

Remember: if they thrust in toward you above with the lance, then go up and set aside the thrust in front of your left hand (with the sword on their left side), and leap toward them and look for the setting upon. If you can't do that, then let your sword fall and look for the wrestling.

If they thrust toward you with the lance when you are in the lower [second] guard, then set aside the thrust in front of your left hand (with the sword on their right side) and look for the setting upon or the wrestling.[22]

10 On setting aside (*absetzen*) with empty hand:

Text
53 What they extend, your left hand beats;
Leap surely in, catch what you meet.
54 Should they again fear harm, and pull
Back from your catch and your control,
55 Then you will find them all exposed,
Your point harassing unopposed.

Gloss When you stand in the lower [second] guard, if they thrust toward you above with the spear and have it held so that the point extends out of their hand,

[22] The spear against sword is the section with the most disagreement between witnesses of the armored fencing gloss. See the reading apparatus in Chapter 5 for the variations.

and they thrust toward you with it above, then strike their spear away to the side with your left hand, and quickly grab your sword in the middle of the blade again with your left hand and leap toward them, and set upon them, etc.

If they thrust underneath toward your groin with the spear, then catch their spear in your left hand and hold firmly with it, and thrust underneath toward their groin with your right. If they pull their spear toward themselves firmly and want to wrench it out of your hand, then release the spear from your hand up above them so they give an exposure. Quickly grab your sword again in the middle of the blade with your left hand and follow after them swiftly, and set upon them.

11 The exposures:

Text

56 And leather, gauntlets, and the eyes:
Toward these exposures you should rise.

Gloss

This is when you want to set upon someone in armor: you should look for the exposures well. First in the face, or under the armpits, or in the palm or rear of the glove, or in the knee, or between the legs, or in all limbs of the armor where it has joints. These are the best for defeating an opponent who faces you, and you should seek the exposures correctly so that you don't attack a farther one when you can get a closer one.

Do this with all weapons that belong to the duel, etc.

12 The forbidden wrestling (*verpoten ringen*):

Text

57 Forbidden arts of wrestling learn
And bring them forth in battle's churn;
58 Find ways to lock and take control,
Surmount the strong to reach your goal.

Gloss

Remember, this is when one of you runs in on the other: immediately let your sword fall and look for the sure wrestling that belongs to the duel. They are forbidden by all wise masters of the sword, that no one should be allowed to teach them nor show them in the *Fechtschule*,[23] because they belong to duel fencing. They are the arm breaks, leg breaks, groin strikes, death strikes, knee strikes, finger dislocations, eye grabs, and more like that.

[23] This probably doesn't refer specifically to tournaments, but rather to the public training days (overseen by a free fencer or fencing master) which shared the same label.

13 Remember the wrestling here:

If someone grabs you above by wrestling and wants to pull you toward them or push you from them with strength, then strike your right arm outside over their left (in front by their hand) and press on your chest with both hands, and leap behind their left foot with your right and throw them off their feet over your knee.

If they grab you with wrestling but don't hold you firmly, then grab their right hand with your right and pull them toward you. Grab their elbow with your left and step in front of their right foot with your left, and pull them over it like this. Or fall on their arm with your chest and break it like this.

Or grab their left arm with your left hand (in front next to their hand) and pull them toward you, and strike your right arm strongly over their left (in the elbow joint) and break their left over their right with your left hand, and leap behind their right foot with your right and throw them over it like this, etc.

When they go through under your right arm with their left (around your body), strike them strongly down from above with your right arm (outside, in their left elbow joint), and turn yourself from them with it.

When they hold you by the arms (and you them in return) and they stand strongly with the foot that they set forward outstretched, bash them on that same knee with the foot, and break their foot like this.

When it's up to you, you can also bash your knee or foot at their genitals. But you should watch out that they don't grab you by the foot, etc.

When they grab at you with open hands or with outstretched fingers, see whether you can grab one of their fingers and break it upward, and move them in the ring with it, and defeat them at the edge [of the ring] and have great gain.

14 A lesson:

Text

59 In every lesson that you learn,
Your point toward the exposure turn.

Gloss

This means that with all three weapons that belong to the duel (as mentioned above), you should always thrust with the point and do nothing else (or you'll be injured).

15 How one should fence with the sword against sword in the duel:

Text

60 When each from scabbard draws their sword
And both face off with one accord,
61 Then you should strengthen in the fight
And bear in mind to cover right.

Gloss

This is if both have thrown their spears and want to fence with swords: above all else, know and remember the four guards with the retracted sword[24] and always thrust toward the upper exposure from them.

If [your opponent] thrusts in simultaneously with you or binds you on the sword, then immediately notice whether they are Hard or Soft on the sword, and when you've sensed this, drive the pieces against them which are proper against the Hard and against the Soft (as you'll see described below).

16 Remember the first guard with the half sword here:

Hold your sword by the handle with your right hand and grab the middle of the blade with your left, and hold it overhead by your right side and let the point angle downward against your opponent's face, etc.

Remember: if they stand against you in the lower [second] guard and want to thrust toward you underneath, then thrust through down from above (between the sword and their forward hand), and press the pommel downward and turn your point in against their right side, below on the sword, and set upon them.

Thrust out of the first guard toward their face. If they defend widely, then pull it or go through with the thrust toward the other side as before, and when you've set upon them, then slam your sword under your right armpit (with the hilt on your chest) and crowd them away from you like this.

Remember another one: thrust toward them as before. If they parry in front of their left hand with the sword, and keep their point in front of your face and want to set upon you, then grab their sword by the point with your left hand and hold it fast, and thrust forcefully toward their groin with your right hand. If they pull their sword to themselves firmly and want to wrench it out of your hand, then suddenly let their sword go and thus they expose themselves. Quickly grab your sword again in the middle of the blade with the left hand and follow after them.

[24] As with the extended sword, the literal translation would be "short sword", but "retracted sword" captures the intended meaning better—the short is 'shortened' (i.e., *verkurzen*) because the hand on the blade prevents it from extending as far.

Another: if you grab their sword and they yours, then throw their sword from your left hand and grab yours again in the middle of blade with it, and turn your point in outside over their left hand and set upon them. Or throw your sword at their feet and grab their left hand with your left, and drive the arm break (or another wrestle), etc.

When you thrust toward them out of the upper [first] guard, if they fall on your sword with their left hand (in between both your hands), then go in over their left hand (inside or outside) with the pommel and wrench on your right side and set upon them.

If it's up to you, you can also strike with your pommel out of the upper [first] guard well.

17 This is the second guard with the retracted sword in the duel:

Remember: hold your sword with both hands and hold it underneath your right side with the handle by your right knee. Your left foot is in front and your point is against your opponent's face.

When you stand in the guard like this: if they stand against you in the upper [first] guard and want to set upon you above, then thrust before they do and set your point in the exposure under their left armpit, or thrust toward the exposure of the palm of their forward hand, or thrust through above their forward hand and sword, and press your pommel down and set upon them toward the other side.

When they thrust above toward you, grab their sword in front of their left hand with yours, and set the hilt of your sword on your chest with your right hand, and set upon them, etc.

18 A counter against the thrusting through:

When you thrust toward them out of the lower [second] guard, if they thrust through at you out of the upper [first] guard (between your forward hand and the sword), then remember: while they press the pommel down, go up into the upper [first] guard and set upon them.

Thrust toward them out of the lower [second] guard. If they go through underneath your sword with the pommel and want to set aside or wrench with it, then keep your point strongly in front of their face. Press their right hand downward while they turn through, and set upon them.

Thrust strongly toward their face out of the lower [second] guard. If they thrust in simultaneously with you, then grab their sword in the middle (against yours) with your inverted left hand and hold both fast together, and go through under your sword with your pommel and pull upward on your right side with your right arm. In this way, you take their sword.

When someone grabs your sword in the middle (against theirs) with their left hand and wants to wrench it away from you, then remember: while they hold your sword firmly against theirs in their left hand, go up into the upper [first] guard and set upon them.

Thrust inside toward their face out of the lower [second] guard. If they parry, then pull and thrust outside toward their face. If they parry further, then go forward over their right shoulder with your pommel, around their neck, and leap behind their left foot with your right and pull them over your leg with your pommel so they fall, etc.

Whoever goes around your neck in front with their pommel and leaps behind your left foot with their right, grab their right arm with your left hand and press it firmly to your chest, and turn yourself away from them on your right side, and hold them on your right hip and throw them in front of you.

You can also strike them with your pommel well out of the lower [second] guard (if it's up to you).

19 This is the third guard with the retracted sword in the duel:

Hold your sword with both hands (as written above) and lay it over your left knee, and counter all their pieces with parries out of it, etc.

If they thrust toward your face out of the upper [first] guard, and you set aside the thrust against their right side with your sword (in front of your left hand), then go up into the upper [first] guard with your sword and set upon them.

Or go up with your sword and parry their upper thrust between both of your hands, and go in over their forward hand with your pommel and pull downward with it, and set upon them.

Or go through below with your pommel, over their forward hand, and pull downward with it and set upon them. You can also change through below with your pommel and set aside their thrust.

You'll also find written later on how to parry any pommel strikes that an opponent strikes out of the third guard.

20 This is the fourth guard with the retracted sword in the duel:

Hold the sword with both hands (as written above) with the handle under your right armpit, and set the hilt in front of you on the right side of your chest so that the point is against your opponent. And you should come into this guard out of all three guards mentioned before.

Thus, when you thrust toward their exposure and your point is stuck in their armor, always turn your hilt onto your chest and crowd them away from you like this, and don't let them escape from your point. In this way, they can neither thrust nor cut nor strike against you.

Remember: if you want to set upon someone, set upon their face, or upon their neck, or upon their left shoulder, or under their left armpit.

When you've set upon them, if they can reach farther than you, then crowd them away from you there (so that your point goes up above and is set into the mail well).

But if they can't reach as far as you, then let your sword sink down to your right hip (with the pommel at your right side) so that your point is upward, and set into their mail as before, and then crowd them in front of you and don't let them escape from your sword.

21 The Before (*Vor*) and After (*Nach*):

Text

62 Before and after: these two things
Learn well to gauge with backward spring.

Gloss

This means that above all else, you should know the Before and the After, since all the art of the duel proceeds from these two things.

Remember to always advance before they do (be it with a strike or with a thrust) and do it in such a way that they must parry, and as quickly as they bind your sword with their parry, nimbly drive a piece so they can't advance with their piece in front of your work, and this is called 'the Before'.

Likewise, remember here what is called 'the After': it is the counter to all pieces that they drive against you.

Learn it like this: if they advance such that you must parry, then with your parry, immediately seek the nearest exposure with your point. In this way, you win the work with your parry, and this is called 'the After'.

Remember here that in duel fencing, there shouldn't be more than one step out and one step in.

Learn it like this: if they rush at you such that you can't come to parry, you should only take one step backward with your left foot, and then see that you set upon them surely by stepping forward with your left foot (or grab them with wrestling).

22 The pursuit (*nachreisen*) with the sword in the duel:

Text

63 Pursue whenever steel meets steel,
And you'll confound the strong with zeal.
64 If they defend, pull back and thrust;
If they defend, rush in you must.
65 If they should fight you long and wide,
Then artfully their fate decide.

Gloss

Remember: this is the pursuit; drive it against the strong ones who fence wide and extended, and otherwise don't hold well to the correct art.

Arrange yourself against them like this: lay yourself against them in a guard. If they also lay themselves in one, then remember precisely: when they pull their sword to themselves and want to thrust, or rise up and want to strike, quickly follow after them toward the nearest exposure with your point before they bring the thrust or the strike. If they see your setting upon and they parry widely with their sword but don't seek your exposure with their point, then pull and go through again, and always drive the pulling as often as they go after your sword with a parry.

With this, you also come to the arm breaks and other pieces, and this is the art against all those that fence wide and extended, toward the sword and not the body.

23 The setting upon (*ansetzen*):

Text

66 If they attack you fierce and strong,
Defeat them when you shoot in long.

Gloss

Remember, this is when someone has set upon you and crowds you: thrust in toward the palm of the hand by which they hold their sword in the middle.

Or, if they've turned that hand around, thrust down from above in toward the same hand again, or thrust at their arm behind at the back of their glove. Run forward when the thrust sticks, so you win on their side and win another great advantage.

Or thrust through down from above, over their forward hand and your sword, and set your hilt on your chest and set upon them, etc.

If they've set upon your left shoulder, then step backward with your left foot so their point goes off and yours sticks. You can also extend your sword when you pull your pommel to your chest with the crowding.

24 How one should parry strikes:

Text

67 The other point can guard and smite,
So meet their sword and don't take fright.
68 With either hand, both left and right,
Turn then your point to piece their sight.

Gloss

Remember: the pommel is the other point of the sword, the point that smites (*schlagenden Ort*).

If they want to overrun you with strong blows with it, then hold your sword over your left knee and stand in the [third] guard.

If they're strong and strike toward your head, then sweep the strike away against their right side with the sword in front of your left hand, and go up on the sword into the upper [first] guard.

Or, if they're weaker than you are, then step in boldly and catch the strike on your sword between both hands, and set the point into their face.

When you catch their pommel strike in the middle of your sword, go over their sword with your pommel (in front by the hilt) and pull downward with it on your right side. In this way, you take their sword.

If they strike toward your left knee with their pommel, then catch the strike on your sword between your hands (so the pommel is angled toward the ground), and go through with your pommel below their sword by the hilt and pull upward on your right side. In this way, you pull their sword from their hands.

Or, if they strike below toward your foot with their pommel, then throw your sword against their strike (with your pommel at your left side toward the ground) and leap toward them with it, and look for the wrestling.

When you hold your sword by your right side in the lower [second] guard, if they strike toward your point with their pommel and let it run around widely, then meanwhile, leap boldly toward them in such a way that they strike the pommel over you so it doesn't harm you, and set upon them.

You can also pursue and set upon them wherever you want while they let their pommel run around widely.

25 On pommel strikes:

Text

69 With strikes, your forward foot defend,
That you may fight on to the end.

Gloss

Remember, this is when you want to pommel strike: aim very precisely at their forward members.

Learn it like this: when you want to strike, hold your sword overhead in the [first] guard and act like you want to set upon their face, but release your sword with your right hand and grab the middle of the blade with it (to help your left hand), and strike your pommel toward their forward foot or forward hand (with which they hold the sword in the middle). You can also strike out of the lower [second] guard from the same right side in this way.

Very precisely foresee when they strike toward your forward knee or forward hand with their pommel, so that in the strike, you parry in such a way that they can't harm you (using the counters written above), etc.

Paul Kal helps a Bavarian duke arm himself, presumably Ludwig IX 'the Rich', Duke of Bayern-Landshut (to whom Kal dedicated his book). The banner reads "Take this sword, gracious lord; you will be protected by the Mother of God and the chivalrous knight Saint George." (*Nemt hin genediger herr das schwert ir wert von der muter gots und riter sand iorgen aller riterschafft gewert*, PKM: 5^{r}.)

4. Manuscript discussion

Michael Chidester

This chapter will review the witnesses used in the translation. There are three texts that might be attributable to Ainring. The first is a gloss of Johannes Liechtenauer's *Record* of the long sword. This attribution is fairly certain because Ainring is credited as the author in the text itself—at the end of the witness in the *Glasgow fight book* (GF) and the beginning of the witnesses in *Joachim Meyer-Rostock* (JMR), and the *Sigmund ain Ringeck fight book* (SR). The other two are anonymous glosses of Liechtenauer's *Record* of mounted and armored (dismounted) fighting or dueling. These can't be definitively ascribed to Ainring, but each one appears alongside the long sword gloss in two of the manuscripts that include Ainring's name—the armored in JMR and SR and the mounted in GF and SR.

While this is not proof of authorship, these glosses are still included in this book because even if they aren't Ainring's work, they're among the few glosses of Liechtenauer's teachings that still lack a comprehensive treatment and analysis, and because adding them to Ainring's gloss of the long sword gives an interpretation of Liechtenauer's complete poem.

No autograph or archetype is known to exist, but there are eight manuscripts containing one or more of these texts. In addition to those mentioned above, two manuscripts include the full text of one section of gloss (armored fencing): *Albrecht Dürer* (AD) and *Willibald Pirckheimer* (WP); the other three include only fragments of one section (the long sword): *Anton Rast* (AR), *Hans Medel* (HM), and *Hans von Speyer* (HS). The nature of these fragmentary inclusions will be discussed below.

A final manuscript worth mentioning is the *Paris fight book* (P), which is mostly textless but includes a series of illustrations depicting the pieces of Liechtenauer's *Record* of the long sword. Among these are five which depict pieces that aren't found in any gloss other than Ainring's. For this reason, P may also be considered a transmission of the text.

Analysis and comparison of these manuscripts has been generally lacking in the literature, and discussion of Ainring has mostly focused exclusively on SR, which has often been treated as the only surviving witness of his work (or one of two, along with HM).

In conjunction with editing SR, Martin Wierschin proposed that it "was written by Ringeck himself or that he supervised its production",[1]

[1] *sie von Ringeck selbst geschrieben oder ihre Herstellung von ihm überwacht wurde*; Wierschin 1965: 43.

Witness	Date	Wi	Hi	Le	Ha
Albrecht Dürer or *Oplodidaskalia* Wien, Albertina, Hs. 26-232	ca. 1600	—	45	9.11	AD
Anton Rast Augsburg, Stadtarchiv, Reichsstadt Nr. 82	1553	—	01	8.1	AR
Glasgow fight book* or *Sigmund Emring **Glasgow, Kelvingrove Museum, E.1939.65.341**	**1508**	—	—	**1.2**	**GF**
Hans Medel or *Sigmund Schining* Augsburg, Universitätsbibliothek, Cod. I.6.2° 5	after 1566	14	06	1.1	HM
Hans von Speyer Salzburg, Universitätsbibliothek, M I 29	1491	32	43	—	HS
Joachim Meyer-Rostock **Rostock, Universitätsbibliothek, Mss. var. 82**	**1570**	—	—	—	**JMR**
Paris fight book or *Cluny fight book* Paris, Musée de Cluny, Cl. 23842	1480–1500	03	14	2.3 6.2	P
Sigmund Ainringck fight book **Dresden, SLUB, Mscr. Dresd. C.487**	**ca. 1510**	**04**	**16**	—	**SR**
Willibald Pirckheimer Nürnberg, Stadtbibliothek, Pirckh. Papp. 353	ca. 1500	—	—	—	WP

Table 1: Known witnesses of Ainring's gloss, including their catalog numbers in WIERSCHIN 1965 (WI), HILS 1985 (HI), LENG 2008 (LE), and the HAGEDORN *Sigla* system (HA) in the bibliography.

and also asserted that Ainring was the sole author of the entire manuscript (apart from Liechtenauer's *Record*). This assumption was based on the fact that Ainring was credited as the author of the gloss of the long sword, while no other section thereafter included any attribution (thus extending the attribution from the single section it specified to the entire manuscript without offering any justification).

HANS-PETER HILS and JAN-DIRK MÜLLER carried this assumption forward unchallenged[2] and RAINER WELLE compounded it, describing SR as "a masterly plagiarism" of the *Peter von Danzig fight book* (PD), which he attributed to the personal authorship of Peter von Danzig.[3] This idea was echoed as late as GRZEGORZ ŻABIŃSKI's 2010 dissertation on the *Goliath fight book* (G).[4]

[2] HILS 1985 and MÜLLER 1992.

[3] *ein meisterhaft angelegtes Plagiat*; WELLE 1993: 65. (Such claims are fortunately absent from his more recent works, such as WELLE 2021.) This was a response to HILS' assertion that Danzig had plagiarized from Ainring; see HILS 1985: 112.

[4] ŻABIŃSKI 2010. He attempted to prove it by comparing SR, N, and PD in the footnotes to his translation of the Pseudo-Danzig gloss of the long sword, though where he transcribed and translated N and PD, he merely cited and summarized WIERSCHIN 1965 and TOBLER 2001 for SR.

These 20th-century researchers were likewise content to classify the gloss in **HM** as a direct copy of **SR** with minor changes,[5] failing to recognize that while it does borrow several passages from Ainring, it diverges from **SR** in many details, and it also contains various unique teachings as well as pieces from the anonymous glossator nicknamed "Nicolaüs".[6]

The first published English translation of an Ainring manuscript was CHRISTIAN HENRY TOBLER's publication of **SR** (based on WIERSCHIN's edition) in 2001, and to his credit, he acknowledged both the lack of evidence both for Ainring's direct involvement in the creation of **SR** and for Ainring's authorship of its unattributed contents.[7] This idea failed to take root, though, and DAVID LINDHOLM's translation in 2003 and 2006 credited Ainring as the sole author yet again.[8] (Coincidentally, both of these publications were extensively illustrated, with black and white photos in TOBLER and drawings by PETER SVÄRD in LINDHOLM, despite the fact that the witnesses referencing illustrations were not yet known.)

The presence of Ainring's glosses in other manuscripts wasn't widely discussed even in the early 21st century.[9]

Though **GF** was mentioned by SYDNEY ANGLO in his 2000 overview of the fencing treatise tradition, including scans of four illustrated folia from its Ainring gloss (ff. 16v–17v & 19v), it wasn't the subject of much research until 2008, when it was included in RAINER LENG's catalog and the long sword gloss was correctly identified as being Ainring's.[10] Around this time, photos from a handheld digital camera were added to the online library of the Academy of European Medieval Martial Arts, and in 2009, it was edited and publicized by DIERK HAGEDORN.[11]

5 See, e.g., WIERSCHIN 1965: 20–21; HILS 1985: 35.

6 See HAGEDORN ET AL. 2021 and HAGEDORN 2023 for two other witnesses of Nicolaüs.

7 TOBLER 2001: ix–x. This was preceded by a translation of most of the manuscript into 'modern' German released online by CHRISTOPH KAINDEL in 1998, a partial translation of the long sword by JÖRG BELLINGHAUSEN in 1999 and continued by DAVID RAWLINGS in 2003 (likewise online), who completed the long sword but stopped partway through the armored fencing. It was common in the early 21st cen. for students of historical fencing to produce transcriptions and translations and release them freely online; these documents (and the websites that hosted them) are largely gone from the internet or preserved only on Wiktenauer, so this section will focus on work that either was published in print or has had a broad and enduring influence.

8 LINDHOLM & SVÄRD 2006: ix–1.

9 The 2002 French translation of the long sword by PHILIPPE ERRARD, ANTOINE FOURNIER, DIDIER DE GRENIER, and MICHAËL HUBER failed to consider any manuscript but SR (and a 2012 expansion that also translated HS failed to recognize the Ainring passages mixed into Lew's gloss); the 2006 Spanish translation by EUGENIO GARCÍA-SALMONES patterned after it shared this omission. A partial Hungarian translation by GÁBOR ERÉNYI and a Swedish translation by ANDREAS ENGSTRÖM were both released in 2009, also referring only to SR.

10 LENG 2008.

11 HAGEDORN 2009B. ANTON KOHUTOVIČ edited ff. 1r–24v that same year.

JMR seems to have first surfaced around this time as well. The first mention in print was WELLE's review of LENG's catalog, in which one of his criticisms was the omission of this manuscript.[12] This criticism could apply equally to HEIDEMARIE BODEMER's dissertation released that same year, an art historical overview of fencing treatises which ignores the art in GF and omits JMR completely.[13] WELLE didn't mention JMR's contents in this publication, but understanding of it among researchers increased as microfilm scans began circulating privately.[14] The first complete edition of the manuscript was produced by HAGEDORN in 2015.[15]

The first work that sought to combine multiple extant witnesses into a reconstruction of Ainring's original text was a translation of the long sword gloss by CHRISTIAN TROSCLAIR in 2015,[16] which he donated to Wiktenauer and which I helped him expand to incorporate all known witnesses—not just GF, JMR, and SR, but also the pieces of HS and AR that I had identified.[17] He translated HM separately that same year.[18] This monumental translation was subsequently published privately in a book distributed to Wiktenauer donors in 2016.[19]

This didn't establish a new trend for Liechtenauer studies, though. THORE WILKENS' biographical sketch of Ainring returned to mentioning only SR and HM as witnesses of his work,[20] and RACHEL KELLETT's analysis of wounding in the text acknowledged only one other (GF) before proceeding to focus purely on SR.[21] STEPHEN CHENEY's 2020 translation of the long sword gloss was likewise based only on GF and SR,[22] and his translation of the mounted fencing only on GF.[23] The recent edition and French translation of HM by BENJAMIN CONAN and HÉLÈNE LEBLANC includes a thorough attempt at comparing the long sword glosses in SR and HM, but doesn't mention any other witnesses.[24]

[12] WELLE 2009: 38.
[13] BODEMER 2008.
[14] Important early work was contributed by KEVIN MAURER and JENS-PETER KLEINAU, who edited and translated portions of it which they circulated online.
[15] HAGEDORN 2015B.
[16] TROSCLAIR 2015A. This has undergone several revisions since this time, but none of the later versions incorporated the breadth of sources used in the 2015 edition. The original version was later translated to Indonesian by PRADANA PANDU MAHARDHIKA in 2017.
[17] AR's Ainring gloss was edited along with other contents by WERNER UEBERSCHÄR in 2012, but he didn't acknowledge the presence of Ainring's text.
[18] TROSCLAIR 2015B.
[19] CHIDESTER, ET AL. 2016.
[20] WILKENS 2016.
[21] KELLETT 2015.
[22] CHENEY 2020A.
[23] CHENEY 2020B.
[24] CONAN & LEBLANC 2024.

Aside from this, most recent work on Ainring has been done as an extension of studies of **AD**, which was discovered to contain Ainring's armored fencing gloss when it was edited by HAGEDORN in 2016[25] (the previous edition by FRIEDRICH DÖRNHÖFFER ignored the unillustrated portions almost completely[26]). Two editions and a translation of **AD** have been published in the last five years, but were (quite reasonably) limited in scope to just that manuscript and only mentioned the others in passing.[27] These studies also lead to the (re)discovery of **WP**, which hasn't yet been the subject of any dedicated publications.

P has likewise received little scholarly attention to date, perhaps due to the restrictive policies of its owning institution, and nowhere in the literature has it been associated with Ainring.

Given this history, a description and comparison of all the known witnesses of Ainring's glosses is overdue. It will begin with thorough descriptions of the three principal manuscripts and then treat the other six manuscripts more briefly.

4.1. GF – GLASGOW FIGHT BOOK

Glasgow, Kelvingrove Museum, *E.1939.65.341* • paper, III + 107 ff. + III • 1508 • Bayern[28]

E.1939.65.341 is a paper manuscript with 107 leaves in quarto format, measuring 155 mm × 206 mm. The page block has five sewing supports, and the edges are painted or speckled red.

Five folia (ff. 17^v–19^v) have paper glued over the original text with new text written on them (in the same hand as the surrounding pages), suggesting that they were added during creation (see Fig. 2).

The manuscript shows evidence of having been trimmed down from its original size, including the older numeration on ff. 35^r–56^r often disappearing off the top of the page and illustrations cut off at the bottom, so it may have been created from *reçute* or chancery sheets (indicating an original page size of 157 mm × 225 mm).[29]

25 HAGEDORN 2016B.
26 DÖRNHÖFFER 2010.
27 HAGEDORN 2021A, WELLE 2021, and HAGEDORN & JAQUET 2022.
28 Some of the information in this section is based on observations and notes that DIERK HAGEDORN and HELEN HAGEDORN made when they examined the manuscript in September 2024, which they provided to me in a private communication. This will be cited as HAGEDORN & HAGEDORN 2024.
29 DANE 2012: 52.

Figure 1: GF's current binding.

It's bound in a 20th-century dark green Morocco cover with gold-foiled ruling around the edges and on the spine, which reads "LIECHTENAUER – FECHT UND RINGERBUCH" at the top and "1508" at the bottom.

4.1.1. Dating and origin

GF is internally dated 1508;[30] Leng's attempt to verify this through watermark analysis was inconclusive.[31] It's written in a Bavarian dialect.[32]

The manuscript appears to consist of three production units: A, ff. 1–33; B: ff. 34–63; C: ff. 64–107.[33] Units A and B were produced by different scribes and artists and show different *mise-en-page*, but they share the same paper; unit B is also numbered by page in a way that the other two are not. The scribe from unit A also worked on unit C, which begins on the same paper as the others but transitions to a new paper stock for quires VI–VII.[34]

Based on this, it seems likely that all three production units were created in the same workshop around the same time. The fact that three units end in multiple blank folia (see Fig. 3) suggests that there was no plan to include these units in the same book until after they were completed, so the next section couldn't begin on those last folia (this also seems to be the case with HS, as explained below).

GF's circumstances of creation are otherwise unknown, as is most of its history.

30 GF: 22r.

31 Leng 2008: 9.

32 Leng 2008: 11.

33 For a discussion of production units and related terminology, see Kwakkel 2012.

34 Hagedorn & Hagedorn 2024 includes descriptions of six watermarks, four of which are common to quires I–V and two of which appear in VI–VII.

4.1.2. HISTORY AND PROVENANCE

In 1933, it was noted as having a binding from Zaehnsdorf Bindery;[35] JOSEPH ZAEHNSDORF (1816–1886) first opened his London bindery in 1844 and it continued as a family business until 1947, but the time of binding can't currently be narrowed down further than the 89 years between 1844 and 1933.

GF was shown in München by EMIL HIRSCH (1866–1954) in 1933,[36] and soon sold to JOHANN ULRICH "Ulrico" HOEPLI (1847–1935). While HIRSCH was a target of Nazi persecution starting in 1933, and beginning that year, his sales are considered questionable by the German Lost Art Foundation; GF has not been listed among his coerced sales. He was formally ordered to liquidate his collection and close his business in 1935 (this was later extended into 1937 as he attempted to appeal the order).[37]

HOEPLI showed GF in Milano that same year, but it's unclear if he still owned it in 1939 when it was purchased by ROBERT LYONS SCOTT (1871–1939). This was shortly before SCOTT's death, at which time his entire collection was bequeathed to the Kelvingrove Art Gallery and Museum, where it now forms the R. L. Scott Library.

4.1.3. CONTENTS

GF contains no title or dedication of any sort; any introduction it once had was lost with the initial quire (see below). It's a miscellany containing fencing and wrestling treatises from masters of the Fellowship of Johannes Liechtenauer[38] (including not only Ainring, but also Martin Huntsfeld, Andre Lignitzer,[39] and Ott Jud[40]), other masters associated with the tradition (including Andres Juden, Hans Döbringer, Jobs von der Nyssen,

35 HOEPLI 1933: 43.
36 HIRSCH 1933: 11–12.
37 See *www.proveana.de/en/link/act20000877.*
38 This list of sixteen fencing masters associated with Johannes Liechtenauer was recorded by Paul Kal in ca. 1470; it's found in PKM, PKB, and PKW.
39 Huntsfeld's teachings are also recorded in AD,* AR,* G, HC, HS,* JMR,* L, LSM, O, PD, PKW, PMD, PMM, PMW, and WP;* Lignitzer's teachings are also recorded in AD,* G, HC, HS,* L, LSA, LSM, O, PD, PMD, PMM, PMW, PKW, SR,* and WP.* (Manuscripts marked with an asterisk (*) are detailed in this chapter.) For resolving disputes about authorship of the works of Martin Huntsfeld, Lew, and Andre Lignitzer, see JAQUET & WALCZAK 2014.
40 Ott's teachings are also recorded in AD,* G, HS,* HTG, HTN, HTGö, HTM3, L, LSM, O, PD, PMD, PMM, PMW, PKW, SR,* WF, and WP.* (Manuscripts marked with an asterisk (*) are detailed in this chapter.)

1^{r} – 22^{r}	Illustrated gloss of Liechtenauer's *Record* of the long sword by Sigmund Ainring *"Hie merck den pruch der wider den õbern twer haẘ..."*
22^{v} – 24^{r}	Anonymous treatise on fencing *"Hie merck wie man fechten sol mit dem langen schwert, aus der huet..."*
24^{v} – 25^{r}	*New Record* of the long sword by Martin Syber *"Hernach geschribn ist ain newe zetl des langen schwerts..."*
25^{v} – 26^{v}	Anonymous treatise on the long knife *"Merck ettliche stuck in dem messer..."*
27^{r} – 29^{v}	Treatise on fencing by Andres Juden, Jobs von der Nyssen, Nicklas Prewsen, and Hans Döbringer *(30^{r}–34^{v} blank)* *"Hie hebt sich an ettlicher maistern gefechte Andres Juden Jobs von der Nyssen..."*
35^{r}	Portrait of a seated master
35^{v} – 56^{r}	Anonymous illustrated treatise on wrestling *(56^{v}–60^{v} blank)* *"Der Standt zům Ringen..."*
61^{r}	A four-line proverb *(61^{v}–63^{v} blank)* *"Wo lieb ist, da ist auch leid..."*
64^{r} – 66^{r}	Treatise on wrestling by Andre Lignitzer (no author listed) *"Hie heben sich an die Ringen..."*
67^{r} – 73^{v}	Treatise on wrestling by Ott Jud *"Das sind die Ringen die der Ott gemacht hat der ain taufter Jüd gewessen ist..."*
74^{r} – 82^{r}	Anonymous gloss of Liechtenauer's *Record* of mounted fencing (currently attributed to Sigmund Ainring) *"1 Jag von deiner prust zw seiner rechten handt..."*
83^{r} – 84^{r}	Anonymous gloss of Liechtenauer's *Record* of armored fencing (fragment) *"Alhye hebt sich an Dye glos vnd die auß legung der ritterlichen kunst des..."*
84^{r} – 95^{v}	Anonymous gloss of Liechtenauer's *Record* of armored fencing (currently attributed to Pseudo-Peter von Danzig) *"Das ist der text vnd die gloß von ringen zu kampf..."*
95^{v} – 97^{v}	Treatise on the dagger by Martin Huntsfeld (no author listed) *"Das ist das fechten Vnd dye ringen mit dem tegen..."*
97^{v} – 100^{r}	Anonymous treatise on the dagger *"Ein vast gutte Lerr vnd ist gutt zu wissen..."*
100^{v} – 104^{v}	Treatise on the short sword by Martin Huntsfeld *"Das ist Maister Merteins hüncz feldes Fechten In harnckh auß den vier huten..."*
105^{rv}	Treatise on the sword and buckler by Andre Lignitzer *(106^{r}–107^{v} blank)* *"Hie heben sich ann die stůgk mit dem Bůgler die Maister andre lignizer..."*

Table 2: Contents of GF. Where the correct author of a section is currently known but the manuscript differs, the attribution from the manuscript is given in parentheses.

Nicklas Prewsen,[41] and Martin Syber[42]), and several anonymous works which are mostly unique.[43]

Unit A covers unarmored fencing with the two-handed sword (illustrated and unillustrated); unit B unarmored wrestling (illustrated); and unit C a variety of other activities: fencing on horseback and on foot, in armor and out of armor, with the spear, the two-handed sword, the one-handed sword with the buckler, the long knife, the dagger, and wrestling.

The two glosses of Liechtenauer's *Record* of armored fencing with the short sword (ff. 83^r–84^r & 84^r–95^v) are particularly interesting. The first is a unique fragment glossing couplets 1–7, while the second is a fragment of the gloss currently attributed to the anonymous author nicknamed Pseudo-Peter von Danzig and covers the remainder (8–29).[44] Since the text of the unique gloss continues directly into the Pseudo-Danzig gloss with no indication that the source of the text had changed, it seems likely that the scribe or client for the manuscript viewed these two commentaries as equivalent or interchangeable.

4.1.4. Structure and foliation

The collation of **GF**, which is composed of 107 folia plus six flyleaves, can be represented according to BRADSHAW's model as follows:[45]

I^{+5} (f. 5) + II^{28} (f. 33) + III^{18} (f. 51) + IV^{14-2} (f. 63) + V^{16} (f. 79) + $VI\text{–}VII^{14}$ (f. 107)

The first quire currently consists of just five cut folia which were inserted into the binding; if it were originally the same size as quire II, then 23 folia were lost from the beginning of the manuscript (not shown in Fig. 2), whereas if it were the same size as quire III, then the loss is only 13 folia. These two quires include the Ainring gloss of the long sword, and comparison to other witnesses shows that the first 33 paragraphs of the gloss are missing; of these missing pieces, there are eight instances in which **JMR**

[41] The text by these four masters is only found in one other manuscript, N.
[42] Syber's teachings are also recorded in HS and JMR, both of which are detailed in this chapter.
[43] The treatise on fencing on ff. 22^v–24^r is also found in SR and the 1516 treatise of Andre Paurenfeyndt. The wrestling treatise on ff. 35^v–56^r is also found in AD* (mixed together with pieces from BF), though that may have been copied directly from GF. The dagger treatise on ff. 97^v–100^r shares a complex relationship with the anonymous dagger treatises in GE, PMD, PMM, PMW, and JMR;* this will be the subject of a forthcoming paper by BART WALCZAK.
[44] In addition to GF, the Pseudo-Peter von Danzig gloss is also recorded in G, LSA, O, PD, PFS, PKW, and SR.* (Manuscripts marked with an asterisk (*) are also detailed in this chapter.)
[45] Collation formula based on diagram from HAGEDORN & HAGEDORN 2024. For a discussion of different methods of symbolizing manuscript structures, see DOROFEEVA 2019.

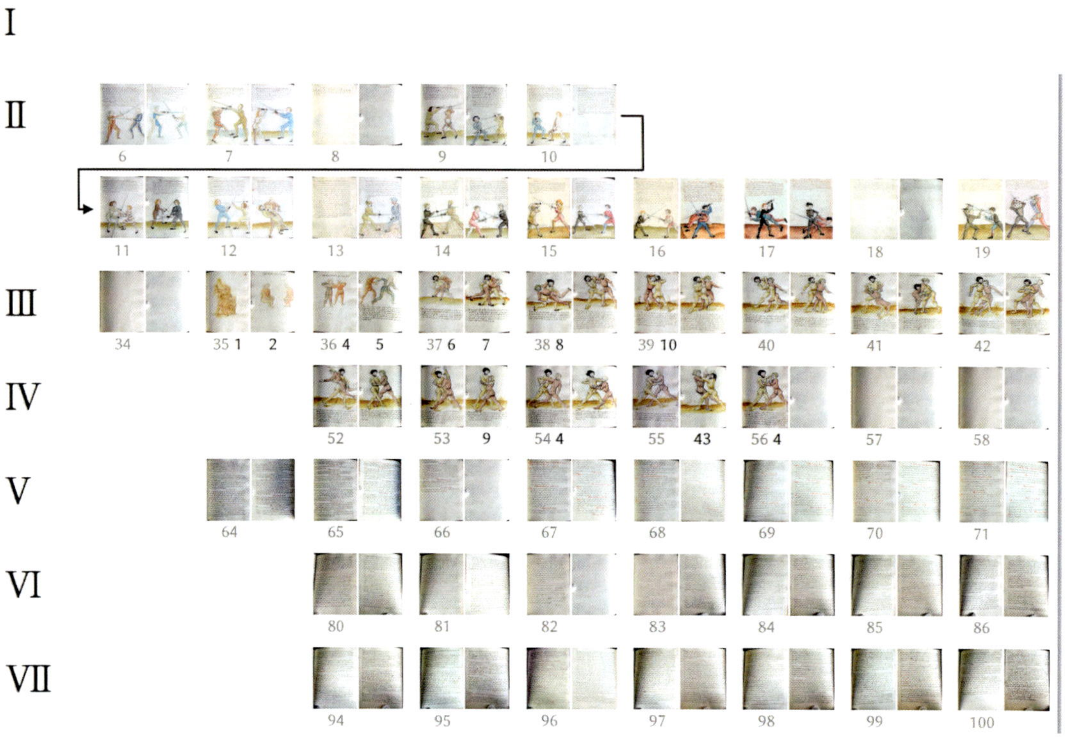

Figure 2: Collation of GF; the black and grey numbers represent the pen and pencil numbers, the heavy gray line and the fold of this book indicate the central stitching, and the light grey line separates the leaves of the first quire, which are single cut sheets glued into the binding that →

refers to illustrations and at least ten pieces which are missing from JMR that might arguably have merited illustration. These 33 paragraphs could have entirely filled the remainder of an 18-folio quire, but if it were a 28-folio quire, then even if they were each written on a separate side, that would still leave several folia whose contents are unaccounted for.[46]

Other material lacunae are seen between ff. 62–63 and 63–64; since these are surrounded by blank folia, it's likely that no text was lost.

The manuscript is foliated continuously in pencil in the upper right corner, which must have occurred after these losses. The remains of an older numbering in brown pen (by page rather than by folio) are usually visible in the upper outside corners of ff. 35^{r}–56^{r}, though part or all of the number is often lost due to trimming. These numbers skip '3' but seem to have otherwise been continuous: 1–2 (f. 35^{rv}) and 4–44 (ff. 36^{r}–56^{r}).

[46] SR places an unglossed witness of Liechtenauer's *Record* before the long sword gloss, as do several other manuscripts (including D, G, and PKW), so that could also have been the case here and account for the rest of the quire.

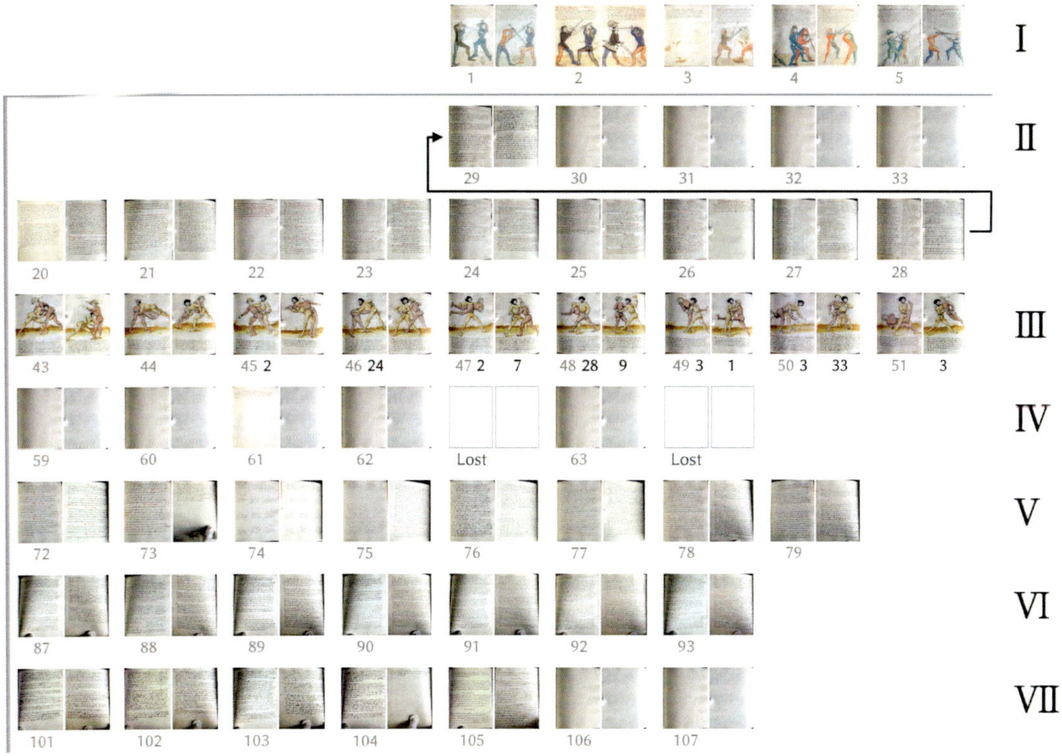

(continued) thus have no stitching, from the others. Quire II is much too wide to display as a continuous sequence on this page, so the outer five paper layers are connected to the rest with black arrows.

4.1.5. Writing and decoration

GF is written by five hands in a non-calligraphic practical bastarda (hand I: ff. 1^{r}–29^{v}, 64^{r}–66^{r}, 75^{r}–101^{r}; hand II: ff. 35^{v}–56^{r}; hand III: ff. 67^{r}–75^{r}; hand IV: ff. 100^{v}–104^{v}; hand V: f. 105^{rv}).[47] A short poem was added to the top of f. 61^{r} in an elaborate cursiva by a sixth hand.[48] On ff. 17^{v}–19^{v}, the original text has been covered up by paper patches glued onto the page;[49] the replacement text was also written by hand I.

In the first illustrated section (ff. 1^{r}–19^{v}), the text is written in a single column of 5–14 lines on the top half of each page, while in the second

47 Leng 2008: 10.

48 Hagedorn & Hagedorn 2024.

49 An attempt to recover the hidden text using backlighting and digital manipulation allowed partial transcription of the underlying text on ff. 17^{v}, 18^{v}, & 19^{r}. The text on 17^{v} appears to be piece 75 (which isn't otherwise in GF), while 18^{v} and 19^{r} are pieces that were rewritten slightly later in the manuscript (pieces 77 and 78, respectively). Transcription provided in Hagedorn & Hagedorn 2024.

illustrated section (ff. 35^{r}–56^{r}), it's written in 4–10 lines on the bottom half of the page (see Fig. 2). The unillustrated pages generally have a single column of 24–32 lines per page, except when blank space was left for an illustration that was never attempted. A few illustrated pages (ff. 35^{r}–36^{r}, 55^{v}) have no accompanying text.

There are no decorative borders or Lombard capitals in GF. The text generally has rubricated titles and keywords but lacks any other decorative flourishes. On ff. 74^{r}–75^{r}, the text is written inside rows of circles (twelve each on 74^{rv} and two on 75^{r}).

4.1.6. ILLUSTRATIONS

There are 74 illustrations in the manuscript, executed in pen and ink with watercolor by six artists. The Ainring gloss employs four: artist I, ff. 1^{r}–2^{v} & 4^{r}; artist II, ff. 3^{v} & 4^{v}–6^{v}; artist III, ff. 7^{r}–14^{r}; and artist IV, ff. 14^{v}–19^{v}. The illustrated wrestling section employs the other two: artist V, ff. 35^{r}–36^{v}; and artist VI, ff. 37^{r}–56^{r}.[50]

All illustrations are frameless, and they typically occupy an area about 150 mm square. In the first illustrated section (ff. 1^{r}–19^{v}), the illustrations appear in the lower part of the page, while in the second illustrated section (ff. 35^{r}–56^{r}), they appear in the upper part. Dimples and perforations in the paper along the black ink lines show that the illustrations were traced from an earlier manuscript rather than being original compositions.

LENG characterizes the artists as follows:

> *Artist I: rough, static figure depictions, clumsy facial drawings, and rough, flat coloring; figures placed on the suggestion of a grass background, also several notable doodles on the grass background (1^{r} toad, 2^{r} face, 4^{r} salamander), regular pencil drawings recognizable;*
>
> *Artist II: figures contoured with a few lines and only a rough suggestion of facial features, coloring flat but shaded with darker mixtures at the edges, figures standing freely in space;*
>
> *Artist III: somewhat more dynamic figures and more precise facial drawings, figures colored flatly but shaded with rich hatching, all figures placed on an ochre-colored grass background; 10^{r}, a notable figure with the letters A–H on his clothing; 13^{v} and 14^{r}, somewhat more careful drawings of people with hints of richly decorated and slashed clothing, both figures here in striped clothing with turban-like headgear placed on a green and red tiled floor;*
>
> *Artist IV: simple, contoured figures, flatly colored, occasional hatching, coarse facial features, monochrome-colored fighters (with*

[50] LENG 2008: 11.

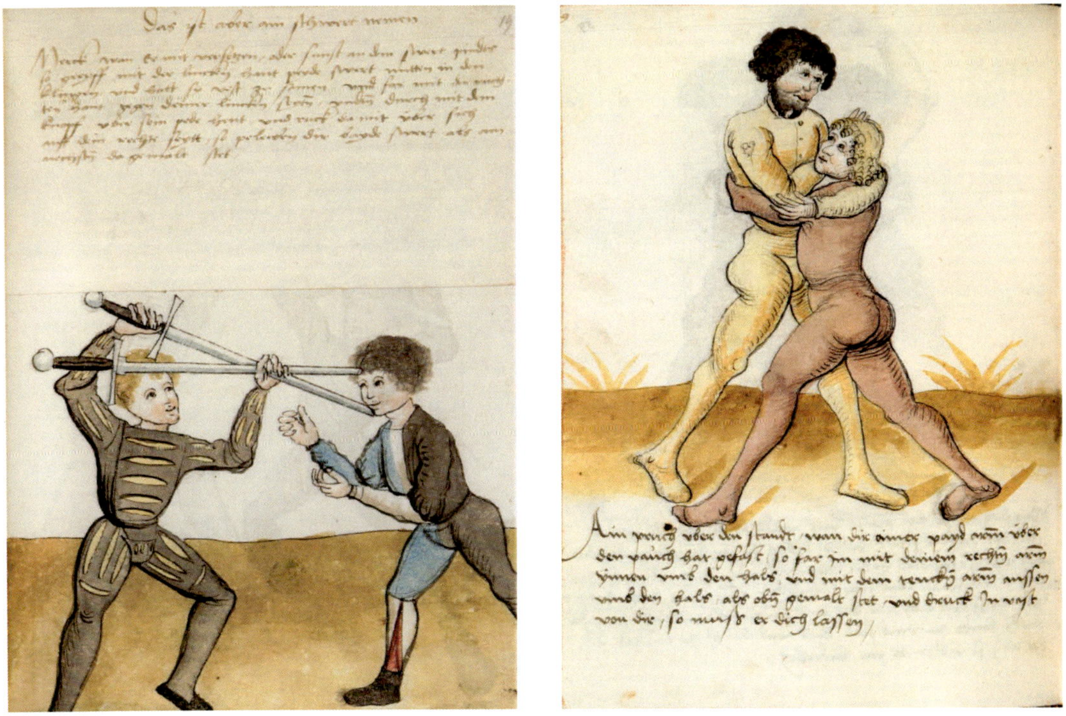

Figure 3: *Mise-en-page* of GF's two illustrated sections (left, f. 19^{r}; right, f. 53^{v}); f. 19^{r} also shows a paper patch covering and replacing the original text.

a few exceptions of contrastingly-colored slashed clothing); all figures placed on a monochrome, ochre-colored grass background;

Artist V: small static figures, colored shading, almost continuously provided with hatching, with shadows cast on an indicated horizon line;

Artist VI: dynamic pairs of wrestlers and a clear sense for the precise representation of the individual grips described in the text; figures placed throughout on a grass background with stylized plant growth, both in tight-fitting clothing, colored shading, with additional use of hatching; fighters are individualized throughout: one figure in yellow clothing with dark hair and beard, the other fighter dressed in red, blond and beardless; the yellow-clad fighter almost always wears a sign on his chest or arm (crown with crossed swords?), which is replaced by an upright B in 42^{r}.[51]

[51] Translated from LENG 2008: 11–12. All translations are my own except where noted.

4.1.7. LITERATURE[52]

Editions: HAGEDORN 2009B; TOBLER 2022 (ff. 35^{v}–56^{r} & 97^{v}–100^{r}); HAGEDORN [forthcoming].

Translations: TOBLER 2010B (ff. 25^{v}–26^{v}); TROSCLAIR 2015A (ff. 1^{r}–22^{r}); CHENEY 2020A (ff. 1^{r}–22^{r}); CHENEY 2020B (ff. 74^{r}–82^{r}); LEE 2022: 150–151 (ff. 25^{v}–26^{v}); TOBLER 2022 (ff. 35^{v}–56^{r} & 97^{v}–100^{r}); HAGEDORN [forthcoming].

Literature: HIRSCH 1933: 11–12; HOEPLI 1933: 43–44; ANGLO 2000: 184–186, XVIII–XXI, XXV–XXVIII; LENG 2008: 9–12; FORGENG 2012: 167–168; JAQUET & WALCZAK 2014; WALCZAK & STARKO 2018.

Scans: *wiktenauer.com/wiki/Category:MS_E.1939.65.341*[53]

4.2. JMR – JOACHIM MEYER-ROSTOCK

Rostock, Universitätsbibliothek, *Mss. var. 82* • paper, I + 128 ff. + I • before 1570 • Strasbourg[54]

Mss. var. 82 is a paper manuscript with 128 leaves in quarto format, measuring 153 mm × 197 mm. The page block has three sewing supports, and the edges are painted or speckled blue.

The manuscript shows evidence of having been trimmed down from its original size, including text frequently disappearing off any of the three edges, so it may have been created from *reçute* or chancery sheets (indicating an original page size of 157 mm × 225 mm).

JMR is bound in a 16th-century white leather cover over pasteboard, with a blind-tooled pattern of diagonal triple-lines forming a lattice on both covers. The cover also includes a leather flap which wraps around the fore-edge to enclose the book. A staple mounted on the front board and a hasp in the closure flap that would allow the manuscript to be held closed with a pin or padlock. An inscription written in brown ink on the cover reads: "F [?] I M 1 5 7 0 S", while "L 41" is carved into the side of the cover flap.

[52] The literature sections of this chapter are intended to provide roadmaps for further reading about these manuscripts; I have attempted to be thorough, especially as regards English-language sources, but in a few cases I have found references to items that I was unable to access, and those are omitted here since I can't verify their relevance. For my purposes here, an 'edition' includes a transcription and discussion of the text but not necessarily a translation.

[53] The museum has never released a complete set of scans of GF; instead, these are photos taken with a handheld digital camera some time in the early 21st cen. (photographer unknown).

[54] Some of the information in this section is based on an unpublished profile of the manuscript prepared for the University Library in 2014 by HEIKE TRÖGER, which was provided to me by ANNIKA BOSTELMANN. This will be cited as TRÖGER 2014.

Figure 4: JMR's current binding, with the cover flap closed (left) and open (right).

4.2.1. DATING AND ORIGIN

JMR appears to be a single extended production unit; it consists of an initial unit (ff. 6^{r}–122^{r}) extended with additional materials at the beginning and end (ff. 1^{r}–5^{v}, 122^{v}–126^{v}).

Watermark analysis suggests that the initial production unit uses a single paper stock produced in Pfreimd in 1559.[55] Given the fact that it was trimmed before receiving its current binding (while the extensions were not), it seems likely that it was produced some time in the 1560s and then acquired by Joachim Meyer in a completed state.

The extensions are on paper produced in Montbéliard in 1564.[56] The date '1570' is inscribed on f. 123^{r} as part of a title page for the final portion, but it isn't clear if that indicates when he acquired the manuscript or simply when he began writing that section. The same date is written on the front cover, suggesting that the manuscript was bound in its current form by that year.

It's unclear if the rapier text on ff. 112^{r}–122^{r} was part of the initial creation, or was added for Meyer on blank sheets at the end of the initial unit.

55 TRÖGER 2014.
56 TRÖGER 2014.

4.2.2. HISTORY AND PROVENANCE

By 1570, **JMR** seems to have taken its current form and been acquired by Strasbourg *Freifechter* Joachim Meyer (1537–1571), who wrote his name on ff. 5ʳ & 123ʳ and possibly also the cover.[57]

In 1570, Meyer accepted a contract as a fencing master to Johann Albrecht (1525–1576), Duke of Mecklenburg, and on 11 January 1571, he began the 500-mile journey to the ducal court in Schwerin, ultimately arriving on 10 February. Sadly, he took sick on this journey and died soon thereafter; his death was ultimately reported to the duke on 24 February.[58]

Anton, brother of Meyer's widow Appolonia, offered the manuscript to Johann Albrecht as a gift as part of negotiations for the return of Meyer's effects and payment for his service,[59] and **JMR** was added to the ducal library in Schwerin and cataloged in 1573 under the signature *L. 41.*[60]

Its subsequent history is that of the entire library. It was still in Schwerin as late as 1769, when it was cataloged as item 27 in a list of manuscripts in the judicial office.[61] After this, the ducal libraries were transferred to the short-lived Friedrichs-Universität Bützow (which had been founded in 1760 by Friedrich (1717–1785), Duke of Mecklenburg).

In 1789, just a few years after Friedrich's death, the university was merged into the Universität Rostock along with its libraries,[62] where **JMR** received a library stamp reading *Ex Bibliotheca Academiae Rostochiensis* (on ff. Iʳ, 1ʳ, & 127ᵛ) and the signature *Mss. var. 82* (written inside the front cover). It remains in that collection to this day.

4.2.3. CONTENTS

JMR's first title page (f. 5ʳ), added by Joachim Meyer, declares it to be a "Fencing book of a quick succession of all sorts of pieces, on horse and on foot and in the duel, recorded by the most ancient and noble masters, etc. Also a short introduction to the rapier by Joachim Meyer".[63]

57 TRÖGER 2014 reads "I M" as Ioachim Meyer.

58 For a biography of Joachim Meyer, see VANSLAMBROUCK 2023.

59 LIAM CLARK provided me with the text of a letter from the duke to Anton (care of the Strasbourg city council) which mentions this gift and also notes a "gift" to her of 30 Thaler that may have been payment for it. Strasbourg, Archives, V14 #141.

60 Rostock, Universitätsbibliothek, Mss. Meckl. J 203: 179ʳ, where it's cataloged along with a copy of Meyer's then-recently-printed book bound in red velvet.

61 Rostock, Universitätsbibliothek, Mss. Meckl. O 116.c: 147ᵛ.

62 Rostock, Universitätsbibliothek, Mss. Meckl. J 81: 404, the catalog of books being transferred.

63 *Fechtbuch von allerhand geschwinden Stucken zu Rosß vnd zu fuß vnd zum Kampf Aufgezeichnet von den fürnemstem alten theichsten meistern etc. auch einen kurtzen grund vom Rapier gestellet durch Jochim Meyer.*

1^{v} – 5^{r}	Additions by Joachim Meyer *(5^{v} blank)* *".A. . . . . . Angesicht linie..."*
6^{r} – 13^{v}	Gloss of Liechtenauer's *Record* of the long sword by Sigmund Ainring *"Hie heben sich an die Zedttel In den geschrieben stedt die Ritterlich kunst..."*
13^{v} – 39^{v}	Anonymous gloss of Liechtenauer's *Record* of the long sword (currently attributed to Lew) *"Hie hebt sich an ein gute gemeine lehr des langen schwerts..."*
39^{v} – 41^{r}	*New Record* of the long sword by Martin Syber *"Das hernach geschribenn ist ein newe tzetel des lanngenn schwerts..."*
41^{r} – 58^{r}	Anonymous gloss of Liechtenauer's *Record* of mounted fencing (currently attributed to Lew) *"1. Jagenn vonn der brust, tzu seiner Rechten seittenn..."*
58^{r} – 63^{v}	Treatise on the short sword by Martin Huntsfeld (author listed as Lew) *"Hie hernach hebt sich an meister lebenn kunst fechtens im harnisch..."*
63^{v} – 67^{r}	Treatise on armored wrestling by Martin Huntsfeld (no author listed) *"Item das sein die vnnterhabenn vnnd aufstehenn im kampf..."*
67^{r} – 74^{v}	Treatise on armored fencing by Andre Lignitzer (author listed as Martin Huntsfeld) *"Das kurtz schwert meister Merteinn huntzveltz in dem Kampf..."*
74^{v} – 75^{r}	*Record* of armored fencing by Johannes Liechtenauer (no author listed) *"Ob du wilt ringenn, hinter pain Recht lernn springenn..."*
75^{v}	Verses on mounted fencing by Martin Huntsfeld (no author listed) *"Die tzetel weist ringenn tzu dem kampf tzu Ros..."*
76^{r} – 85^{v}	Anonymous treatise the dagger *"Degenn vechtenn mit lerer hannd des ersten..."*
86^{r} – 96^{v}	Anonymous techniques for various weapons *"Mit dem schwert ein stuck, leg dich fur mitt dem schwert als in die eisnen..."*
96^{v} – 98^{r}	Treatise on the sword and buckler by Andre Lignitzer (no author listed) *"Das erst stuck mit dem Puckler..."*
98^{r} – 110^{r}	Anonymous gloss of Liechtenauer's *Record* of armored fencing (currently attributed to Sigmund Ainring) *(110^{v} blank)* *"Hie hebt sich ann meister Joannes Liechtenawers vechtenn Im harnasch..."*
111^{r}	Index of mounted fencing pieces *(111^{v} blank)* *"Reister vom Ros Fechten bis hieher..."*
112^{r} – 122^{r}	Treatise on the rapier, potentially by Joachim Meyer *"Fechten Im Rapier, Wiltu fechten zu einer hand..."*
122^{v} – 127^{r}	Treatise on the rapier by Joachim Meyer *(127^{v}–128^{v} blank)* *"Düs vor geschrüben Fechten hab üch aus den alten zedlen gezogen..."*

Table 3: Contents of JMR.

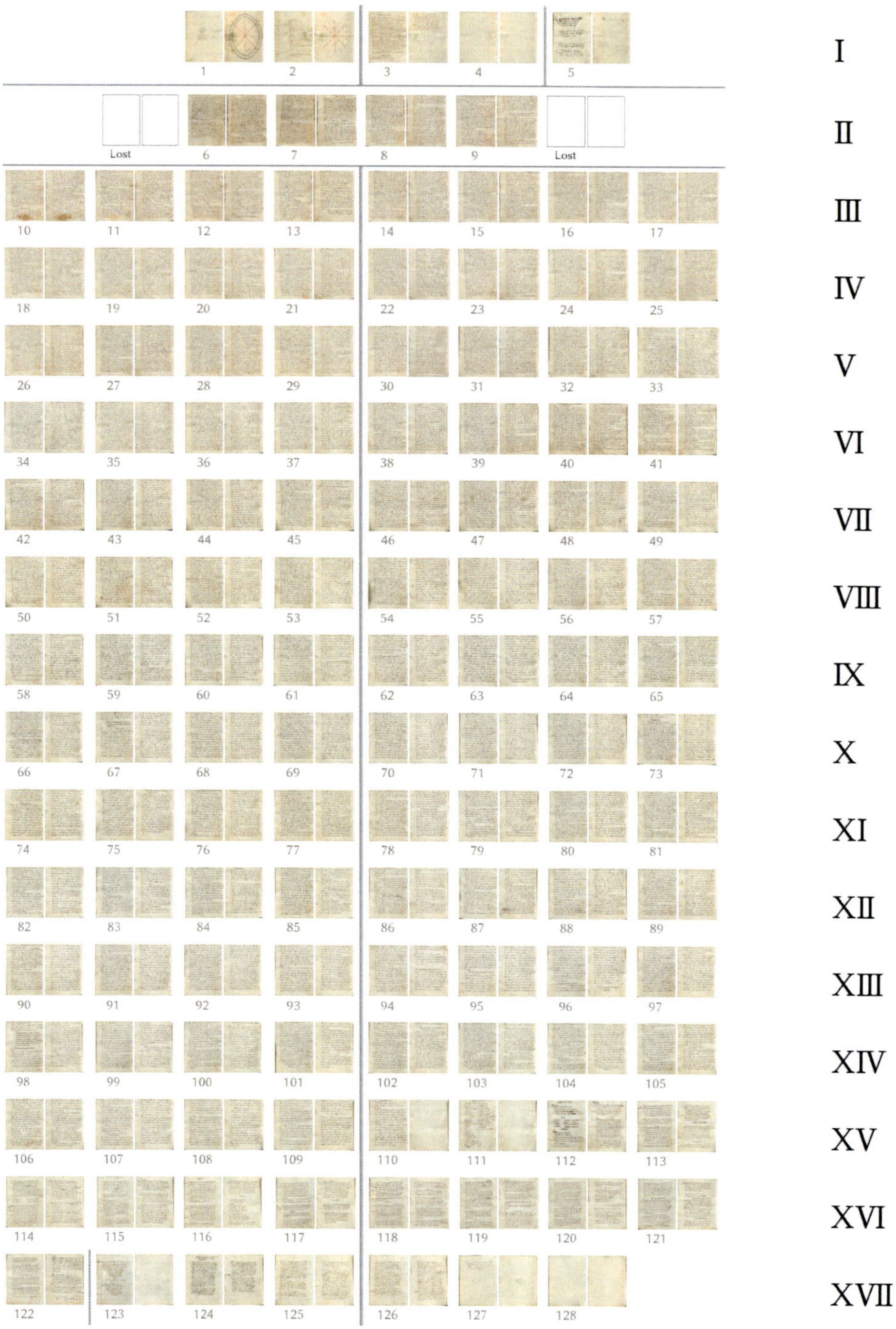

Figure 5: Collation of JMR; the heavy gray line indicates the fold, and the lighter lines indicate folia added to the manuscript; since the older numeration is largely cut off, it's not represented here. Quire II consists of four cuts leaves that were glued into the spine together and thus have no stitching.

The initial production unit is a miscellany containing treatises from masters of the Fellowship of Johannes Liechtenauer (including Ainring and also Martin Huntsfeld, Andre Lignitzer, and Hans Pegnitzer[64]), other masters associated with the tradition (including Lew[65] and Martin Syber), and several anonymous works which are mostly unique.[66] Its constituent texts cover fencing on horseback and on foot, in armor and out of armor, with the spear, the two-handed sword, the one-handed sword with the buckler, the dagger, and wrestling. Meyer's extensions consist primarily of rapier fencing lessons.

JMR's witness of the Ainring gloss on the long sword is incomplete, including only about 40% of the total pieces. The pieces it includes are mostly from the beginning of each major section and more from earlier in the treatise rather than later; this probably wasn't due to the scribe stopping short of the end since it includes portions of every major section.

The gloss of the armored fencing is complete, but heavily disorganized (this is discussed more below).

4.2.4. Structure and foliation

The collation of JMR, which is composed of 128 folia plus two flyleaves, can be represented according to BRADSHAW's model as follows:[67]

I^{4+1} (f. 5) + II^{+4} (f. 9) + $III–XVI^{8}$ (f. 121) + XVII^{6+1} (f. 128)

In quire I, f. 5 is a single leaf glued to the back of f. 4; this quire contains Meyer's additions to the beginning of the manuscript. Quire II consists of four cut leaves glued together, and based on the very conspicuous missing segment of the Ainring gloss in between the last line of f. 9^{v} and the first of f. 10^{r}, it seems likely that these are the remains of a longer quire that was cut out of the book and lost some of its leaves. It's possible that f. 5 was initially attached to the missing folio between ff. 9 & 10.

Meyer added more writings on ff. 122^{v}–127^{r} in quire XVII; the paper used in the initial production unit ends at f. 122, which is glued to the beginning of quire XVII, and Meyer seems to have begun writing on the verso side.

64 This isn't a treatise, merely a single piece on f. 94^{r} which is attributed to Pegnitzer. However, it's the only teaching from this master known to survive.

65 Sometimes erroneously referred to as 'Jud[e] Lew' in the literature, a name that doesn't appear in any primary source, due to the *explicit* in L which implies that he was Jewish. Lew's teachings also appear in AD,* HC, HS,* L, LSM, PMD, PMM, and PMW. (Manuscripts marked with an asterisk (*) are also detailed in this chapter.)

66 Note that JMR is part of a group of manuscripts that wrongly attribute the teachings of Huntsfeld to Lew and the teachings of Lignitzer to Huntsfeld. See JAQUET & WALCZAK 2014.

67 Collation formula based on TRÖGER 2014.

(As mentioned before, it's unclear whether the text on ff. 112^r–122^r was also added for Meyer.) There's a second title page on f. 123^r which notes that the remainder of JMR contains the teachings of Stephan Heinrich (1543–1613), Count of Eberstein, who Meyer probably met and fenced with at the Diet of Speyer in 1570.[68]

JMR received two separate sets of folio numbers. The older set is frequently cut off at the top right corner, indicating it was applied before the trimming and therefore before the current binding. The first clearly-visible number is '8' on f. 12^r. Assuming a lost leaf between ff. 9–10, the foliation began on 6^r, the first surviving folio of the initial production unit.[69]

A more recent foliation was added in pencil at the bottom center of each page, and this is continuous throughout the current state of the manuscript (from ff. 1–128).

4.2.5. Writing and decoration

JMR is written by four hands in cursiva with occasional titles in a bold Fraktur (hand I: ff. 6^r–110^r; hand II: ff. 1^v–3^v; hand III: ff. 111^r–122^r; hand IV: ff. 5^r, 122^v–127^r).[70] There's also minor marginal notes from several hands which may or may not be among these four. Hand IV probably belonged to Joachim Meyer,[71] but the other writers are unknown.

In the initial production unit, the text is written in a single column of 15–30 lines (sometimes cut off at the top, bottom, or side). The extensions are similar, but with more variation in line density. The initial production unit has a large right-side margin on both recto and verso sides.

There are no borders, Lombard capitals, or other decorations apart from the occasional Fraktur title.

4.2.6. Illustrations

Though the text makes frequent reference to illustrations (especially in the Ainring gloss), none of these are included in JMR.

The only illustrations in the manuscript are two geometric designs on ff. 1^v & 2^v. Each is a bounding shape with a complex series of intersecting heavy red, light yellow, and plain brown lines. The diagram on f. 1^v is an oval with heavy blue bounding lines, while the diagram on f. 2^v is a circle with plain brown bounding lines.

[68] VanSlambrouck 2023: lii.

[69] If the entire outer bifolium of this quire were lost, then that would mean that it was originally six folia and the first leaf was unnumbered, which wouldn't be unusual if it had no text on it.

[70] Tröger 2014.

[71] Forgeng 2016: 167.

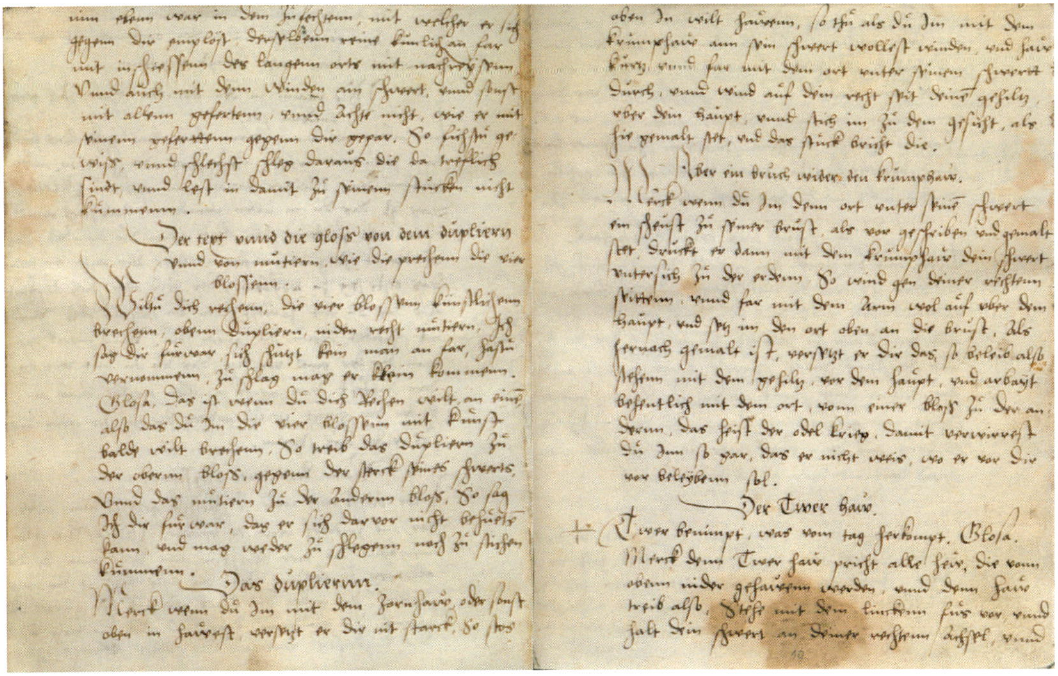

Figure 6: *Mise-en-page* of JMR and the site of a potential missing leaf (ff. 9^{v}–10^{r}).

4.2.7. LITERATURE

Editions: HAGEDORN 2015B.

Translations: FORGENG 2015: 167–170 (ff. 122^{v}–127^{r}); TROSCLAIR 2015A (ff. 6^{r}–13^{v}).

Literature: JAQUET & WALCZAK 2014; FORGENG 2015: 8–10; VAN-SLAMBROUCK 2023: lii–lv.

Scans: *rosdok.uni-rostock.de/resolve/id/rosdok_document_0000005792*

4.3. SR – SIGMUND AIN RINGECK FIGHT BOOK

Dresden, Sächsische Landesbibliothek – Staats- und Universitätsbibliothek, *Mscr. Dresd. C. 487* • paper, 149 ff. • ca. 1510 • Schwaben

Mscr. Dresd. C.487 is a paper manuscript with 148 leaves in quarto format, measuring 110 mm × 150 mm. The page block is sewn directly to the cover in the long-stitch fashion (leaving exposed thread at the top and bottom of the spine).

SR is bound in a 16th century limp parchment cover (without boards); the parchment was recycled from a 12th century missal (folded down to

Figure 7: SR's current binding, with the cover flap tucked under the front cover.

the size of the manuscript), and the outside is painted red. The cover also includes a parchment flap which wraps around the fore-edge to enclose the book. The tie that would have held the book shut is missing.

The outer spine has a sticker marked with the current signature: [printed:] "Msc. Dresd." [written:] "C487". Opposite this, on the flap enclosing the fore-edge, the title or description "Johan Liechtnawers Fechtbuch geschriebenn" is recorded in brown ink at the top, and a torn sticker at the bottom seems to include the former signature, *C 241*.

The manuscript is quite worn, with the cover heavily damaged in places and the pages frequently showing smears and liquid damage.

4.3.1. Dating and origin

Dating **SR** has proven problematic.

WIERSCHIN proposed a date in the first half of the 15th century (or the very end of the 14th) based on a paleographic analysis, and believed the watermarks he identified were coherent with this time period.[72] HILS accepted this date range and narrowed it to the 1440s by assuming that

[72] WIERSCHIN 1965: 79–84.

Ainring's text was produced after Albrecht III became Duke of Bayern-München in 1438 and that it was the source for PD in 1452 and therefore must have been created before it.[73]

This date range continued to be accepted[74] until 2010 when HOFFMANN prepared a new watermark study for his catalog of the library's manuscripts; his analysis assigned a much later date—the first years of the 16th century—and he further identified a manuscript from 1510–12 with the same constellation of watermarks.[75]

The place of origin is less disputed: while the original texts reflect a Bavarian dialect, the versions in SR show clear Swabian influence, indicating that it was produced in Schweben.

4.3.2. HISTORY AND PROVENANCE

The early history of SR is unknown.

Its first folio bears a simple heraldic device: a shield containing a representation of a *globus cruciger* (a circle with a second circle inside and four spokes inside it, surmounted by a cross) and the letters ·C· and ·e·, which WIERSCHIN reads as an abbreviation for 'Christi episcopus' and sees it as an indication that it was owned by a cleric.[76] The title on the fore-edge flap is in a 16th or 17th century hand.

SR doesn't appear in historical records for over two hundred years after its creation, by which time it was already in the Königlich Sächsischen Bibliothek. In the great cataloging project of 1755, it was listed as *Nr 815.*, which is inscribed in the upper left corner of f. 1v.

Another signature, *C. 241.*, is written on both the outside cover on a sticker and at the bottom of f. 1r in red ink. WIERSCHIN suggests that this was an earlier signature from the same library,[77] but HOFFMANN only confirms that it seems to be older.[78]

When Sachsen became a kingdom in 1806, the library became the Bibliotheca Regia Dresdensis ("Royal Dresden Library"), and library stamps with this name are found on ff. 2r, 3r, & 125v. This then became the Sächsische Landesbibliothek ("Saxon State Library") after Germany became a republic in 1919.

In 1996, it merged with the Universitätsbibliothek Dresden ("Dresden University library") to form the current Sächsische Landesbibliothek – Staats-

[73] HILS 1985: 55.
[74] See, e.g., TOBLER 2001: ix and BODEMER 2008: 103.
[75] See HOFFMANN 2022: 319 for the latest version.
[76] WIERSCHIN 1965: 84.
[77] WIERSCHIN 1965: 84.
[78] HOFFMANN 2022: 320.

1^{rv}	*Front matter*
2^{r}	First title *(2^{v} blank)* *"Hie hept sich an die Ritterlich kunst deß langen schwerts"*
3^{r} – 9^{v}	*Record* of the long sword by Johannes Liechtenauer *(10^{r} blank)* *"IN sant Jörgen namen höbt an die kunst deß fechtens…"*
10^{v} – 48^{v}	Gloss of Liechtenauer's *Record* of the long sword by Sigmund Ainring *"Hie hept sich an die vßlegung der zedel…"*
49^{r} – 54^{r}	Anonymous treatise on fencing *"Hie mörck vß den nebenhůtten Daß ist auch den streÿchen zů fechten…"*
54^{r} – 55^{v}	Treatise on the sword and buckler by Andre Lignitzer (no author listed) *"Her nach stond geschriben die stuck mitt dem buckler…"*
55^{v} – 57^{r}	*Record* of the long sword by Johannes Liechtenauer (fragment; no author listed) *"Wer dir aber haẘet Zornhaẘ Im drawet…"*
57^{r} – 59^{v}	Anonymous gloss of Liechtenauer's *Record* of the long sword (fragment) *(60^{r}–65^{v} blank)* *"Wer dir aber hawet, Zor^{n}haẘ ort im drawet…"*
66^{r} – 77^{v}	Anonymous treatise on wrestling *"Hie heben sich an gůtte ringen vnd ander gůtt brich…"*
78^{r} – 84^{r}	Treatise on wrestling by Ott Jud (no author listed) *"Hie heben sich an andere gůtte ringen vnd brůch…"*
84^{r} – 86^{v}	Anonymous treatise on wrestling *(87^{rv} blank)* *"Hie hebent sich an die zůlauffenden ringen…"*
88^{r}	Second title *(88^{v} blank)* *"In sant Jorgen namen höpt an die kunst"*
88^{r} – 108^{r}	Anonymous gloss of Liechtenauer's *Record* of armored fencing (currently attributed to Sigmund Ainring) *(108^{v} blank)* *"Hie höpt sich an der ernstlich kampff zů roß vnd fůß, Alhie hept sich ann…"*
109^{r} – 110^{v}	Anonymous gloss of Liechtenauer's *Record* of mounted fencing (fragment; currently attributed to Sigmund Ainring) *"Hie hept sich an die ritterlich kunst das fechten zuo roß…"*
110^{v} – 121^{v}	Anonymous gloss of Liechtenauer's *Record* of mounted fencing (fragment; currently attributed to Pseudo-Peter von Danzig) *(115^{r} blank)* *"Ain stuck mitt der kurczen Glefen gegen der langen zů roß…"*
122^{r} – 125^{v}	Four displaced pages *(126^{r}–148^{v} blank)* *"die ~~wil~~ wÿl er durch windt vnd secz im an…"*

Table 4: Contents of SR.

und Universitätsbibliothek Dresden ("Saxon State and University Library Dresden"). SR's current signature at the SLUB, *Mscr. Dresd. C. 487*, is recorded inside the back cover and on the spine.

4.3.3. Contents

SR has consistently been described in the literature as a work with a single author: Sigmund Ainring.[79] This is typically because he is credited as the author of the second section, the gloss beginning on f. 10r, while the subsequent treatises in the manuscript (including the other sections of gloss) have no attribution.

In fact, SR is a miscellany containing treatises from masters of the Fellowship of Johannes Liechtenauer[80] (including not only Ainring, but also Andre Lignitzer and Ott Jud) as well as several anonymous works which are mostly unique.[81] That the scribes omitted the names of Lignitzer and Ott is not uncommon in the tradition.

Its constituent texts cover fencing on horseback and on foot, in armor and out of armor, with the spear, the two-handed sword, the one-handed sword with the buckler, and wrestling.

It can be roughly divided into thirds, with the Ainring glosses forming bookends for shorter segments of teaching by other authors; both Ainring sections have title pages of a sort: f. 2r states "Here begins the chivalric art of the long sword" in large red text,[82] while f. 88r states "In Saint George's name, here begins the art" in large black text with pen flourishes.[83]

The anonymous gloss of Liechtenauer's *Record* of fencing with the long sword (ff. 57r–59v) is of particular note because it's otherwise unique in the corpus; it only addresses three of Liechtenauer's couplets (27–28 & 43), but the six blank folia immediately after it might indicate that the scribe planned to copy a longer text and left sufficient space for it, but never finished copying.

The two glosses of Liechtenauer's *Record* of mounted fencing (ff. 109r–110v, 110v–121v) are also interesting. The first is a fragment of the Ainring gloss (only the text glossing couplets 1–2) while the second is a fragment of the gloss currently attributed to the anonymous author nicknamed Pseudo-Peter von Danzig and covers the remainder (couplets 3–40). Since

79 See, e.g., Wierschin 1965: 43; Hils 1985: 55–57; Welle 1993: 40–41, 50–66, 92–103; Bodemer 2008: 103; and Hoffmann 2022: 319–323.

80 This group and its members are detailed in the description of GF.

81 The wrestling treatise on ff. 84r–86v is also found in WF.

82 *Hie hept sich an die Ritterlich kunst deß langen schwerts.*

83 *In sant Jorgen namen höpt an die kunst.*

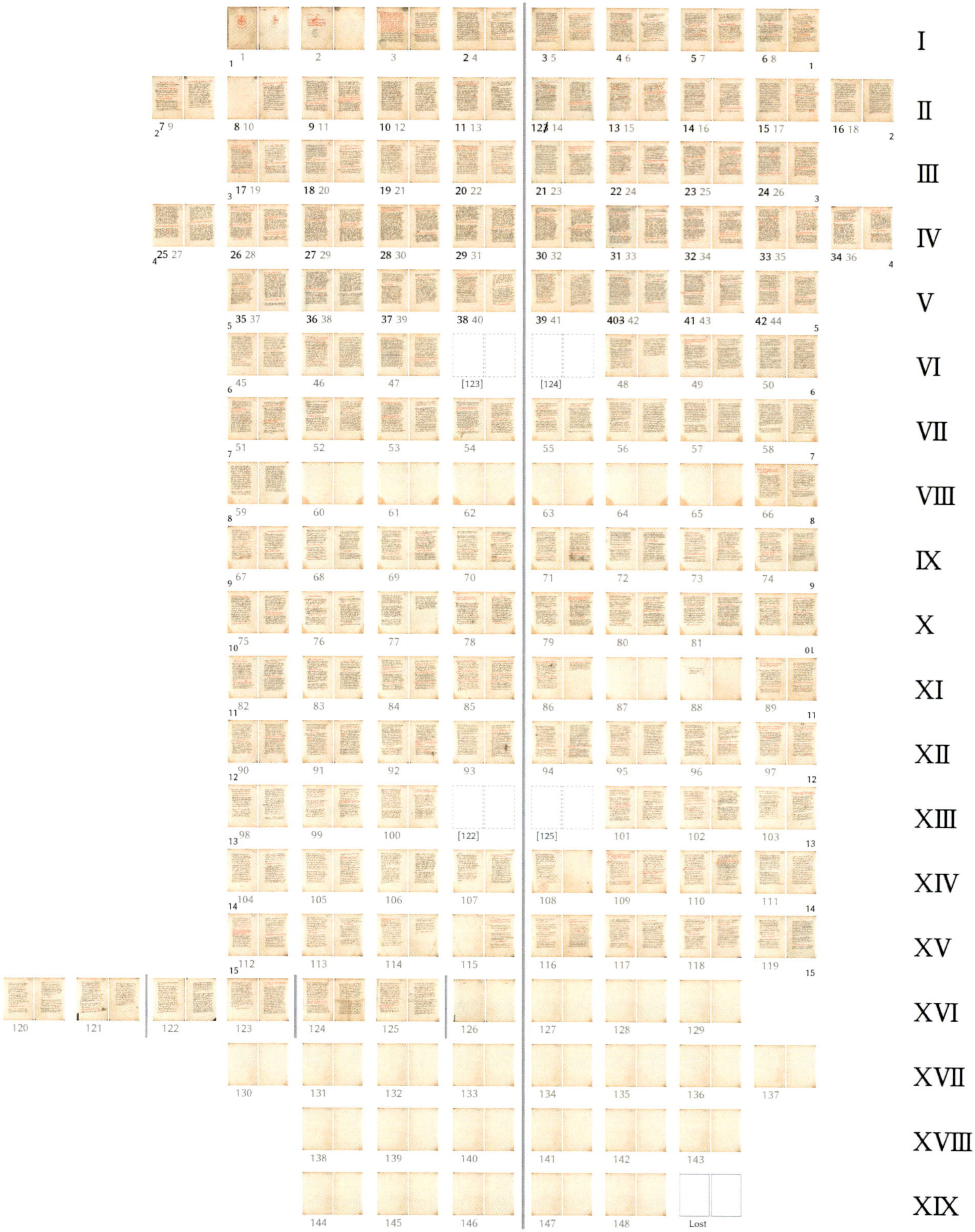

Figure 8: Collation of SR; the black and grey numbers represent the pen and pencil foliations, and the quire numeration is displayed in subscript; the heavy gray line indicates a fold.

the text of the Ainring gloss continues directly into the Pseudo-Danzig gloss with no indication that the source of the text had changed, it seems likely that the scribe or client for the manuscript viewed these two commentaries as equivalent or interchangeable.

4.3.4. STRUCTURE AND FOLIATION

The collation of SR, which is composed of 148 folia without flyleaves, can be represented according to BRADSHAW's model as follows:[84]

I^{8} (f. 8) + II^{10} (f. 18) + III^{8} (f. 26) + IV^{10} (f. 36) + V^{8} (f. 44) + VI^{8-2} (f. 50) + $VII{-}XII^{8}$ (f. 97) + XIII^{8-2} (f. 103) + $XIV{-}XV^{8}$ (f. 119) + XVI^{6+4} (f. 129) + XVII^{8} (f. 137) + XVIII^{6} (f. 143) + XIX^{6-1} (f. 148)

While the quires are generally quaternions, there are sporadic ternions and quinternions.[85] The center bifolium was separated from quires VI and XIII, and later sewn back in toward the end of the manuscript. They were nested together, with the bifolium from XIII outside the bifolium from VI, and sewn in between (current) ff. 121 & 126 in quire XVI; whoever did this repair apparently considered it more important to place them alongside the final written folio (f. 120) rather than in the center of the quire.

SR has two sets of folio numbers. The first begins on the third leaf (the first with writing) and continues to the end of quire V. It's written in brown ink in a 16th century hand, and may have been added by the scribe or original owner. This hand also added quire numbers to the first fifteen quires, in the bottom center of the first recto and last verso of each one.[86]

The second foliation is written in pencil in a more recent hand and runs through the whole manuscript—but skips a folio at the end of quire X, running 1–81 and 82–148 (these are the folio numbers used in the present edition).[87]

4.3.5. WRITING AND DECORATION

SR is written by two hands in a loose bastarda (hand I: ff. 2^{r}–26^{v}, 37^{r}, 39^{r}–125^{v}; hand II: ff. 27^{r}–36^{v}, 37^{v}–38^{v}) with frequent deletions, insertions, and

[84] Collation formula and diagram based on HOFFMANN 2022: 319; WIERSCHIN 1965: 78 agrees but omits the structure of quires XVI–XIX.
[85] Gatherings of eight, six, and ten folia, respectively.
[86] HOFFMANN 2022: 319.
[87] Rather than impose a third numbering scheme to further confuse the issue, this unnumbered folio is usually just labeled '81a'.

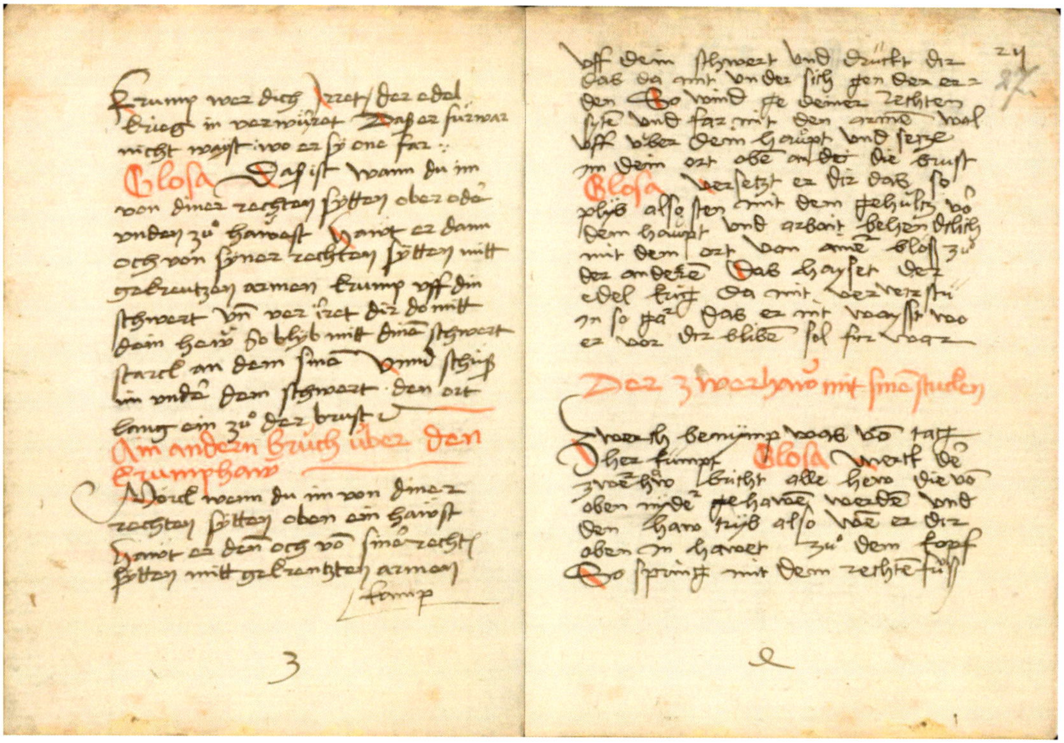

Figure 9: *Mise-en-page* of SR (ff. 26v–27r).

other attempts to correct apparent copying errors. Both hands are reminiscent of the first half of the 15th century;[88] given the current accepted date of the manuscript, these scribes must have been very old (or trained in a very old style).[89]

The text is written in a single column of 15–22 lines per page. There are no decorative borders or Lombards, but some sections begin with elaborate decorative Bastarda capitals and there are frequent red and black pen flourishes at the end of paragraphs. The titles are generally rubricated, as is the 'gloss' label, and there is other minor rubrication in the form of red slashing and underlining. The rubrics are generally written larger than the other text, as are certain keywords indicating the structure of the text (e.g., *merck*, *item*, *auch*, *das vernym also*, etc.).

4.3.6. ILLUSTRATIONS

SR is unillustrated.

[88] WIERSCHIN 1965: 83.

[89] For a detailed analysis of the characteristics of each hand, see WIERSCHIN 1965: 80–83.

4.3.7. LITERATURE

Editions: WIERSCHIN 1965; HAGEDORN 2008B; VERELST 2016 (ff. 10^v–48^v).

Translations: TOBLER 2001; LINDHOLD & SVÄRD 2003 (ff. 10^v–54^r); LINDHOLD & SVÄRD 2006 (ff. 54^r–55^v, 66^r–108^r); ŻABIŃSKI 2010: 67–80, 141–470, 515–548 (ff. 1^r–48^v); TROSCLAIR 2015A (ff. 10^r–48^v); CHENEY 2020A (ff. 10^r–48^v).

Literature: HILS 1985: 54–57; MÜLLER 1992: 267–270; WELLE 1993: 40–41, 50–66, 92–103; BODEMER 2008: 103–109; TOBLER 2010A: 6–8; JAQUET & WALCZAK 2014; KELLETT 2015; TALAGA & TALAGA 2018; SMITH 2021; HOFFMANN 2022: 319–323.

Scans: *digital.slub-dresden.de/id280775717*

4.4. OTHER MANUSCRIPTS

The remaining six manuscripts are profiled more briefly in this section.

4.4.1. AD – ALBRECHT DÜRER

Wien, Albertina, *Hs. 26-232* • ca. 1600 • Nürnberg

Description: Paper, 220×310 mm, I+126 ff.; 6 sewing supports; red leather binding with gold-foiled decorations and the title "OPVS ALBERTI DURERI", probably from ca. 1600; apparently composed of four production units: A, ff. 1–11; B, ff. 12–95; C, ff. 96–103; D, ff. 104–128; pages show signs of trimming on all three edges, including loss of text.

Origin: The oldest part of the ms. is D, probably created around 1500 by or for Willibald Pirckheimer (1470–1530); B and C were likely produced around the same time in the early 1500s under the supervision of Albrecht Dürer (1471–1528); ff. 53, 55, 57, & 69 were removed from the ms. and replaced by new blank leaves soon after creation; the interleaving of guard leaves was likely done in the latter part of the 16th cen.; A was produced last, around 1600 when the current ms. was compiled (the epitaph on f. 125^r was also added at this time); A is written on different paper from the rest of the ms.; WELLE concludes that A was created when the now-lost ms. ADB was copied from AD, and this unit was swapped between the two.[90]

[90] WELLE 2021: 111.

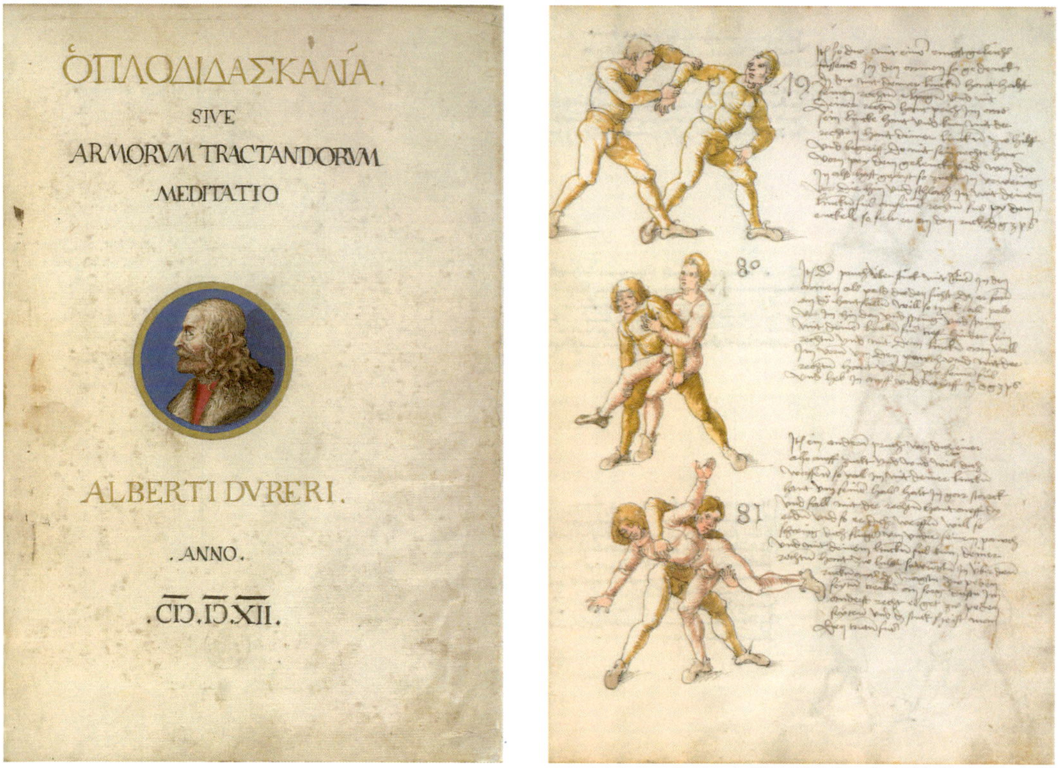

Figure 10: *Mise-en-page* of AD, parts A (left, f. 1ʳ) and B (right, f. 40ʳ).

Provenance: Owned by a member of the Flemish circle of Abraham Ortelius (1527–1598) in the 16th cen., who organized its current binding;[91] used as the exemplar for the lost ms. ADB in ca. 1600; listed in the 1607/1611 library catalog of Rudolf II (1552–1612), Holy Roman Emperor, in Prague;[92] possibly gifted to Sebastian Cressel by Ferdinand II, Holy Roman Emperor, in the 1630s and taken to Graz;[93] owned by Jacob Stahl in 1657;[94] rediscovered in Graz in 1823 by VINCENZ WEINTRITT and sold to the Imperial Library in Wien in 1833;[95] transferred to the Albertina when it was established in 1920.[96]

Contents:

1ʳ, title;

2ʳ, poems by Johannes Vivianus;

91 WELLE 2021: 94–110.
92 WELLE 2021: 90–91.
93 WELLE 2021: 117–118.
94 Who left an owner's mark inside the front cover.
95 HAGEDORN & JAQUET 2022: 15.
96 LENG 2008: 131.

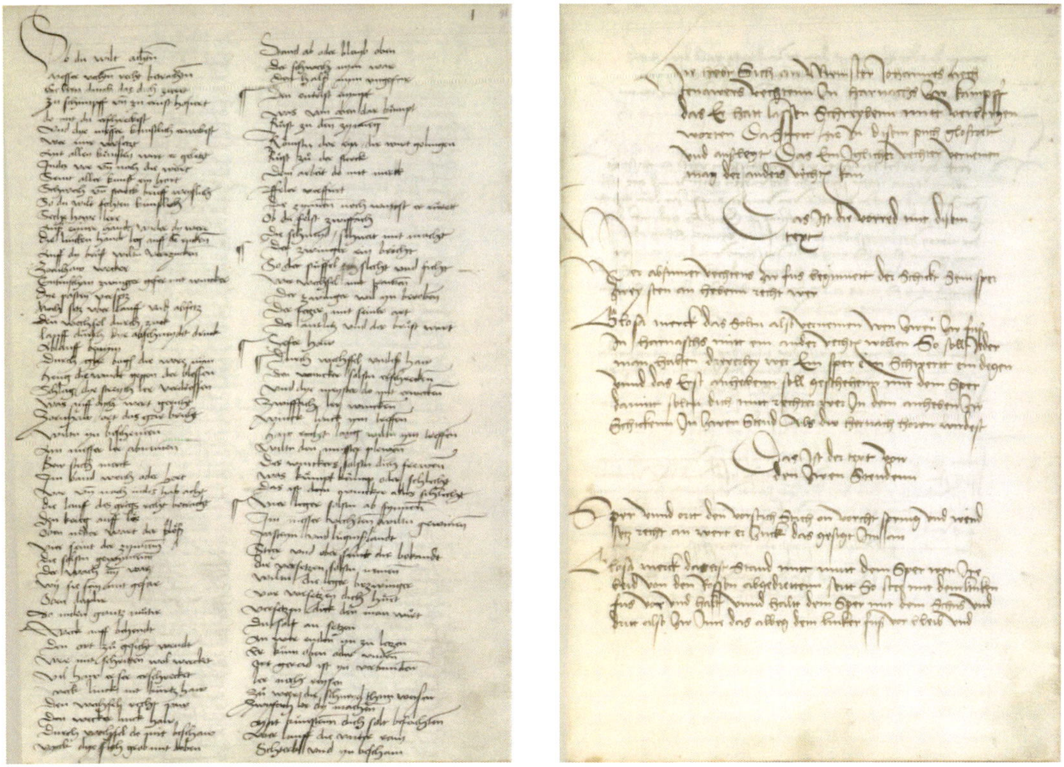

Figure 11: *Mise-en-page* of AD (continued), parts C (right, f. 96^{r}) and D (left, f. 105^{r}).

3^{r}, poem by Thomas Petit du Claux-Hardy;

13^{r}–52^{v}, anonymous illustrated wrestling (currently attributed to Albrecht Dürer);

60^{r}–66^{r}, anonymous illustrations of long sword (currently attributed to Albrecht Dürer);

66^{v}–67^{v}, anonymous illustrations of dagger (currently attributed to Albrecht Dürer);

74^{r}–91^{v}, anonymous illustrations of long knife (currently attributed to Albrecht Dürer);

96^{r}_{A}–100^{v}_{A}, *Record* and gloss of the long knife by Johannes Lecküchner (fragment; no author listed);

100^{v}_{A}, poem on mounted fencing by Martin Huntsfeld (no author listed);

100^{v}_{A}, *Record* of armored fencing by Johannes Liechtenauer (fragment; no author listed);

100^{v}_{AB}, wrestling by Ott Jud (fragment; no author listed);

101^{r}_{A}, anonymous sword;

101^{r}_{B}, anonymous dagger;

101^{r}_{B}–101^{v}_{A}, armored fencing by Andre Lignitzer (fragment; no author listed);

105^{r}–112^{v}, anonymous gloss of Liechtenauer's *Record* of armored fencing (currently attributed to Sigmund Ainring; disordered);

112^{v}–116^{v}, anonymous dagger;

116^{v}–124^{r}, anonymous gloss of Liechtenauer's *Record* of mounted fencing (currently attributed to Lew);

125^{r}–127^{r}, epitaph.

Collation:[97] I^{12} (f. 11) + II^{18-2} (f. 27) + III^{18} (f. 45) + IV–VI^{18-4} (f. 87) + VII–XI^{8} (f. 126); in quires II–VII, the original bifolia are interleaved with blank guard sheets, increasing their size from 8 folia to 18 before some guard leaves were lost; missing guard sheets between ff. 13–14 & 27–28.

Foliation: Folia numbered continuously in pencil (ff. 1–128) in a modern hand; illustrations numbered individually: wrestling, 1–120; long sword, 1–13; dagger, 14–22; long knife 1–58; folia in unit C numbered in pen (1–8?), mostly lost to trimming.

Writing: Written in a Bavarian dialect in 1 (A, B, & D) or 2 (C) columns by 10 hands; I: ff. 1^{r}–3^{r}, humanist antiqua; II: ff. 13^{r}–29^{v} & 33^{r}–42^{v}, cursiva by Dürer (also the numbering in the wrestling, sword, and long knife 1–46); III: ff. 29^{v}–31^{v}, cursiva by Pirckheimer; IV: inserted missing bits of text and fixed losses due to trimming in the previous; V: ff. 44^{r}–52^{v}, bastarda with cursiva elements; VI: numbering 14–22 in dagger and 47–58 in long knife; VII: ff. 96^{r}–12^{v}, bastarda and cursiva; VIII: ff. 105^{v}–124^{r}, bastarda (same as hand I in WP); IX: f. 124^{rv}, textualis and cursiva; X: 125^{r}, humanist antiqua;[98] writing in gold ink or pigment on ff. 1^{r} & 125^{r}; no decorative capitals or borders, but occasional use of larger plain capitals to begin sections in C & D.

Illustration: 203 pen and watercolor illustrations from the workshop of Albrecht Dürer (17 half-page and 186 third-page) showing pairs of fencers and wrestlers in clothing typical of the turn of the 16th cen.;[99] 2 later illustrations: a profile portrait of Dürer on f. 1^{r}, possibly by Melchior Lorck or

97 HAGEDORN & JAQUET 2022: 290–297. Assuming the current state of quire III is how quires II–VII were all expanded, there would also be single missing guard leaves between ff. 45–46, 59–60, 73–74, & 87–88 and double guard leaves missing between ff. 52–53, 66–67, & 80–81. WELLE 2021: 36–39 & 89 offers a slightly different collation, attaching f. 27 to the outside of quire III and also proposing that rather than missing two folia from quire II, ff. 13 & 27 were originally counterparts in the outer bifolium of quire II and there was simply no outer wrapper of blank sheets in any quire but III.

98 Breakdown of the hands from HAGEDORN & JAQUET 2022: 290, with details of the script supplied from LENG 2008: 131.

99 The extent of Dürer's personal involvement is difficult to determine, but WELLE 2021: 51–83 makes an extensive and detailed argument that he did the bulk of the work.

Abraham Ortelius,[100] and a sketch of a skull on f. 125^{r}, probably by Georg Hoefnagel.[101]

Literature: Wassmannsdorff 1888; Strauss 1974: 2663–2701; Hils 1985: 116–119; Welle 1993: 110–128; Anglo 2000: 183–186; Bodemer 2008: 161–171; Leng 2008: 131–134; Welle 2014: 13–16, 127–131; Widauer 2017; Welle 2020.

Editions: Wassmannsdorff 1870 (ff. 13^{r}–52^{r}); Dörnhöffer 1910 (ff. 13^{r}–91^{v});[102] Hagedorn 2016b; Hagedorn 2021a; Welle 2021; Hagedorn & Jaquet 2022.

Translations: Hagedorn & Jaquet 2022.

Scans: *sammlungenonline.albertina.at/?query=search=/record/objectnumbersearch=[26232/1]&showtype=record*

4.4.2. AR – Anton Rast

Augsburg, Stadtarchiv, *Reichsstadt Nr. 82* • 1553 • Augsburg

Description: Paper, 210×310 mm, 110 ff.; modern brown leather binding bearing Augsburg city heraldry.[103]

Origin: Augsburg, 1553; created based on an exemplar (currently lost) that Paul Hektor Mayr (1517–1579) purchased from the estate of *Marxbruder* and sword polisher Anton Rast († 1549);[104] Rast's work was probably based directly on **BF**[105] as well as other unknown sources.

Provenance: After Mayr's execution in 1579, his assets were seized by the Augsburg city council; **AR** wasn't sold with the rest of his fencing treatises and has been owned by the city ever since.[106]

Contents:

1^{r}–3^{v}, frontmatter by Paul Hektor Mayr;

4^{r}–16^{r}, anonymous gloss of Liechtenauer's *Record* of the long sword (currently attributed to Nicolaüs);

16^{r}–17^{v}, gloss of Liechtenauer's *Record* of the long sword by Sigmund Ainring (fragment; no author listed);

18^{r}–22^{v}, *Record* by Johannes Liechtenauer;

22^{v}–23^{v}, short sword by Martin Huntsfeld (fragment);

24^{r}–33^{v}, anonymous illustrated wrestling;

36^{r}–42^{v}, anonymous illustrated dagger;

100 Welle 2021: 96.

101 Welle 2021: 105.

102 Versions of this treatise were also released in 1907 and 1909, but this is the edition I accessed.

103 Roth 1917: 41.

104 As described on AR: 2^{r}.

105 Welle 2014: 127–131.

106 Leng 2008: 93.

Figure 12: *Mise-en-page* of AR, parts A (left, f. 13^{r}) and B (right, f. 45^{r}/#49).

43^{r}–47^{v}, anonymous illustrated long knife;
48^{r}–59^{v}, anonymous illustrated long sword;
60^{r}–67^{v}, armored fencing by Anton Rast(?);
68^{r}–71^{v}, staff by Anton Rast(?);
72^{r}–83^{v}, mounted fencing by Anton Rast(?);
84^{r}–100^{r}, illustrated descriptions of various types bits for horses;
101^{r}–107^{v}, *Horse pharmacopeia* by Albrant;
109^{v}–110^{r}, illustration of a horseman.

Collation: Not available.

Foliation: Old numeration in pen at top of leaf in three units (skipping first 3 leaves); A: 1–20 by folio (ff. 4–23); B: 1–154 by page (ff. 24^{r}–100^{v}); C: 155–164 by folio (ff. 101–110), final two numbers in pencil; newer numeration in pencil at bottom by folio throughout (ff. 1–110).

Writing: Written in a Bavarian dialect in a single column by three hands; I: 1^{r}–2^{r} & 107^{v}, cursiva, probably Paul Hektor Mayr; II: 2^{r}–107^{r}, calligraphic bastarda; III: 109^{v}, bastarda;[107] no decorative capitals or borders; simple rubrication in the form of red-ink titles.

107 LENG 2008: 94.

Illustration: 154 half-page pen and watercolor illustrations; 120 are multi-colored depictions of pairs of fencers (24^r–83^v), 33 are sketches of horse bits with grey shading (84^r–100^r), and 1 is a depiction of a single armored horseman (110^r); all except 110^r by Heinrich Vogtherr the Younger (1513–1568).[108]

Literature: ROTH 1917: cxliii–cxliv; HILS 1985: 24–25; WELLE 1993: 184–186; LENG 2008: 93–95; WELLE 2014: 127–131; WALCZAK 2022.

Editions: UEBERSCHÄR 2018.

Translations: TROSCLAIR 2015A (ff. 16^r–17^v); LEE 2022: 151–152 (ff. 25^v–26^v).

Scans: *commons.wikimedia.org/wiki/Category:Reichsstadt_"Schätze"_Nr._82*

4.4.3. HM – HANS MEDEL

Augsburg, Universitätsbibliothek, *Cod. I.6.2° 5* • after 1566 • upper Germany

Description: A convolution of four different works in the same binding (A, ff. 1–6; B, ff. 7–20; C, ff. 21–42; & D, ff. 43–49). Paper (A, B, & D) and parchment (C), 223×290 mm, I+49 ff.+I; parchment folia widened with paper strips glued on the fore-edge to match the width of the paper folia (paper folia in IV similarly widened with much smaller strips); modern tan leather binding with gold-foiled Oettingen-Wallerstein heraldry in the center; red and green ribbon bookmarks.

Origin: Various works of different ages; fencing regulations (A & B) recorded from 1491–1566;[109] treatise by Hans Medel (C) written before 1539;[110] prints (D) collected in 1552.[111] A & B likely bound as a single usage unit prior to being bought by Paul Hektor Mayr; all four units bound together not later than 1579.

Provenance: Confiscated by the city with Mayr's other possessions in 1579; 26 "fencing and other books" auctioned to Markus Fugger (1529–1597) in 1580;[112] 13 fencing books inherited by Marquart Fugger (1595–1655); these were sold to his brother-in-law Ernst II (1594–1670), Count of Oettingen-Wallerstein, in 1653;[113] signatures 'i.c.i9' and 'I.6.fol.5.' added to f. I^v in the 19th cen.; cataloged by WILHELM FRIEDRICH LÖFFELHOLZ

[108] As noted on AR: 110^v.

[109] Dates recorded on HM: 1^r, 20^r, etc.

[110] See HM: 21^r.

[111] See HM: 43^r.

[112] WELLE 2014: 22. Since this manuscript shares its provenance with the others in Mayr's library from this point, I rely on RAINER WELLE's thorough research into BF's provenance.

[113] WELLE 2014: 23. Library stamp "F. Öttingen Wallerstein'sche Bibliothek" inside both covers.

Figure 13: *Mise-en-page* of HM part C (ff. 24^{v}–25^{r}).

(1809–1891), Baron of Kolberg and chief archivist of the Oettingen-Wallerstein Library, in 1870;[114] purchased by the state of Bayern in 1980 and donated to the Universitätsbibliothek Augsburg.

Contents:

1^{r}–5^{v}, Augsburg fencing regulations;

7^{r}–20^{r}, chronicles and regulations of the Marxbrüder; 20^{v}, notes by Paul Hektor Mayr;

21^{r}–36^{v}, Gloss of Liechtenauer's *Record* of the long sword by Hans Medel von Salzburg;

37^{r}–42^{v}, 'seven stances' by Hans Medel von Salzburg;

43^{r}–49^{r}, 12 engravings by Maarten van Heemskerck.

Collation: Not available.

Foliation: Oldest numeration in pen at top of page in B (1–14) *&* C (1–22); foliated in pen at top of page in A *&* B (1–7); newest numeration in pencil at top of page throughout (ff. 1–49).

Writing: Written in Bavarian (A *&* C) and Franconian (B) dialects by three hands (one per section) plus incidental writing from others; A *&* B, written in a single column in cursiva with headings in textualis; C, written

[114] WELLE 2014: 23.

in a single column in bastarda; Mayr added annotations and short notes scattered throughout;[115] no decorative capitals or borders, but in C, space was left for 3-line capitals on 21^{r}, 22^{v}, 23^{r}, 25^{v}, 27^{r}, 30^{v}, & 37^{r}.

Illustration: A, sigil used by Paul Hektor Mayr in pen on f. 1^{r};[116] B, unillustrated; C, 17 pen and wash illustrations (on ff. 23^{r}, 24^{v}, 25^{v}, 26^{r}, 27^{r}, 29^{r}, 30^{v}, 31^{r}, 31^{v}, 34^{r}, 37^{v}, 38^{v}, 39^{r}, 39^{v}, 40^{r}, 41^{r}, 42^{r}) by one artist showing pairs of fencers in 16th cen. clothing standing on a shaded horizon line; D, 12 intaglio prints attributed to Maarten van Heemskerck (artist) and Dirck Volckertszoon Coornhert and Cornelis Bos (engravers) showing pairs of muscular nude fencers and wrestlers.[117]

Literature: WIERSCHIN 1965: 20–21; HILS 1985: 34–36; LENG 2008: 7–8.

Editions: DUPUIS 2009 (ff. 1^{r}–20^{v}); ENGSTRÖM 2009 (ff. 20^{v}–42^{v}); CONAN & LEBLANC 2024 (ff. 21^{r}–42^{v}).

Translations: TROSCLAIR 2015B (ff. 21^{r}–42^{v}); CONAN & LEBLANC 2024 (French, ff. 21^{r}–42^{v}).

Scans: *www.nbn-resolving.de/urn:nbn:de:bvb:384-uba002007-6*

4.4.4. HS – HANS VON SPEYER

Salzburg, Universitätsbibliothek, *M I 29* • 1491 • southwest Germany

Description: Paper, 157×214 mm, I+161 ff.; 3 sewing supports; 15th cen. blind-tooled red leather binding with two clasps; trimmed down from its original size (catchwords in some quires are cut off); apparently composed of four production units: A, ff. 1–9; B, ff. 10–45; C, ff. 46–117; D, ff. 118–161; beginning of each section marked with red tabs on the fore-edge.

Origin: Created in 1491 by Hans von Speyer;[118] 2 other names were recorded by Speyer in the text alongside his own signatures: 'Johannes de zeynissen de Novaciuitate' on f. 44^{r} and 'Johannes zeynissen de novacuutate' on f. 117^{r}; their relationship to the ms. is unknown.

Provenance: Uncertain; date of accession unknown; former owner's marks: on f. 1^{r}, (pen) 'Ecclesia Berchtolsgadis Can Reg 1665', (pencil) 'V. 1. H 159', (library stamp) 'K. K. Studien. Bibliothek Salzburg'; inside cover in different hands, (pencil) '~~III 2 J.~~', (pen) '~~278~~', (pen) '~~Mj: j54~~', (pencil) 'V. 1. H 159', (pencil) '159/121', (pencil) 'M I 29'.

115 LENG 2008: 8.

116 Thanks to JOHANNA HOPFGARTNER for pointing out that while this sigil doesn't appear in any of the other fight books owned by Mayr, it's common in other books from his library.

117 LENG 2008: 8.

118 Signed on HS: 44^{r} & 117^{r}; signed and dated on HS: 158^{r}.

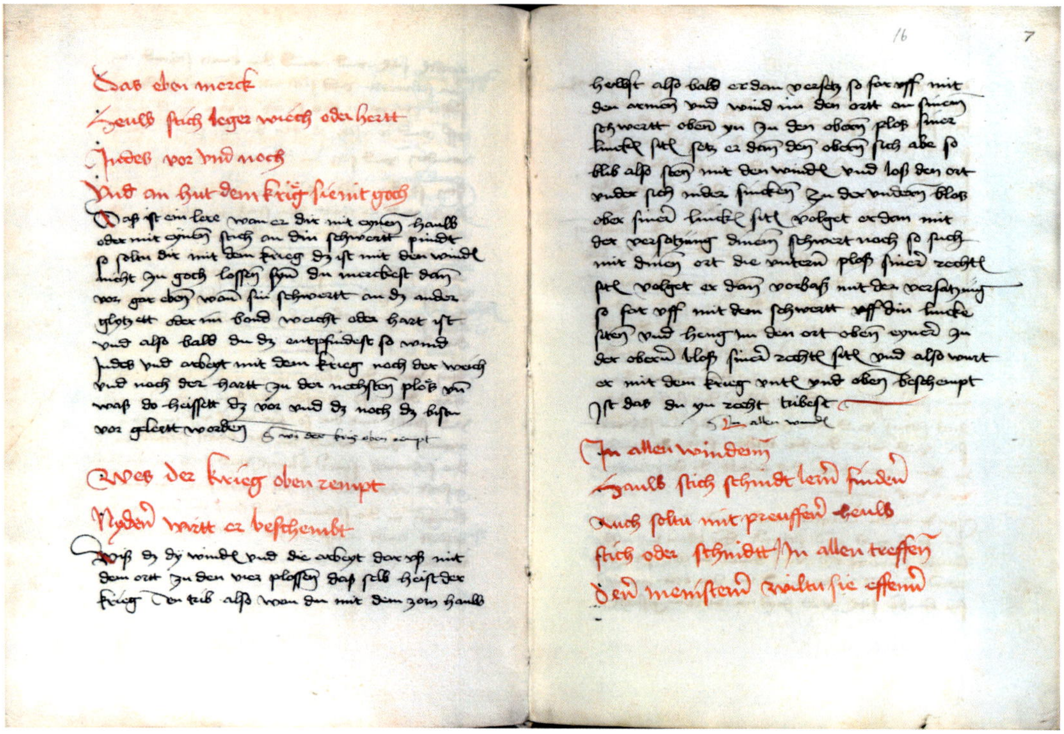

Figure 14: *Mise-en-page* of HM (ff. 15^{r}–16^{v}).

Contents:

1^{r}–2^{v}, *New Record* of the long sword by Martin Syber;

3^{r}, anonymous poem on fencing;

5^{r}–7^{r}, comparison of terminology by Andreas;

10^{r}–44^{r}, anonymous gloss of Liechtenauer's *Record* of the long sword (currently attributed to Lew), with inclusions from other authors including 11 paragraphs from Sigmund Ainring;

46^{r}–117^{r}, *Record* and gloss of the long knife by Johannes Lecküchner (no author listed);

119^{r}–126^{v}, wrestling by Ott Jud (no author listed);

129^{r}–130^{r}, *Record* of armored fencing by Johannes Liechtenauer (no author listed);

130^{r}–134^{v}, short sword by Martin Huntsfeld (author listed as Lew);

134^{v}–136^{v}, armored wrestling by Martin Huntsfeld (no author listed);

137^{r}–141^{r}, armored fencing by Andre Lignitzer (author listed as Martin Huntsfeld);

143^{r}–158^{r}, anonymous gloss of Liechtenauer's *Record* of mounted fencing (currently attributed to Lew).

Collation:[119] I–III^{10} (f. 29) + IV–V^{8} (f. 45) + VI–VIII^{10} (f. 75) + IX^{12} (f. 87) + X–XIII^{10} (f. 127) + XIV^{14} (f. 141) + XV–XVI^{10} (f. 161).

Foliation: Old numeration in pen in the first three units: A, 1–9; B, 1–36; C, 1–72; new numeration in pencil throughout (ff. 1–161).

Writing: Written in an Alemannic dialect in a single column by one hand, a loose bastarda,[120] identified as Hans von Speyer based on the signatures at the end of units B, C, & D (ff. 44^{r}, 117^{r}, & 158^{r}); no decorative capitals or borders, though space was left for a 3-line capital on f. 46^{r}; simple rubrication in the form of red-ink titles, verses of poetry, and frequent pen flourishes.

Illustration: None.

Literature: WIERSCHIN 1965: 35; HILS 1985: 113–115; LENG 2008: 93–95.

Editions: HAGEDORN 2009; TROSCLAIR [forthcoming].

Translations: TROSCLAIR 2015A (ff. 10^{r}–44^{r}); FABIAN 2024 (ff. 1^{r}–117^{r}); TROSCLAIR [forthcoming].

Scans: *wiktenauer.com/wiki/Codex_Speyer_(MS_M.I.29)*

4.4.5. P – PARIS FIGHT BOOK

Paris, Musée de Cluny, *Cl. 23842* • 15th cen.$^{4/4}$ • southern Germany

Description: Paper, 158×218 mm, XI+201 ff.+XI; 3 sewing supports; 15th cen. blind-tooled brown half-leather binding over wood with remnants of two clasps; pages show signs of trimming on bottom and fore-edges, including loss of text and illustration.

Origin: Produced in the last quarter, 15th cen., with art apparently based in part on the treatises of *Blume des Kampfes*, *Gladiatoria*, Peter Falkner, Paul Kal, and Hans Talhoffer, as well as including unique art (or using other unknown resources).[121]

Provenance: Potentially used by Jörg Wilhalm as an exemplar for his treatise on the long sword in ca. 1522;[122] "Ain Ander wendelß" written on f. 195^{r} in drypoint; inside front cover: (pencil) "37.5." and "6110"; acquired by the Fürstenberg family prior to 1865 and housed in Castle Donaueschingen under signature Cod. 862 (library stamp on ff. 1^{r} & 223^{r}, "F. F. BIBLIOTHEK DONAUESCHINENS");[123] sold to IAN WOODNER (1903–1990)

119 JUNGREITHMAYR 1988: 5.

120 JUNGREITHMAYR 1988.

121 Attempting to identify which of the extant manuscripts from each of these textual traditions was employed as an exemplar, if any, is beyond the scope of this work.

122 HAGEDORN, ET AL. 2021: 25, 40–41.

123 When the collection was cataloged by KARL AUGUST BARACK.

Figure 15: *Mise-en-page* of P (left, f. 17^r; right, f. 17^v).

of New York in a private sale in the 1980s;[124] acquired by Dr. Jörn Günther Rare Books of London by 2003;[125] sold to the Musée de Cluny in Paris in 2008.[126]

Contents:

12^r–57^v, illustrations of long sword;
60^r–69^v, illustrations of long knife;
72^r–87^v, illustrations of dagger;
95^r–114^v, illustrations of wrestling;
118^r–131^v, illustrations of sword and buckler;
133^v–146^r, illustrations of mounted fencing;
146^v–148^r, illustrations of mounted vs. foot;
149^r–178^v, illustrations of armored fencing;
180^r–181^v, illustrations of long shield;
183^r–186^v, illustrations of pole weapons;
187^r, illustrations of long shield;

124 According to a 2005 Sotheby's description that has since been removed from the internet, previously available at *search.sothebys.com/jsps/live/lot/LotDetail.jsp?&lot_id=4GDGX*

125 LENG 2008: 27.

126 According to the museum's website; see *www.musee-moyenage.fr/collection/oeuvre/traite-de-combat.html*

187^{v}–189^{r}, illustrations of pole weapons;
190^{r}–191^{v}, illustrations of spear;
194^{r}, illustration of curved dagger;
194^{v}, illustration of a duel between a man and a woman;
195^{r}–212^{v}, illustrations of armored fencing.

Collation: ^{11} (f. 11) + I–II^{12} (f. 35) + III^{12-1} (f. 46) + IV^{12} (f. 58) + V^{12+1} (f. 71) + VI^{12} (f. 83) +VII^{12-1} (f. 94) + VIII^{10+2} (f. 106) + IX^{12-1} (f. 117) + X^{8} (f. 125) + XI^{12} (f. 137) XII^{12+1} (f. 150) + XIV^{12} (f. 176) + XV^{10} (f. 186) + XVI^{6} (f. 192) + XVII^{2} (f. 194) + XVIII^{12} (f. 206) + XIX^{6} (f. 212) + ^{11} (f. 223); 11 flyleaves on both sides of page block; ff. 59, 95, 100, & 145; sheets missing between ff. 35–36, 90–91, & 113–114.

Foliation: Numbered continuously in the upper right corners, including flyleaves (ff. 1–223).

Writing: Scattered incidental notes from several hands (mostly focused on the long sword section) present on ff. 12^{r}–20^{v}, 21^{v}–23^{r}, 27^{r}–30^{r}, 32^{r}–34^{r}, 35^{v}, 36^{r}–37^{v}, 38^{v}–47^{r}, 48^{r}–57^{v}, 83^{r}, 170^{v}, 171^{v}, & 195^{r}.

Illustration: 358 half-page watercolor illustrations by two artists: I, 12^{r}–194^{v}; II, 195^{r}–212^{v}; illustrations depict fencers and wrestlers with various weapons, including in armor and on horseback.

Literature: WIERSCHIN 1965: 13–14; HILS 1985: 46–50; WELLE 1993: 140–142; LENG 2008: 27–29, 76–77; HAGEDORN, ET AL. 2021: 25, 31, 40–41.

Editions: *None.*

Translations: *None.*

Scans: *www.photo.rmn.fr/C.aspx?VP3=SearchResult&VBID=2CO5PCQFH3MF*

4.4.6. WP – WILLIBALD PIRCKHEIMER

Nürnberg, Stadtbibliothek, *PP. 353* • ca. 1500 • Nürnberg

Description: Paper, 163×223 mm, 68 ff.; unbound, held together by a single long stitch with a seal.[127]

Origin: Uncertain; apparently received by Willibald Pirckheimer in a partially-finished state and then expanded by him with additional texts.

Provenance: Part of a cache of documents by discovered during the renovation of Pirckheimer's former house in 1757;[128] planned for individual auction in 1861, but approximately 1,700 items from the collection purchased as a group by Nürnberg prior to the auction;[129] kept in the city archives to the present.

127 WELLE 2021: 48.
128 WELLE 2021: 49.
129 SCHEIBLE & HOLZBERG 2012: 2, 3.

Contents:

1^r–3^v, anonymous gloss of Liechtenauer's *Record* of armored fencing (currently attributed to Sigmund Ainring; disordered);

4^r, anonymous dagger;

4^v–5^v, anonymous gloss of Liechtenauer's *Record* of mounted fencing (currently attributed to Lew);

6^r–13^v, anonymous gloss of Liechtenauer's *Record* of armored fencing (currently attributed to Sigmund Ainring) continued;

14^r–17^v, anonymous dagger (continued);

18^r–27^r, anonymous gloss of Liechtenauer's *Record* of mounted fencing (currently attributed to Lew) (continued);

27^r–32^v, armored fencing by Martin Huntsfeld (author listed as Lew);

32^v–37^v, armored fencing by Andre Lignitzer (author listed as Martin Huntsfeld);

37^{rv}, *Record* of armored fencing by Johannes Liechtenauer;

37^v, *Record* of the long sword by Johannes Liechtenauer (fragment);

39^r–62^r, anonymous gloss of Liechtenauer's *Record* of the long sword (currently attributed to Lew);

63^r–65^v, anonymous mounted fencing.[130]

Collation: Single quire; there's evidence that some leaves were bound out of order.[131]

Foliation: Numbered continuously (ff. 1–68).

Writing: Written in three hands; III, which wrote ff. 39^r–65^v and made corrections to the other two, can be identified as Pirckheimer.[132]

Illustration: None.

Literature: SCHEIBLE & HOLZBERG 2012; WELLE 2021: 48–49; HAGEDORN & JAQUET 2022: 14–15.

Editions: *None.*

Translations: *None.*

Scans: *Not available.*

4.5. COMPARISON AND ANALYSIS

As mentioned at the start of this chapter, there are nine manuscripts that contain some witness of Ainring's teachings. Some of these have been analyzed in detail by scholars, while others have received little attention.

[130] Contents provided by DIERK HAGEDORN.

[131] WELLE 2021: 49.

[132] WELLE 2021: 49.

	Witness	Date	LS	MF	AF
AD	*Albrecht Dürer* or *Oplodidaskalia* Wien, Albertina, Hs. 26-232	ca. 1600			×
AR	*Anton Rast* Augsburg, Stadtarchiv, Reichsstadt Nr. 82	1553	/		
GF	***Glasgow fight book* or *Sigmund Emring*** **Glasgow, Kelvingrove Museum, E.1939.65.341**	**1508**	×	×	
HM	*Hans Medel* or *Sigmund Schining* Augsburg, Universitätsbibliothek, Cod. I.6.2° 5	after 1566	/		
HS	*Hans von Speyer* Salzburg, Universitätsbibliothek, M I 29	1491	/		
JMR	***Joachim Meyer-Rostock*** **Rostock, Universitätsbibliothek, Mss. var. 82**	**1570**	×		×
P	*Paris fight book* or *Cluny fight book* Paris, Musée de Cluny, Cl. 23842	1480–1500	/		
SR	***Sigmund Ainringck fight book*** **Dresden, SLUB, Mscr. Dresd. C.487**	**ca. 1510**	×	/	×
WP	*Willibald Pirckheimer* Nürnberg, Stadtbibliothek, Pirckh. Papp. 353	ca. 1500			×

Table 5: Known witnesses of Ainring's gloss along with which sections of gloss are present in each one. An × indicates the presence of an entire section (or a substantial majority), while a / indicates that only one or more small fragments are present.

It's easy to declare that texts are or are not connected based on external factors or a glance at their contents, but as RAINER WELLE has pointed out, the only correct way to establish these relationships is through the use of a thorough reading apparatus comparing the witnesses on a word-by-word basis.[133]

DANIEL JAQUET and BARTŁOMIEJ WALCZAK further emphasize that such assessments must be made separately for each section, treating each one as a potentially independent unit of transmission,[134] and thus allow for the possibility that a manuscript was copied from multiple separate sources rather than being a faithful copy of a single one. Once all of the sections of a manuscript have thus been evaluated, the results may be combined and the manuscript may finally be placed correctly in a comprehensive stemma.

133 WELLE 1993: 55.
134 JAQUET & WALCZAK 2014: 123.

Such efforts at creating individual stemmata have been done previously for the writings of Martin Huntsfeld,[135] Lew,[136] Andre Lignitzer,[137] and Ott Jud,[138] but this analysis is lacking for the writings of Sigmund Ainring.[139]

The complete reading apparatus may be viewed in Chapter 5, but a survey of the findings is included here.

4.5.1. LONG SWORD

The gloss of the long sword is present in seven manuscripts (**AR**, **GF**, **HM**, **HS**, **JMR**, **P**, and **SR**) and may be divided into an introduction and 90 pieces.

It's attributed to Sigmund Emring in **GF**, Sigmund Einring in **JMR**, Sigmund ain ringeck in **SR**, and Sigmund Schining (or Schninig) in **HM**. The other witnesses lack the introductory section and have no attribution.

SR includes the most pieces, with 87 of the 90 (though it's missing substantial segments of a few more). **GF** has only 50 pieces, while **JMR** has only 35 out of 90 and is missing most of the latter part of the gloss. Both of these manuscripts are damaged in their current states, and **GF** may have lost up to 32 pieces and **JMR** 5. The other three have only fragments of Ainring's gloss inserted into a different text: **HM** has 21 pieces (scatter through Hans Medel's gloss), **HS** has 11 (scattered through Lew's gloss), and **AR** has 8 (appended to Nicolaüs' gloss).

GF and **JMR** both make frequent reference to illustrations, but only **GF** actually has any of them.[140] **SR** lacks these illustrations and almost all referenced to them, but that seems to be the product of an effort to carefully eliminate all such references rather than evidence that they were a later invention, because one reference remains intact[141] while two others were mangled but not removed completely.[142] The other three textual witnesses make no reference to illustrations.

Interestingly, while these references to illustrations appear in the same position in all three witnesses, the phrasing tends to vary. For example, where **GF**: 1ʳ has *als am nãchstn hernach gemalt stett*, **JMR**: 11ʳ has only

135 JAQUET & WALCZAK 2014; also discussed in HAGEDORN ET AL. 2017: 72–75.

136 HAGEDORN ET AL. 2017: 60–86.

137 JAQUET & WALCZAK 2014; also discussed in HAGEDORN ET AL. 2017: 75–76.

138 WELLE 1993: 39–83; also discussed in HAGEDORN ET AL. 2017: 77–78.

139 Though a brief discussion of the differences between the major Liechtenauer glosses appears in HAGEDORN, ET AL. 2017: 69–72 & 78–79.

140 Though three folia (ff. 8ʳ & 18ʳᵛ) mention illustrations that aren't present in the manuscript.

141 *alß am nesten gemelt ist* on SR: 31ʳ.

142 SR: 20ʳ has *alß vor stett* and SR: 27ʳ replaces it with the word *gloss*.

als hie stet gemalt. Given how precise the copyists are in preserving the wording of the text and gloss otherwise, this difference seems noteworthy. Perhaps, like the rubricated titles (which also change considerably between witnesses), these references were considered to be of secondary importance—a part of the apparatus rather than the authoritative text—and therefore the phrasing was unimportant. This might also explain the decision to silently remove them from the unillustrated SR.

None of these witnesses seems to be directly based on any other extant one. This can be easily demonstrated because even apart from small differences in word order and phrases absent or present, each surviving witness includes one or more entire pieces that are missing from any witness that they might have been copied from.

For example, the second piece in the current configuration of GF ("the counter against the lower crosswise strike"), piece 35 in the present book, is missing from all witnesses except JMR. JMR has sometimes been suggestedto be a late copy of GF, but the presence of piece 84 ("the slice over") in JMR[143] and not GF invalidates that supposition.

The three fragmentary witnesses are poor candidates to be the source of any of the three longer ones (at a minimum, they could only be one source among other lost ones), but HS does help demonstrate the distance of HM from the others. There's one piece shared by both and missing in all other witnesses: piece 33 ("the counter to the laying-on"). Since there's no other evidence that Hans Medel had access to HS and he certainly didn't copy any of the manuscript's other unique passages, this strongly suggests that he was working from a non-extant Ainring manuscript.

Finally, at first glance, AR could conceivably have been derived from GF by omitting the titles and dropping a few phrases here and there. However, it includes the phrase *zu der andern seiten. zu der plöß* in piece 88, which matches the phrasing in SR—but not GF, which has *zu der anderen plös oder sytten*.

Based on this, I conclude that all surviving textual witnesses likely descended independently from Ainring's original text. While it's true that AR, GF, and JMR are "closer" to each other than any of the others are, there's no evidence requiring that they descend from the same intermediate copy, and their shared distinctiveness from the other three might as easily indicate that the others simply diverged more from the archetype.

It's hard to convincingly answer the question of whether the original Ainring gloss was illustrated. The three principal witnesses all refer to illustrations, but they could conceivably derive from an intermediate manuscript which added illustrations to a previously unillustrated

143 Including a reference to an illustration, unlike the version in SR.

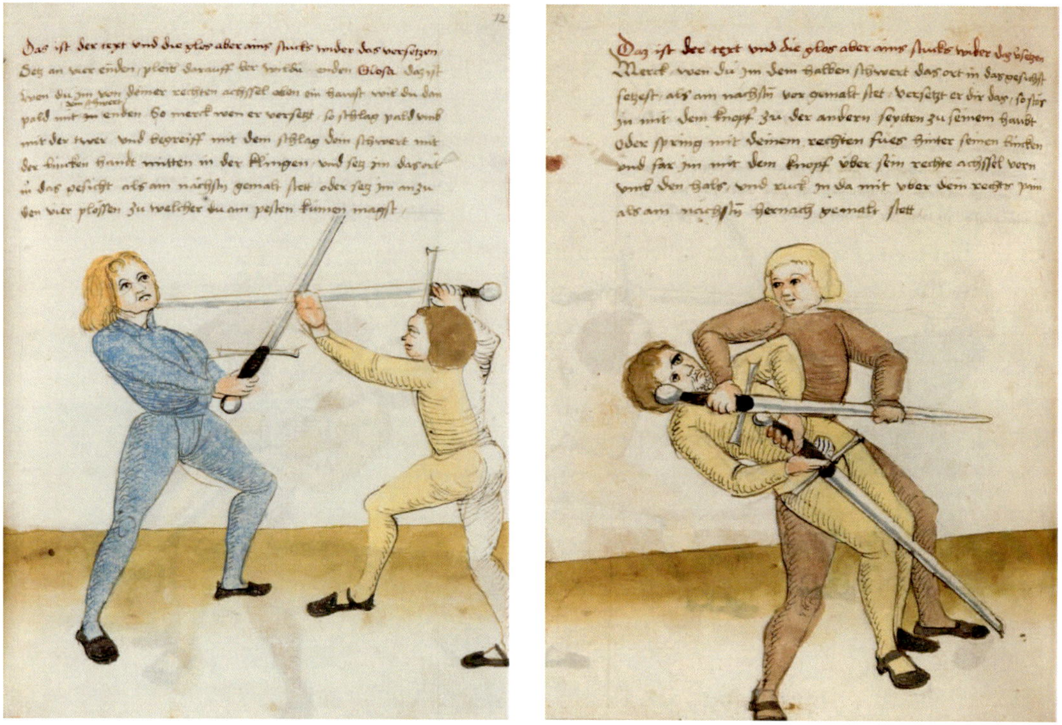

Figure 16: Pieces 60–61 of the long sword according to GF (left, f. 12^{r}; right, f. 12^{v}).

text.[144] However, the level of differentiation between the witnesses (including between SR and GF) is extensive and can't be explained fully by such an intermediate copy, so absent more conclusive evidence, the simpler explanation for the present is that the original included illustrations, presumably similar to those in GF.

Finally, lacking any substantial text, the position of P is hardest to pin down. It includes at least five pieces unique to Ainring's gloss:[145] f. 17^{r} shows piece 22,[146] and ff. 30^{r}–33^{v} show pieces 58–61.[147]

Of these, piece 61 is particularly noteworthy. The text in GF reads:

> *Remember, when you set the point in the face with the half sword (as is pictured nearby previously), if he parries that, bash in toward the other side of his head with the pommel. Or leap with your right*

[144] This was the case with G, for example, a collection of mostly 15th cen. treatises which were illustrated by Georg Lemberger in the 1530s. See WELLE 2017.

[145] It's possible that there are other illustrations in the manuscript also intended to accompany Ainring's text, but in cases where Ainring's descriptions agree with one or more other glossators, there's no basis on which to attribute the illustrations to Ainring's text specifically.

[146] Also, P: 17^{v} seems to show a counter to piece 22 not described by Ainring.

[147] With additional 'extra' counters (including one, P: 33^{v}, which is only included in HM: 33^{v}).

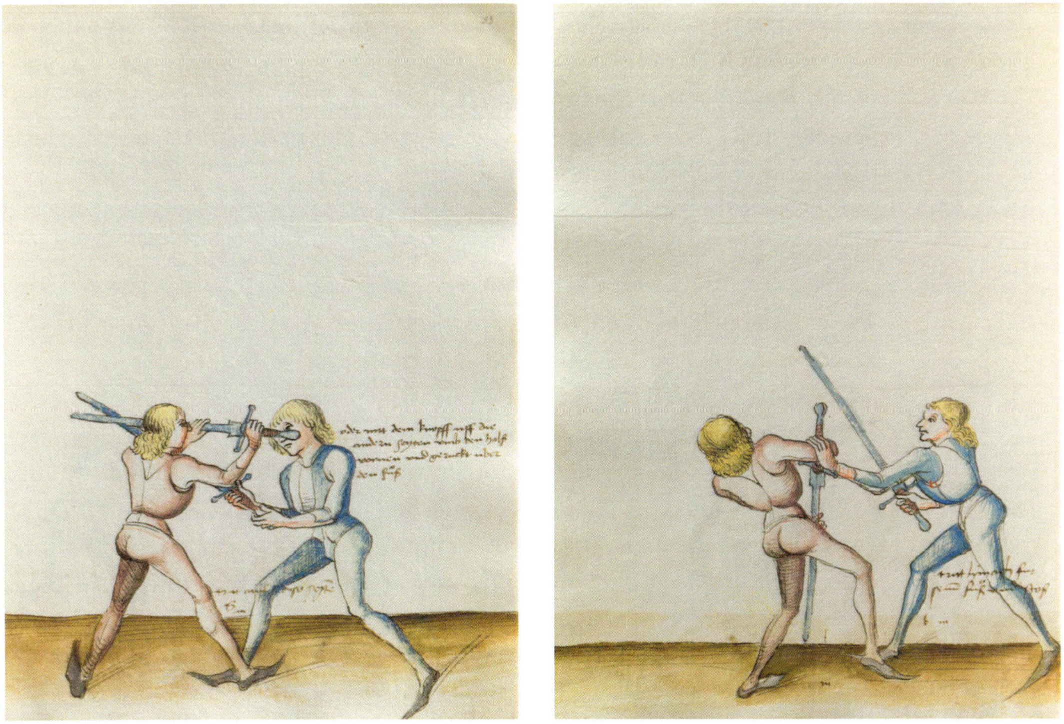

Figure 17: Piece 61 of the long sword according to P (left, f. 33^{r}; right, f. 33^{v}).

> *foot behind his left and go in with the pommel over his right shoulder, around the neck in front, and pull him over your right leg with it (as is pictured nearby hereafter).*[148]

To this text, **HM** adds: "Counter: take the elbow."[149]

Thus, while it mentions two potential outcomes, it specifies that the illustration will depict a pommel-assisted throw over the fencer's right leg. **P**, on the other hand, shows the pommel bash to the opponent's head (f. 33^{r}) before proceeding to Medel's counter. What's more, the caption, which is one of the longer in **P**, states "Or with the pommel on the other side around the neck in front, step to the side, and pull over the foot",[150] indicating that the creator was also familiar with the other option in the piece.

[148] *Merck, wen du im dem halben schwert das ort in das gesichst setzest, als am nächsten vor gemalt stet, versetzt er dir das, so stōs in mit dem knopf zu der andern seytten zu seinem haubt, Oder spring mit deinem rechten fües hinter seinen lincken vnd far im mit dem knopf v̈ber sein rechte achssel vorn vmb den hals, vnd ruck in da mit vber dein rechts pain als am nächsten hernach gemalt stett.* GF: 12^{r}.

[149] *Bruch nym den elenpogen.* HM: 33^{v}.

[150] *oder mit dem knopff uff die andern seyten vmb den halß vornen vnd geruckt uber den fuß tritt auff dise seyten.* P: 33^{r}.

This indicates that the textual source for these pictures was probably a copy from which the references to illustrations had been removed (though it's also possible that the artist or client merely liked the outcome with the pommel strike more and chose to include that illustration instead; that question could only be settled by finding the source).

4.5.2. Armored fencing

The gloss of armored fencing is present in four manuscripts (AD, JMR, SR, and WP) and may be divided into an introduction and 25 pieces.

This gloss is anonymous in all witnesses.

These witnesses are far more uniform than in either of the other two sections; SR includes some portion of all 25 pieces, while AD and WP are missing all of piece 17, and JMR is missing all of piece 20.

SR's independence from the other three can be established based on differences in the text and especially in piece 9 ("How one should arrange oneself when the spear is against the sword"), in which SR includes this instruction that's missing from the other three:

> *Likewise, if he stabs toward you with the glaive when you are in the lower guard, then set aside the thrust in front of your left hand, with the sword on his right side, and look for the setting upon or the wrestling.* [151]

Whereas the other three witnesses include an instruction in piece 9 that's missing from SR:

> *Or likewise, if he stabs toward you underneath, then set aside the thrust in front of your left hand and leap with the right behind his left, and go in with the pommel in front over his right shoulder around the neck, and throw him over the knee.* [152]

AD, JMR, and WP are much more homogenous (apart from the large gap in each one mentioned above). However, though AD and WP have the same pieces and the same gap, piece 19 ("the third guard with the short sword) includes one instruction in WP that's missing from AD:

[151] *Item sticht er dir zů mitt der glefen wann du stäst in der vndern hůt So secz im den stich ab mitt dem schwert vor diner lincken hand vff sin rechte sÿtten vnd wart deß anseczents oder der Ringen.* SR: 93^{r}.3–93^{v}.1

[152] *Item oder sticht er dir vnden zw so setz den stich ab vor deiner linckenn handt vnd spring mit dem Rechten fuß hinter sein lynckenn vnnd far Im mit dem knopff fornn vber sein rechte achsel vmb den hals vnnd wurff In vbers knye* WP: 8^{v}.3 (see also AD: 109^{r}.2 & JMR: 104^{r}.3).

> *Or likewise, go through below with the pommel, over his forward hand, and pull downward with it and set upon. You can also change through below with the pommel and set aside his thrust.*[153]

Given Willibald Pirckheimer's known involvement in the creation of both of these manuscripts (detailed above), it's likely that AD was directly copied from WP.

AD, JMR, and WP are heavily disorganized compared to both SR and Liechtenauer's *Record*. At multiple points in the gloss, all three witnesses simultaneously jump to a different part of the text, e.g. *sticht er denn || zu Kampf* (jumping from the middle of piece 4 to the middle of piece 21 at the ||).[154] This suggests that both JMR and WP were copied from the same source, and that this source had some of its pages out of order. Since JMR includes the text missing from WP (and AD), it couldn't be a copy of either manuscript and must have proceeded from that source independently.

4.5.3. Mounted fencing

The gloss of the mounted fencing is present in two manuscripts (GF and SR) and may be divided into an introduction and 28 pieces. It's also accompanied by Liechtenauer's '26 Figures' in GF.

Like the armored fencing, it's anonymous in both witnesses.

SR is only a brief fragment including the introduction and three pieces. This fragment is largely identical with GF, but it includes an extra rubric for the introduction (*Hie hept sich an die ritterlich kunst das fechten zuo roß*). The introductory paragraph also reads *Vnnd stat her nach*, whereas GF has *das stett alhÿe in diesem puech*.

In piece 1, SR refers to the weapon used at one point as *dem sper* rather than *der gleffen* like GF.

The other differences are limited to short words, but this seems sufficient to suggest that these two witnesses are independent.

4.5.4. Conclusion

The broad consistency in wording, phrasing, and sequence of instructions shown in the reading apparatus in Chapter 5, even where not required by the structure of the *Record*, indicates that all of these witnesses proceeded

153 *Item Oder far vnden durch mit dem knopff vber sein vorgesatzte hant vnd ruck damit vntersich vnnd setze an Auch magstu vnden durch wechselnn mit dem knopff vnd Im denn stich absetzenn* WP: 13r.4

154 AD: 105v.2, JMR: 98v.3, & WP: 1v.2.

from a single written archetype of each gloss rather than, for example, an oral tradition.

Using this analysis, a *stemma* of the known witnesses can be constructed by combining the findings for each manuscript. It isn't possible to determine from the current evidence whether the archetypes for all three sections of gloss originated from the same author, but I'll assume for the moment that they originated in a single manuscript. In doing this, I must acknowledge that the hypothetical "original Ainring" could be three separate manuscripts, one for each gloss.

Since none of the witnesses of the long sword or mounted fencing show signs of direct copying (despite the strong similarity between AR, GF, and JMR), their *stemmata* are quite simple. Only the armored fencing presents any complication, and I posit a hypothetical intermediate manuscript that JMR and WP (and therefore also AD) are based on which introduced their chaotic structure.

As an interesting final note, the *stemma* of Huntsfeld, Lew, and Lignitzer's works by JAQUET & WALCZAK connects HS and JMR through a hypothetical earlier manuscript they call 'Lew 3',[155] which could be the same as 'Ainring 2' on this *stemma*.

[155] JAQUET & WALCZAK 2014: 122.

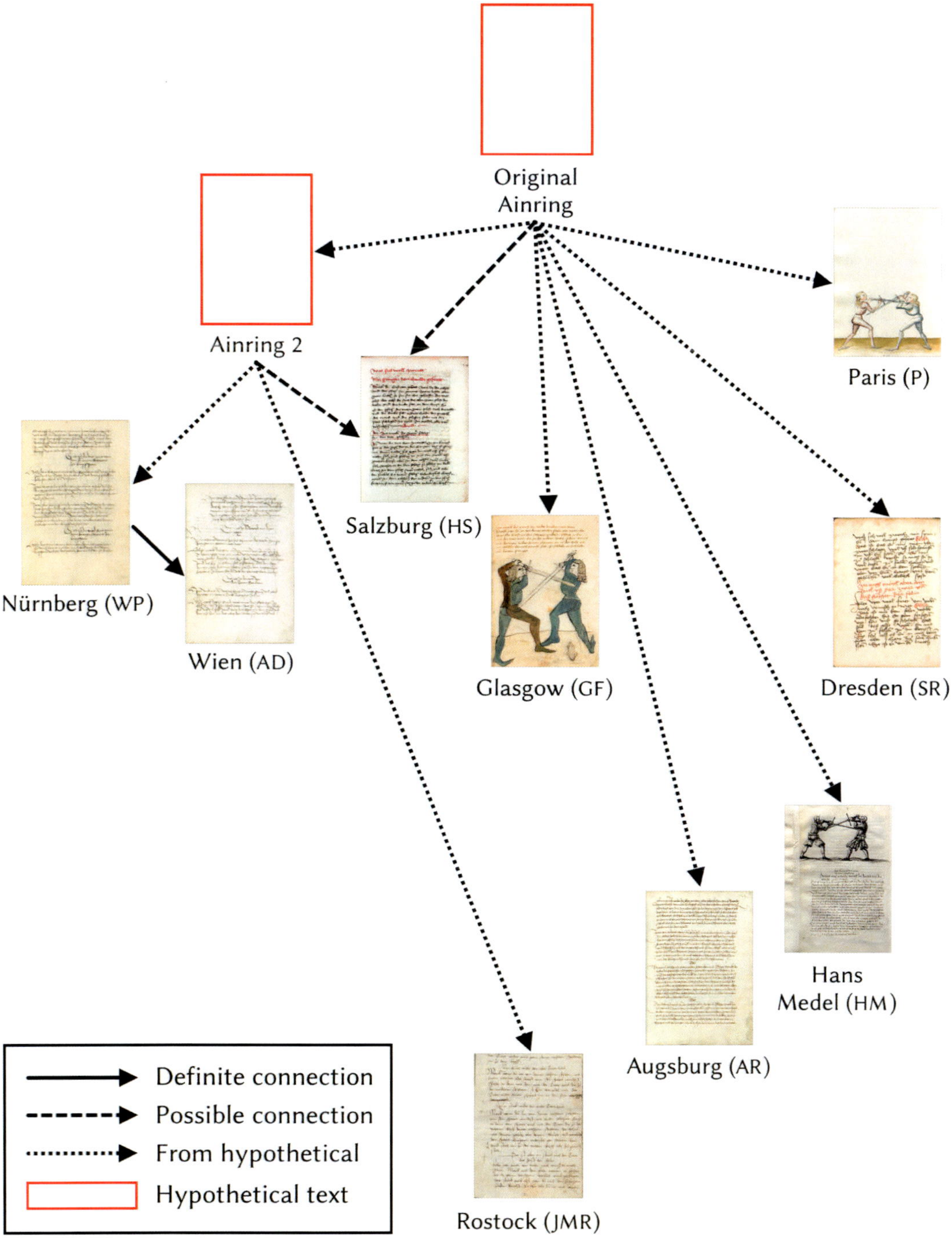

Figure 18: Combined *stemma* of the nine witnesses to Ainring's gloss (and/or the other two associated glosses). Vertical placement represents an approximate timeline, with the earliest witness (P) dating to the fourth quarter of the 15th cen. and the latest witness (JMR) dating to the third quarter of the 16th.

5. Reading apparatus

This chapter contains all the witnesses to Ainring's glosses laid out side by side, including original transcriptions by DIERK HAGEDORN and original English translations by MICHAEL CHIDESTER.

It begins with the gloss of the unarmored fencing with the extended sword (which concludes with the illustrations from GF for reference), and this is followed by the gloss of the armored fencing with the 'short' sword and then the gloss of the mounted fencing.[1]

5.1. Transcription notes

MARTIN WIERSCHIN previously offered a complete transcription of SR in his book *Meister Johann Liechtenauers Kunst des Fechtens* in 1965, although he relied on another source, namely the *Nicolas Pol Hausbuch* (N), to create it. From this manuscript, he corrected and supplemented many of the careless mistakes and missing parts in SR. He also slightly adjusted the orthography, introduced (modern) punctuation and removed the numerous abbreviations.

The present transcription of SR dispenses with all of this and follows the original as far as possible. This results in a less fluid readability compared to WIERSCHIN's edition, but it does offer a virtually unadulterated view of the manuscript. SR is written far less carefully than, for example, the *Peter von Danzig fight book* (PD) or the *Lew fight book* (L). There are significantly more abbreviations, but also crossed out and corrected passages. The scribe also made numerous other errors, which remained uncorrected.

The only other Ainring witness which has been published in a comprehensive edition is AD, which I published in 2021 and RAINER WELLE also published that same year. Otherwise, these texts are being published here for the first time.

The transcriptions here are based as faithfully as possible on the original. The letter "v" is not resolved into "u" or "v". Abbreviations have been expanded where possible, and the added letters indicated using underlining.

The manuscripts use numerous ligatures that are no longer common in modern typesetting. These ligatures are broken down into their individual

[1] This isn't how Liechtenauer's *Record* is typically presented and might obscure the fact that these two parts of the *Record* form a single teaching in which the fighters dismount their horses partway through to fight on foot. However, SR organizes the three sections this way, and it's the only witness that contains all three glosses (though the mounted is only a tiny fragment). This sequence also avoids any confusion when the layout changes for the mounted fencing.

letters. Several different forms of the "s" were used in the manuscripts. The corresponding ligatures are also resolved, as is the letter combination of long and round "s" at the end of the word. All that remains is a "sz" ligature, which is represented by "ß".

The capitalization reflects the original, though upper- and lowercase are, by modern standards, quite arbitrary. Occasionally, a word is suddenly highlighted by a capital letter in the middle of a sentence. In many cases the difference between the capital and the lowercase is so small that it was only possible to guess what was originally meant.

Lastly, letters, words, or passages that were struck out in the original are represented by ~~crossed out~~ text. The use of red ink generally follows the source manuscripts, and words that are larger than their surrounding text are rendered in **bold** text. The first letter in each clause is often struck with red ink in some of the manuscripts (including GF, HS, and SR), and this is represented by a red pipe placed |next to the character; the pen flourishes common in these manuscripts are represented by the ~ character in the corresponding color.

5.2. Translation notes

The translations in this chapter will bear some resemblance to that in Chapter 3, but I approached them a little differently.

The aim of this chapter is to explore the differences between the witnesses, and because of this, I chose to stick more closely to the original text, including oddities of word order and phrasing so that where there are differences, they will be more obvious. This should also facilitate comparing the translation to the transcription. Where the scribes make obvious errors, I make no effort to correct them and instead translate the text at face value if at all possible, indicating the error with a [!] symbol.

I also did less polishing and changed fewer words for the sake of readability. In this chapter, fencers *hew* their cuts rather than *cutting* them and they *wind* their swords instead of *turning* them. I also preserved the masculine pronouns that Ainring assigned to his fencers, both for the sake of the original voice and since removing them sometimes required more extensive rephrasing.

When marginal notes are inserted into the main text, this is indicated by the given insertion symbol (such as '//') followed by the translation of the insertion in {curly brackets}.

Dictionary definitions referenced in the footnotes are based on GRIMM for Early New High German (ENHG) and LEXER for Middle High German (MHG). References to Paul Hektor Mayr's Latin translation of Lew's gloss are based on a translation that KENDRA BROWN and REBECCA GARBER are preparing for eventual publication.

Finally, it is a custom in our community to leave certain magic words untranslated. I despise this practice and believe that the translator's job is to translate, and that while a reader is free to choose whether to use German or English terms in conversation, a translator who refuses to translate a term deprives the reader of that choice. Furthermore, these 'special' terms often don't appear in the text as nominative nouns or infinitive verbs, but rather in grammatical forms that aren't easily untangled. Turning "set upon him" (*setz im an*) into "perform an *Ansetzen* against him" distorts the language to an extreme degree. For these reasons, I haven't left any words untranslated in my translations. Given the close proximity of the German and English texts in the following pages, it should be simple for the reader to identify the German equivalent to any particular term.

5.3. LAYOUT NOTES

The apparatus in this chapter presents up to four witnesses side-by-side, with English in the upper register and German in the lower register. Witnesses are placed in roughly chronological order, and in all cases, the identity of each witness is indicated by a flag at the top of each column, displaying both the city where the manuscript resides and the HAGEDORN *Siglum* in blue letters. In the few cases where a witness appears outside its usual column, the flag is instead **white text on a blue background** to make that clear and obvious.

In order to minimize confusion, the three principal witnesses occupy the same column wherever applicable: **SR** in the far left, **GF** in the center left, and **JMR** in the far right. If there is no text from a particular witness, the blank space in its column is preserved to indicate this.

In the extended sword gloss, the fourth column (center right) is shared between the other three textual witnesses: **AR**, **HM**, and **HS**. (For **AR** and **HM**, this places them in chronological order with the others, but **HS** is the oldest witness and must still be placed here for convenience and clarity.) In rare cases, **HS** includes text for the same piece as **AR** or **HM**, and when that happens, **HS** is given priority of placement and the other is shunted to whichever column is empty.[2] The illustrations from **GF** are not included inline, and instead there is a note in the transcription wherever they occur. They appear in their own section at the end of the extended sword.

[2] In one case, there is no empty column available so the HM text is presented in an appendix.

In the armored fencing gloss, there are only four witnesses, so SR and JMR retain their positions from the extended sword (far left and far right, respectively), while WP is placed in the center left column and AD in the center right.

Finally, there are only two witnesses in the mounted fencing gloss, so SR and GF take their usual positions (left and right, respectively) and the other two columns are eliminated, meaning SR and GF appear on every page. (When the SR text runs out, its empty column is retained.)

This apparatus numbers the pieces in the same fashion as Chapter 3. To conserve space and avoid clutter, these numbers only appear in the far-left column (even when SR has no text to contribute).

In order to assist in comparing the texts, all differences in wording are highlighted; this does not extend to spelling differences, unless the differences change one word into a different recognizable word. Blue highlighting indicates words that are not present in all witnesses. Yellow highlighting indicates words that are present in all witnesses but arranged differently in at least one. Green text indicates pairs of words that are different between witnesses but easily explained through scribal error (e.g., *auf* vs. *aus*, *dein* vs. *sein*, etc.).

Footnotes that apply equally to multiple witnesses have their number replicated in all relevant comments (this sometimes means that the note attached to a number may appear on the opposite page from the column you're looking at).

Finally, part of this project included comparing the Ainring text to other similar treatises from this time period (including both the anonymous glosses assigned to Pseudo-Peter von Danzig, Lew, and Nicolaüs and the armored fencing treatise of Martin Huntsfeld), and many pieces show clear influence between these texts. Some are identified as "close matches", meaning virtually every word in one text is present in the other, and others are identified as "similar", meaning that many phrases are shared between the two but there are also substantial differences. In both of these categories, the passages tend to be similar enough to be comparable to the differences between accepted witnesses of Ainring.

These matches are identified in footnotes, including a citation to a witness of the text in question. In order to minimize the number of separate manuscripts being invoked, whenever possible the citations are to one of the manuscripts detailed in Chapter 4: AR for the Nicolaüs extended sword gloss, GF for the Pseudo-Peter von Danzig and Huntsfeld armored teachings, SR for the Pseudo-Peter von Danzig mounted gloss, and JMR for the Lew extended sword and mounted glosses. When no other source for the Pseudo-Danzig gloss is available, PD is used. Editions of all of these texts can be viewed on Wiktenauer.

Hie hept sich an die
Ritterlich kunst deß
langen schwerts

DRESDEN – SR

GLASGOW – GF

The chivalric art of the long sword begins here

1 | The explanation of the *Record* begins here, 6
In which is written the chivalric art of the 'long' sword,[3] which was composed and created by Johannes Liechtenauer, who was a grand master of the art, may God grant him grace. He allowed the *Record* to be written with obscure and cryptic words so that the art should not become common.

And so Master Sigmund Ainringeck,[4] currently fencing master[5] to the highborn prince and noble Lord Albrecht, Count Palatine of the Rhein and Duke in Bayern, had these same obscure and cryptic words interpreted and glossed here in this booklet, as is written hereafter, so that any fencer can well learn and understand them (if he otherwise knows how to fence), etc.

LONG SWORD

2r | Hie hept sich an die Ritterlich kunst deß langen schwerts ~

10v | Hie hept sich an die vßlegung der zedel |in der geschriben stett die |Ritterlich kunst des langes schwerts Die gedicht vnd gemacht hat |Johannes lichtenawer der ain grosser maister in der kunst gewesen ist dem gott genedig sÿ der hatt die zedel laußen schrÿben mitt verborgen vnd verdeckten worten |Darumb dz die kunst nitt gemain solt werden |Und die selbigen verborgnenn vnd verdeckte wort hatt maister |11r| |Sigmund ain ringeck der zÿt des hochgebornen fürsten vnd herrenn hernn aulbrecht pfalczgrauen bÿ |Rin vnd herczog |in baÿern schirmaister |Glosieret vnd außgelegt alß hie in disem biechlin her nach geschrÿben stät dz sÿ ain ÿeder fechter wol verömen vnd vestan mag der da anderst fechten kan, etc ~

[3] The literal translation here would be "long sword", but since it isn't the sword that's long and instead it's holding the sword with both hands on the grip that 'lengthens' it, "extended sword" seems clearer. Compare "retracted sword" in the dueling lessons, which refers to placing one hand on the blade.
[4] The many names of Master Sigmund are discussed in the Introduction and Chapter 1.
[5] A *Schirmmeister* is a fencing teacher, using the late medieval term for fencing (*schirmen* rather than *fechten*). A *Schirrmeißter* is a lord's ßtablemaßter, or a logißtics officer in a military setting in charge of animals and anything pulled by animals (wagons, cannon, etc.). '*Schirmeißter*' could be a spelling of either one; Hans Medel reads it as the former.

The *Record* begins here,
In which is written the chivalric art of the 'long' sword,[3] which was composed and created by Johannes Liechtenauer, who was a grand master of the art, may God grant grace. He had the *Record* composed and written with obscure and cryptic words so that the art should not become common.

And so Master Sigmund Schining,[4] who at the time was fencing master[5] to the highborn prince and noble Lord Albrecht, Count Palatine of the Rhein and Duke in Bayern, thus had these same obscure and cryptic words of the *Record* interpreted and glossed, as is written hereafter in this book, so that any fencer can well learn and understand them (if he otherwise knows how to fence correctly).

And thereafter, [this gloss was] also marked and refined by other masters, and particularly through Hans Medel von Salzburg, which then follows hereafter.

The *Record* begins here,
In which is written the chivalric art of the 'long' sword,[3] which was composed and created by Johannes Liechtenauer, who was a grand master of the art, may God grant him grace. He allowed the *Record* to be written with obscure and cryptic words so that the art should not become common.

And so Master Sigmund Ainring,[4] who at the time was fencing master[5] to the highborn prince and noble Lord Albrecht, Count Palatine of the Rhein and Duke in Bayern, thus had these same obscure and cryptic words of the *Record* interpreted and glossed, as is written and pictured hereafter in this booklet, so that any fencer can well learn and understand them (if he otherwise knows how to fence correctly).

21r | h · Ie[7] hebt sich an die Zetl in der geschribenn stet die ritterliche kunst des langen schwertes die geticht vnd gemacht hat Johannes liechtenawer der ain grosserr̲ maister in der kunst ist gewesenn̲ dem got genad der hat die Zetl geschriben vnd geticht mit verporgenn̲ verdecktenn̲ wortenn darumb das die kunst nit gemain soll werden vnd die selbenn̲ verporgen vnd verdecktenn̲ wort der zetl hat maister Sigmund Schining[8] der diser zeit des hochgeborenn̲ Fursten vnd herrenn̲ Herrnn̲ Albrechtz pfaltzgraue bey Rein vnd hertzog in Baiernn̲ schirmmaister gewesen ist Also glosirt vnd außgelegt Als dann in disem buch hernach geschribenn̲ steet das ain Jeder vechter̲ wol vernemen vnd versteen mag der anderst recht vechtenn kan Und darnach auch von anderen̲ maisternn̲ gemert vnd gebessert vnd besunder durch maister Hansen Medel[9] von Saltzburg wie dann hye nachuolget

6r | Hie heben̲ sich an die Zedttel In den geschrieben stedt die Ritterlich kunst, des langen schwerdts die gedicht vnnd gemacht hat Johannes Liechtennaur der ein großer meister In der kunst gewesen ist. Dem gott gnedig sey, der hatt die Zedtel lasen schreiben mitt verborgen vnd verdackten wortten, darumb das die kunst nicht gemeyns solt werden, vnd dieselbigen verborgen vnd verdagkten worten der Zedtel, Die hat meister Sigmundt Einring zu derselbigenn zeit des hochgeboren fursten vnd herren herren Albrechts Pfaltzgraff bey Rein vnd hertzog in Bayernn schiermeister gewesen ist, also glossirt vnnd ausgelegtt, als sie den in diesem buchlein hernach geschrieben, vnd gemalt stehnn, das sie ein Jeder fechter wol vernemen vnd verstehen mag, der da anderß recht fechten kan.

6 GF's conclusion seems like a shortened version of these introductions; see the end of this section (paragraph 93).
7 An "h" is written in small form as a placeholder. Apparently, an initial spanning three lines was originally intended.
8 Due to the peculiarity of the writer, the letter "i" is sometimes not dotted at all or only very hesitantly, so other readings are possible: *Schninig*, *Schming*, *Schnung*.
9 See previous note; it can alternatively be read *Niedel*.

2| The foreword of the *Record*:

1 Young knight learn
To have love for God, honor women,
2 Thus waxes your honor.

3 The art that adorns you,
courts honor in wars.
4 Wrestling's good grips,
Glaive, spear, sword, and knife,
5 Wield manfully.

6 Hew here, run there;
Charge in, meet or move on.
7 Thus the knowledgeable
Hate the man who sees prizes.
8 Understand this:
All art has measure and moderation.

Die vor red der zedel ~

|Jungk ritter lere |Got liebhaben fröwen ia ere |so wöchse dein ~~are~~ ere |Kunst die dich ziert In kriegen zu ern hoffiert |11^{v}| |Ringes gütt fesset |Glefen sper schwert vnd messer, manlich bederben |Haw drin hart dar |Rausch hin triff oder las faren daß in die wÿsen, |hassen den man sicht brÿsen |Daruff dich fasse, alle kunst haben lenge vnd masse ~ ·:· ~

HANS MEDEL – HM

This is the foreword:
Young knight learn
To have love for God, also honor women,
Thus waxes your honor
Over[!] chivalry, and learn
The art that adorns you,
And courts honor in wars.
Wrestling's good grips,
Glaive, spear, sword, and knife,
Wield manfully,
And ruin in others' hands.
Hew here and run;
Charge in, meet or move on.
Thus the knowledgeable
Hate the man who sees prizes.
They understand this:
All art has measure and moderation.

ROSTOCK – JMR

This is the foreword:
Young knight learn
To have love for God, honor women thus,
Thus waxes your honor
Over[!] chivalry, and learn
The art that adorns you,
And courts honor in wars.
Wrestling's good grips,
Glaive, spear, sword, and knife,
Wield manfully,
And ruin in others' hands.
Hew here and guard there;
Charge in, meet or move on
Thus the knowledgeable
Hate the man who sees prizes.
Understand this:
All art has measure and moderation.

Das ist[10] Die vorrede
Junckritter lere, got liebhabenn frawen[b] Joch[a] ere, So wechst dein ere, Über ritterschafft vnd lere, kunst die dich zieret, vnd in kriegen zw erenn hof-fieret, Ringes gute fesser, Glefney sper swert vnd messer Manlich bederben, vnd in anderen henden verderbenn Haw drein vnd hurt rausch hyn triff oder lass faren das In die weisen hassen den man sy*cht*[11] preisen darauf sy fassen, all kunst haben leng vnd massenn

Das Ist die vorredt.
Junckh Ritter lere got lieb haben, frauen so ehre so wechst dein Ehre vber Ritterschafft vnd lehre kunst, die dich ziertt vnd In kriegen Zw ehren hoffirt, Ringens gut Fesser Gleffen Spern schwerdt vnd messer manlich bederben, vnnd In andern henden verderben, Haw drein, vnnd huet dar rausch hin trieffe oder laß farenn das in die weissen hasen, den man sicht preisn darauff dich fasse, alle kunst hab ein leng vnd masse.

LONG SWORD

[10] The first two words of the line appear to have been written (subsequently?) by another hand.
[11] *cht* inserted above the line by another hand.

3| This is the text of many good common lessons of the long sword:

9 If you want to see art,
Go left and right with hewing,
10 And left with right
Is how you most strongly fence.

Gloss: Remember, this is the first lesson of the extended sword: that you should learn to hew the cuts correctly from both sides, if you want to otherwise fence strongly and correctly. Learn it like this: when you want to hew from the right side, see that your left foot is forward, and when you want to hew from the left side, see that your right foot is forward. If you then hew a cut from over your right side, accompany the cut with your right foot. If you do not do this, then the cut is false and incorrect; since your right side remains behind, the cut is shortened and cannot have its correct path downward toward the correct other side in front of the left foot. Similarly, when you hew from your left side but do not accompany the cut with the left foot, then the cut is also false...

Das ist der text von vil gutter gemainer lere des langen schwerts
|Willtu kunst schowen |Sich linck gen vnd recht mitt hawen Und linck mitt rechtem, ist dz du starck gerst fechten :·
Glosa |Merck dz ist die erst lere des |12r| langes schwercz dz du die hew von baÿden sÿtten recht solt lernen hawen |Ist dz du annders starck vnd gerecht fechten wilt |Dz ver nÿm allso |Wenn du wilt howen von der rechten sÿtten |So sich dz dein ~~k~~ lincker fuß vor stee Und wenn du wilt howen von der lincken sÿtten so sich dz dein rechter fuß vor stee ~~Haw~~ |Häustu dann den ober haw von der rechten sÿtten so folg dem haw nach mitt dem rechten fuß tust du dz nicht, so ist der how falsch vnd vngerecht |wann dein |12v| rechte sÿten pleibpt dahinden |Darum ist der haw zu kurcz vnd mag sein rechten gang vndersich zu der rechten ~~sÿten~~ anderen sÿtten vor dem lincken fuß nicht gehaben |Des glÿchen wenn du hawst von der lyncken sÿtten vnd dem haw nicht nachfolgest mitt dem lincken fuß so ist der haw och falsch

This is the text of many good common lessons of the long sword:

If you want to see art,
Go left and right with hewing,
And left with right
Is how you most strongly approach to fence.

Gloss: Remember, this is the first lesson of the extended sword: that you should learn to hew the cuts correctly from both sides, if you want to otherwise fence strongly and correctly. Learn it like this: when you want to hew from the right side, see that your left foot is forward. If you then hew a cut from over your right side, accompany the cut with your right foot. If you do not do this, then the cut is false and incorrect; since your right side remains behind, the cut is shortened and cannot have its correct path downward toward the other side in front of the left foot. Similarly, when you hew from your left side but do not accompany the cut with the left foot, then the cut is also false...

This is the text of many good common lessons of the long sword:

If you want to see art,
Go left and right with hewing,
And left with right
Is how you most strongly fence.

Gloss: Remember, this is the first lesson of the extended sword: that you should learn to hew the cuts correctly from both sides, if you want to otherwise fence strongly and correctly. Learn it like this: when you want to hew from the right side, see that your left foot is forward, and when you want to hew from the left side, see that your right foot is forward. If you then hew a cut from over your right side, accompany the cut with your right foot. If you do not do this, then the cut is false and incorrect; since your right side remains behind, the cut is shortened and cannot have its correct path downward toward the other side in front of the left foot. Similarly, when you hew from your left side but do not accompany the cut with the left foot, then the cut is also false...

Das ist der text von vil guter gemainer lere des langen schwertz

Wiltu kunst schawen, so biß glinck gen vnd recht mit hawen, Und glinck mit rechtem, ist das dw starck begerst zefechtenn. Glosa

Merck das ist die erst lere des langenn schwertz das dw die hew von baiden seitten recht solt lernen hawen ist das dw anderst starck vnd recht vechten wilt das vernym also Wann dw wilt hawen von der rechten seitten So sich das dein lincker fueß vor stee Hawstu dann den oberhaw von der rechten seitten so volg dem haw nach mit dem rechten fueß Thustu das nit |21ᵛ| So ist der haw falsch vnd vngerecht Wann dein rechte seittenn pleibt da hinden so ist der haw zekurtz vnd mag sein rechtenn gang vndersich zw der anderen seitten vor dem lincken fueß nit gehaben Desgleichen wann dw hawst von der glincken seittenn vnd dein haw nit nachuolget mit dem glincken fueß so ist der haw auch falsch

6ᵛ| **Das ist der Textt von viel gutten gemeinen lere des langen schwerdts.**

Wiltu kunst schauen sich lingck gehen vnd recht mit hauen vnd ling mitt rechten ist das du strackh begerst fechten.

Glosa. Merck das ist die erste lere des langen schwerdts, das du die hew von beyden seitten recht solst lernen hauen, Ist das du anderst starck vnnd gerecht fechten wilt, das vernym also wen du wilt hauen, von der rechten seiten, so sich das dein lingker fueß vorstee, vnnd wen du wilt hauen, von der lincken seytten so sich das dein rechter fuß vorstee, hastu dan den vberhaw von der rechten seiten, so volge dem haw nach, mit dem rechten fuß, thustu das nicht so ist der haw falsch vnd vnrecht, wenn dein rechte seiten bleibet dahinden, darumb ist der haw zw kurtz vnd mag seinen rechten ganng vnder sich zw der andernn seiten vor dem lincken fueß nicht gehaben, deßgleichen wen du haust von der lincken seiten, vnd dem haw nicht nach volgest mit dem lincken fueß, So Ist der haw auch falsch,

...Remember, therefore: from whichever side you hew, accompany the cut with the same foot so you can drive all your pieces with strength, and all other cuts should also be hewn like this.

4| Again, the text of a lesson:

11 Whoever chases after cuts,
He allows himself little joy in his art.
12 Hew close to whatever you want:
No change comes into your shield;
13 Toward the head, toward the body,
Do not eschew the skirmish.[12]
14 With the entire body
Fence whatever you want to drive strongly.

Gloss: When you approach to fence with him, you should not watch for his cuts nor wait for how he will drive them against you, because all fencers that watch and wait for another's cuts and want to do nothing other than displace, they allow themselves little joy in such art since they often become struck by it.[13]

|Darum so merck von welcher sÿtten du haust, dz du mitt dem selbigenn fuß haw nachfolgest so magstu mitt sterck alle dein stuck gerecht trÿbenn |Unnd also süllen alle andere hew |13r| och gehawen werden ~:·~

Der text aber von ainer lere

|Wer nach gat hawen, der darff sich kunst wenig fröwen Haw nachent waß du wilt kain wechsel kumpt in dein schilt |Zu koppff zu lÿbe, die zeck nicht vermÿde, mitt ganczem lÿb ficht waß du starck gerst zu trÿben :·

Glosa |Wenn du mitt dem zufechten zu im kumpst so solt du vff sein heẅ nicht sechen noch warten wie er die gegen dir trÿbt wann alle fechter die do sechen vnd warten vff aines anderen heẅ |Unnd wellend anderß ~~nichtß~~ nicht~ thon |13v| dann verseczen die durffen sich söllicher kunst wenig fröwen wann sÿ werden do bÿ offt geschlagen

[12] *Zecken* is typically translated with a variety of words suggesting minor strikes (and *Zeck* also means "tick", leading some people to read it as something like "bug bites"), but the *Lexer* gives an alternative reading of "skirmish" or harassing actions. *Rühren* is "to stir up", "to cause something to move", and "to touch or hit" (including "to land a blow in fencing"); I summarize these senses as "harrying". *Zeckrühr* doesn't appear in *Grimm* or the *Lexer*, so I read it as a compound of these two words and render it as "harrying →

...Remember, therefore: from whichever side you hew, accompany the cut with the same foot so you can drive all your pieces with strength, and all other cuts should also be hewn like this.

Again, a text of a lesson:
Whoever chases after cuts,
He allows himself little joy in his art.
Hew close to whatever you want:
No change through comes onto your shield;
Toward the head, toward the body,
Do not eschew the skirmish.[12]
With the entire body
Fence whatever you want to drive strongly.

Gloss: When you approach to fence with someone, you should not watch for his cuts nor wait for how he will drive them against you, because all fencers that watch and wait for another's cuts and want to do nothing other than displace, they should not be very free in such art since they often become struck by it. Therefore, strike and stab the exposure.[13]

...Remember, therefore: from whichever side you hew, accompany the cut with the same foot so you can drive all your pieces with strength, and all other cuts should also be hewn like this.

Again, the text of a lesson:
Whoever chases after cuts,
He allows himself no joy in his art.
Hew close to whatever you want:
No changer comes onto the shield;
Toward the head, toward the body,
Do not eschew the harrying strikes.[12]
With the whole body
Fence whatever you want to drive strongly.

Gloss: Remember, this is when you approach to fence with him: you should not watch for his cuts, nor also wait for how he will drive them against you, because all fencers that watch and wait for another's cuts and want to do nothing other than displace, they need little joy in such art since they often become struck by it.[13]

Darumb merck von welcher seitten dw hawest das dw mit dem selben fueß dem haw nachuolgest so magstu mit sterck alle deine stuck gerecht treibenn Und also sollen auch all ander hew gehawenn werdenn

Aber[14] Ain text von ainer lere
Wer nachgeet hawen, der bedarff sich kunst wenig frawen, Haw nachent was dw wilt, kain durchwechsel kumpt an deinen schilt, zw kopf zw leib, die zeck nit vermeid, Mit gantzem leibe, vicht was dw starck gerest treiben Glosa,
Wann dw mit dem zuuechten zw ainem kumbst so soltu auff seinen haw nit sehenn noch warten wie er die gegen dir treibt Wann all vechter die da sehenn vnd warten auff aines anderen haw vnd wöllent anderst nicht thun dan versetzen die sollen sich solcher kunst wenig freyen Wann die werden da bey offt geschlagen Darumb schlach vnd stich die plösse

Darumb so merck von welcher seitten du hawest, das du mit dem selbigen fueß, dem haw nachvolgest, so magstu mit stercke alle deine stucke gerecht treiben, vnd also sollen all ander hew auch gehauen werdenn.

Der Textt aber von einer Lehre
Wer nach gehet hauen der darff sich kunst nicht frewen Haw nahet was du wilt, kein wechseler kombt an den schildt zw kopff zw leib die zeckrur nicht vermeydt, mit gantzem leyb ficht was du starck begerest zw treybenn.
|7^r| Glosa. Mercke das ist wen du mit dem zufechten zw im kumbst, so soltu auff sein hew nicht sehen auch noch wartten, wie ehr die gegen dir treybtt wenn alle fechter die da sehen vnd wartten, vff eines andern hew, vnnd wöllen anderst nicht thun, wen versetzen die bedurffen sich solcher kunst wenig freuen wann sie werden dabey offtt geschlagen.

strikes" to express the idea of harrassing during a skirmish. (Thanks to CHRISTIAN TROSCLAIR for digging up the *zecken* lemma.)
[13] This is similar to paragraphs 6 of the Pseudo-Danzig gloss, 5 of the Lew gloss, and 4 of the Nicolaüs gloss; see PD: 10^v.2, JMR: 14^r.3–14^v.1, & AR: 1^r.4. HM is a close match to Nicolaüs.
[14] *Aber* seems to have been written (subsequently?) by another hand.

Likewise, you should remember that everything you want to fence, drive it with the strength of your whole body, and hew in close to his head and body, so he cannot change through in front of your point. And with the cut, in the binding of swords, you should not eschew harrying strikes toward the nearest exposure (which will be delineated to you hereafter in the five cuts and in other pieces).[15]

5| Again, a lesson:

15 Hear what is simple:[16]
Do not fence left from above if you are right,
16 And if you are left,
It is also very clumsy on the right.

Gloss: Remember, this lesson touches on two people, one left and one right. Learn it like this: when you approach to fence with him, if you are right and intend to strike toward the man, then do not hew the first cut from the left side, because it is weak and with it, [you] cannot resist when someone binds strongly on you. Therefore hew the right side, so you can work whatever you want on the sword strongly [and] with art.

|Item du solt mercken alles dz du fechten wilt dz trüb mitt ganczer störck deines lÿbs Unnd haw im do mitt nahent ein zu kopff vnd zu lÿb so mag er vor dinem ort nicht durch wechslen |Und mitt dem haw solt du im den anbinden des schwerts der ~~zek~~ zeckrure nicht vermyden zu der nächsten blöß di dir hernach in den fünff hewen vnd in anderen stucken vßgericht |14r| werden ~:~

aber ain lere

|Höre waß du schlecht ist, ficht nitt oben linck, so du recht bist Und ob du linck bist Im rechten auch ser hinckest ·:

Glosa |Mörck die lere trifft an zwu personen ainen lincken vnd ain grechten, ~~vnd mainest den man zu schlagen So haw den ersten haw~~ |Das vernÿm also. |Wann du mitt zu fechten zu im kumpst Bist du dann gerecht vnd mainest den man zu schlachen |So haẅ den ersten haw nicht von der lingen sÿtten |Wann der ist schwach vnd magst damitt nicht |14v| wider gehalten wann man dir starck daruff bindt Darum so haẅ der rechten sÿtten, |so magst du starck am schwert mitt kunst arbaÿten waß du wilt

[15] The terms '*Zecken*=skirmish' and '*Zeckrühr*=harrying strikes' are not used again in Ainring's gloss, so it's hard to be sure which pieces are being referred to here. There are hints, though: Hans Medel's gloss repeats verse 13 when it covers the take-away, the misser is described as *rühren*, and some more specific pieces are given in the gloss of this section of the *Record* in Pseudo-Peter von Danzig and Lew (see PD: 10v.3–11r.2 →

Likewise, remember also that everything you want to fence, drive it with the strength of your whole body, and hew in close to head and body, so he cannot change through in front of your point. And with the cut, in the binding of swords, you should not eschew the skirmish toward the nearest exposure (which will be delineated hereafter in the five cuts and other pieces).[15]

Again, a lesson:
Hear what is simple:[16]
Do not fence left from above if you are below right,
And if you are left,
It is also very clumsy on the right.

Gloss: This lesson touches on two people, one left and one right. Learn it like this: when you approach to fence with him, if you are right and think to strike toward the man, then do not hew the first cut from the left side, because it is weak and with it, [you] cannot resist when someone binds strongly on you. Therefore hew from the right side, so you can work whatever you want on the sword strongly [and] with art.

Item auch merck Alles das dw vechten wilt das treib mit gantzer sterck deines leibs vnd haw damit nahent ein zw kopf vnd zw leib So mag er vor deinem ort nit durchwechseln Und mit dem haw solt in dem anpinden des swertz der zeck nit vermeiden zw der nechsten plösse Die hernach in den fünf hewenn vnd anderen stucken ausgericht warde

Aber ain text ainer lere
Hör was da schlecht ist, ficht nit oben linck so dw vnden recht bist, vnd ob dw glinck bist, Im rechten auch ser hinckest Glosa
Die lere trifft an zwo person ainen linckenn vnd ainenn rechtenn das vernym also Wann dw mit ainem zuuechten kummest bistu gerecht vnd vermainst den man zeschlagen so haw den ersten haw nit von der lincken seitten Wann der ist schwach vnd magst domit nit widerhalten Wann man dir starck darauff |22r| Spindet Darumb haw von der rechtenn seitten So magstu starck an dem swert mit kunst arbaitten was dw wilt

Likewise, remember that everything you want to fence, drive it with the strength of your whole body, and hew in close, toward the head and toward the body, so he cannot change through in front of your point. And with the cut, in the binding of swords, you should not eschew purposeful strikes toward the nearest exposure (which will be delineated to you hereafter in the five cuts and in other pieces).[15]

Again, a lesson:
Hear what is simple:[16]
Do not fence left from above if you are right,
And if you are left,
It is also very clumsy on the right.

Gloss: Remember, this lesson touches on two people, one left and one right. Learn it like this: when you approach to fence with him, if you are right and intend to strike toward the man, then do not hew the first cut from the left side, because honor is weak and with it, [you] cannot resist when someone binds strongly on you. Therefore hew from the right side, so you can work whatever you want on the sword strongly [and] with art.

Item merck alles was du fechten wilt, das treib mitt gantzer sterck deines leybes, vnnd haw im damit nahent zw dem kopff, vnd zu dem leyb, So mag ehr von deinem ort nicht durch gewechselnn vnd mit dem haw soltu in dem anbinden des schwerdts der Zweck rur nicht vermeydenn zẅ der negsten blöße die dir hernach in denn funff hewenn vnd In andern stucken ausgericht werdenn.

Aber ein lehre.
Hor was da schlecht ist ficht nit oben lingck so du recht bist, vnd ob du lingck bist im rechten auch sere hinckest.
Glosa. Merck die lehre trifft an zwo person einem lingcken vnd einem rechten, vnd das vernym also wen du mit zufechten zw im kumbst, bistu den gerecht, vnd meinst, den man zu schlagen, so haw den ersten haw nicht, von der lincken seyten wan ehr ist schwach vnd magst damit nit wyderhalten wan man dir starck darauff bindt darumb so haw von der rechten seiten so magstu starck am schwerd mit kunst arbeiten was du wilt,

& JMR: 14v.2). Based on these, it seems to be the term for actions that leave the bind and strike to a new exposure, and in this way, there is a nice contrast between the skirmish and the war (*Krieg*).

[16] *Schlecht* often means "bad", but it can also mean "straight, direct, simple", and that makes more sense here. (Thanks to STEPHEN CHENEY for pointing this out.)

Similarly, do not hew from the right side if you are left, since the art is quite awkward for someone who is left to drive from the right side, just as it is also for right from the left side.

6 | This is the text and a lesson on Before and After:

17 Before and after, these two things[17]
Are the sole wellspring of all art.
18 Weakness and strength;
"Within"[18]—remember that word with it.
19 Thus you can learn
To work and defend with art.
20 If you are easily spooked,
Never learn to fence.

Gloss: Remember, this is that above all else, all you should understand the Before and the After, since these two things are the wellspring from which all the art of fencing flows.

|**Des gelichen** |Bist du linck so haw och nitt von der rechtten wenn die kunst ist gar wild ainem lincken ze triben von der rechten sÿtten Des glich ist es och ainem rechten von der lincken sÿtten ~ :

Daß ist der text vnd ~~lere~~ ain lere von vor und nach
|Vor vnd nach die zwaÿ dinck, sÿnd aller kunst ain vrsprimg |Schwöch vnd störck, /In/des/ daß wort domitt mörck |So magst | 15r | du leren mitt kunst arbaiten vnd werenn |Er schrickstu gern, kain fechten nimmer ~~lern~~ geleren ·:
Glosa |Merck dz ist dz du vor allen sachen wol solt verston daß vor und daß nach, wann die zwaÿ ding sind ain vrsprung do alle kunst des fechtenß außgät

[17] This line is the same as the first line of dueling couplet 62.
[18] The meaning of *Indes* changes significantly from Middle High German to Early New High German. It may be translated "within" or "inside" in both languages, but in MHG, *Indes* was a spatial adverb (i.e., within a place or location) and in ENHG it became a temporal adverb (i.e., within a time or event). Liech- →

Similarly, do not hew from the right if you are left, since the art is quite awkward for someone who is left to drive from the right side, just as it is also for right from the left side.

Similarly, do not hew from the right side if you are left, since the art is quite awkward for someone who is left to drive before the right side, just as she is also for right from the left hand.

A text, a lesson on the Before and After:
Before and after, these two things[17]
Are the sole wellspring of all art.
Weakness and strength;
"Within"[18]**—remember that word with it.**
Thus you can learn
To work and defend with art.
If you are easily spooked,
Never learn to fence.

Gloss: This is that above all else, you should correctly learn and understand two things—that is, the Before and the After—and thereafter Weakness and Strength and the word "Within", the wellspring from which comes the entire foundation of the art of fencing. When you learn these things correctly—and above all the word "Within"—and do not forget them in all pieces that you drive, you are a good master.

Desgleichen bistu linck so haw auch nit von der rechten seitten Wann die kunst ist gar wild ainem lincken zetreiben von der rechten seitten desgleich ist auch ainem rechten von der lincken seitten

desgleichen bistu linck So haw auch nicht von der rechten seiten, wann die kunst ist gar wildt einem lincken zu treyben, vor der rechten seiten desgleichen ist sie auch einen rechtem von der lincken hanndt.

Ain text ainer lere von dem vor vnd nach
Vor vnd nach die zway ding, Send aller kunst ain vrsprüng Schwech vnd stercke, Indeß das wort domit mercke So magstu leren mit kunst arbaiten vnd werenn, Erschrickst dw gerne, kain vechten nymmer lerne Glosa
Das ist das dw vor allen sachenn solt recht versteen vnd vernemen die zwai ding das ist das vor vnd das nach Darnach swech vnd sterck des swertz vnd des wortz indeß darauß kumpt der gantz grundt vnd vrsprung aller kunst des vechtens wann dw die ding recht vernimbst vnd vorauß das wort Indeß nit vergist in allen stucken die dw treibst bistu wo ein guter maister

tenauer's *Record* seems to have been written during this transition and straddles these senses: as will be described more fully below, the word 'Within' is used to describe actions the instant (time) when you have felt the pressure of the bind (place) and must choose a response.

Learn it like this: The Before, this is that you should always come forth with a cut or with a thrust toward his exposure such that he must displace you (prior to when he does this to you). Then in the displacement, work nimbly in front of yourself with the sword, from one exposure to another, so he cannot come to his pieces before your work. But if he runs in toward you, then come Before him with the wrestling.

7 | Here remember what is called 'the After':
Remember, if you cannot come Before, then wait until After. These are the counters for all pieces which he drives against you. **Learn it like this:** when he comes forth such that you must displace him, then Within the displacement, work nimbly toward the nearest expo-sure with your sword. In this way, you meet him ere he brings his piece.

|**Daß vernÿm also** |Daß ~~for das~~ vor daß ist dz du all weg solt vorkummen mitt ainem haw oder mitt ainem ~~sch~~ stich Im zu der blöß |Ee wann er dir zu der deinen so muß er dir verseczen, so arbaÿt in der versachung behentlich für dich mitt dem schwert |15^v| von ainer ~~be~~ blöß zu der anderen so mag er vor deiner arbaÿt zu seinen stucken nicht kommen |Aber laufft er dir ~~ein~~ eÿnn |So komme ~~fo~~ vor mitt dem ringen ~~~~~~~~~~~

Hie mörck was da haÿsst daß nach
|Mörck magstu zu dem vor nitt kommen |So wart uff dz nach dz sÿnd die brüch uff allen stuck die er vff dich trÿbt |**Das vernÿm** |**also** Wann er vorkumpt daß du ihm verseczen must |So arbait mitt der versäczung, |Indes behentlich für dich zu der nächsten blöß |So triffestdu in ee |Wann |16^r| er sein stuck verbringtt

The Before

The Before, this is that you should always come forth {if you want} with a cut or with a thrust toward his exposure such that he must displace you. Then in the displacement, work nimbly in front of yourself with the sword, from one exposure to another, so he cannot come to his pieces before your work. But if he runs in toward you, then come forth with the wrestling or running point.[19]

The After:

The After is when you cannot come Before {or otherwise will not take it}, so you wait until After. These are the counters for all pieces which he drives against you. Learn it like this: when he comes forth such that you must displace him, then Within the displacement, work nimbly toward the nearest exposure with your sword. In this way, you meet him ere he brings his piece.

Das vor

Das vor ist das dw albeg vor kum̲men solt, ob du wilt,[20] mit ainem̲ haw oder mit ainem̲ stich Im zw der plösse ee wann̲ er dir zw deiner plösse So muß er dir versetzen̲ So arbait in der versatzung bhentlich für dich mit dem swert von ainer plössen zw der anderen̲ So mag er vor deiner arbait zw seiner nicht kum̲men Aber laufft er dir ein so khum vor mit dem ringen̲ oder̲ ort lauffen̲

Das nach

Das nach ist so dw zw dem vor nit kum̲men magst, // {oder sunst nit geren̲[?] nemenn̲ wöllest[?],}[21] so wart auf das nach das send die prüch auff alle stuck die er auf dich treibt das vernim also wann̲ er vor kumbt das dw Im versetzen̲ must So arbait mit der versatzung Indeß behentlich für dich zw der nächsten̲ plösse So triffstu In ee wann̲ er sein stuck verbringt

[19] "Running point" isn't explained anywhere in the text (nor in the larger Liechtenauer corpus). It may have been described in HM's (lost) gloss on running through (wrestling).
[20] *ob du wilt* added by another hand above the line.
[21] *oder... wöllest* added in lighter ink by the same hand from the previous note (slightly blurred in places).

If you have won the Before like this and he remains After, then in the Before and After, you should also remember how you should work with word "Within" in accordance with the Weakness and in accordance with the Strength of his sword. **And learn it like this:** the sword has its Strength from the sword's hilt to the middle of the blade; with it, you can resist well when someone binds you on it. And the sword has its Weakness further on, from the middle to the point; there you cannot resist.[22]

And when you understand these things correctly, you can work well with art and protect yourself with it, and also teach princes and lords so that they may well succeed with the same art, in play and in earnest; but if you spook easily, you should not learn the art of fencing, because it does no good when a heart becomes fragile or despondent when struck by art.

|**Also** gewinstu aber dz f vor Und er blÿpt nach |Auch soltu in dem vor vnd nach mörcken wie du mitt wort /in des/ arbaitten solt nach der schwech vnd nach der störck seines schwertß |**Und das vernÿm Also** Von dem gehulcze des schwerts ~~by~~ biß in die mitten der clingen |Hatt dz schwert ~~fin~~ sin störcke dar mitt du wol magst wider gehalten wann man dir dar an bindt |Und fürbaß von der mitt biß an den ort hat es sein schwöch da magst ~~nitt~~ nicht wider |16ᵛ| gehalten |Und wenn du die ding recht verstest |So magstu mitt kunst wol arbaitten vnd dich darmitt wören vnd fürbaß lernen fürsten vnd herrenn dz sÿ mitt der selbigen kunst wol mügen besten In schimmpff vnd in ernst |Aber erschrckstu gern so saltu die kunst des fechtens nitt lernen |Wann ain blöds verzags hercz dz tut kain gut wann es wirt bÿ aller kunst geschlagen ~~~~ ~

[22] Other translators divide this section differently, placing the first clause of the second paragraph at the end of the first paragraph. The large *Also* in SR clearly indicates where the division should be, though. →

If you have won the Before like this and he remains After, then in the After and in the Before, you should also remember how you should work with the word "Within" in accordance with Weakness and in accordance with Strength of his sword. And learn it like this: the sword has its Strength from the sword's hilt to the middle of the blade; with it, you can resist well when someone binds you on it. And the sword has its Weakness further on, from the middle to the point; there you cannot resist.[22]

And when you understand these things correctly, you can work well with art and protect yourself with it, and also teach princes and lords so that they may well succeed with this art, in play and in earnest; but if you spook easily, you should never learn the art of fencing because you will be struck by all art. Therefore, you should not learn it, because a fragile, desponded heart does no good in fencing.

Also gewinstu aber das vor vnd er pleibt nach Auch soltu in dem nach vnd in dem vor mercken wie dw mit dem wort Indeß arbaiten solt nach swech vnd nach sterck seines swertes vnd das vernym also Von dem gehiltz des swertz biß in die mitte der klingen hat das swert sein stercke damit dw wol magst widerhaltenn wann man dir darein pindet Und fürbaß von der mitten biß an das ort hat es sein schwech da magstu |22^{v}| nicht widergehaltenn Und wann dw die ding recht versteest So magstu mit kunst wol arbaiten vnd dich damit weren vnd fürbaß lerenn fursten vnd herren das sy mit diser kunst wol mugen besteen in schimpf vnd in ernnst Aber erschrickstu geren so soltu die kunst vom vechten nymmer leren wann dw wurdest bey aller kunst geschlagen darumb soltu sy nicht lerenn Wann ain plöd vertzagt hertz thut kain gut in dem vechtenn

This is a significant problem because placing it at the end of the description of the After rather than the beginning of the Strength and Weakness is how the untextual idea of "retaking the Before" was invented.

8 | The text on the five cuts:

21 Learn five cuts
From the right hand. Whomever they defend,
22 Him we vow
To reward well in these arts.

Remember, the *Record* sets down five obscure cuts. Many masters of the sword know nothing to say about this: that you should not learn to hew other cuts from the right side against him that stands against you in defense, and you can meet him with the first strike if you select one of the five cuts.

Whoever can counter you without injury will be avowed by the masters of the *Record*, because his art will be more successful for him than another fencer who cannot fence against the five cuts.

And how you should hew with the five cuts, you will find that written hereafter in the same five cuts.

Der text von den fünf hewen
|Fünff hew lere von der |17r| rechten hand Wer dz wäre dem wir geloben, in kunsten gern zu lonen ~:·

|Merck die zedel seczt fünff verborgne hew |Da von vil maister des schwerts nicht wissen ~~zuo~~ zu sagen Die soltu anders nicht lernen hawen wann von der rechten sÿtten gegen dem der sich gegen dir stöllet zu der were |Und versuch öb du mitt ainem haẅ vsß den fünffen den man mitt dem ersten schlag mügest treffen |Wer dir die brechenn kan on seinen schaden, so wirt im gelopt |Von dem maister der zedeln |daß im siner kunst |17v| bas gelonet soll werden dann ainer andern fechtern der wÿder die funff heẅ nicht fechten kann |Und wie du die fünff heẅ howen solt, |dz fündest du in den selbigen funff heẅen her nach geschriben.

The text on the five cuts:
Learn five cuts
From the right hand against the defense.
That we vow
To reward well toward these arts.

Gloss: Remember, the *Record* sets down five obscure cuts. Many who call themselves "masters" know nothing to say about this: that you should not learn to hew other cuts from the right side against him that stands against you in defense, and you can meet him with the first strike if you select one of the five cuts.

And whoever can counter the cuts without injury (and particularly what goes after them with other work) will be avowed by the masters of the *Record*, because his art will be more successful when another fencer cannot fence against the five cuts.

And how you should hew with the five cuts, you will find that written and taught hereafter in the same cuts in the *Record*.

This is the text on the five cuts:
Learn five cuts
From the right hand against the defense.
Him we vow
To reward well in these arts.

Gloss: Remember, the *Record* sets down obscure cuts. Many masters of the sword know nothing to say about this: that you should not learn to hew other cuts from the right side against him that stands against you in defense, and you can meet him with the first strike if you select one of the five cuts.

Whoever can counter you without injury will be avowed by the masters of the *Record*, because his art will be more successful when another fencer cannot fence against the five cuts.

And how you should hew with the five cuts, you will find that written hereafter.

Der text von den funff hewenn
Fünff häw lere, von der gerechten hand wider die were, den wir geloben, zw künsten gern zelonen
Glosa
Merck die zetl setzt fünff verporgen hew, da von vil die sich maister nennent nit wissen zesagen Die soltu anderst nit leren hawenn wann von der rechten seitten gegen dem der sich gegen dir stelt zw der were vnd versuch ob dw mit ainem haw auß den fünfen den man mit dem ersten schlag magst treffen Und wer dir die hew prechenn kan an sein schaden vnd besunder was darauf geet mit Jetlicherr arbait der wirt gelobet von den maisternn der zetl das Im sein kunst pas gelonet soll werdenn wann ainem anderen vechter der wider die funff hew nit vechten kan Und wie dw die funff hew hawenn solt das findstu in den selben hewenn in der zetl hye nachgeschriben vnd gelert

7ᵛ | **Das ist der Text von denn funff hewenn.**
Funff hewe lehre von der rechten hanndt wyde[r][23] die wehre dem wir geloben in kunsten geren zu lohnn.
Glosa. Merck die zedtel setz verbo[r]gen hew daruon viel meister des schwerdts nit wissen zu sagen die soltu anderst nicht lerne[n] hauen wan von der rechten seytten gegen de[n] der sich gegen dir stellet zur wehre vnnd ve[r]such ob du mitt einem haw aus denn funff[?] den man mit dem ersten schlag mugst treffe[n] wer dir die brechen kan, an sein schaden den ~~soll~~ wirt gelobett vonn dem meister der zetel da[s] im sein kunst böß gelonet sol werden, wen einem anndernn fechter der wider die funff haw nit fechtenn kann, vnnd wie du die funf hew hawen solt, das vindestu hernach geschribenn.

[23] On this page, the text comes quite close to the binding and isn't always visible in the scans.

Dresden – SR

Glasgow – GF

9| This is the one on the pieces of the *Record*:

23 Wrathful cut, curved, crosswise
Has cockeye with parter.
24 The foolish one parries,
Pursue, overrun, set the cut,
25 Change through, pull,
Run through, cut off, press the hands.
26 Expose him with hanging and
Strike, catch, sweep, stab with pushing.

Gloss: Remember, the main pieces of the art of the long sword are correctly named here; each is individually called by its name so that you can understand them better. There are seventeen in number and they begin at the five cuts.

Das ist der [24] von den stucken der zedeln

|**Zorn** haw · krump · zwerch hat |Schiller, mitt schaittler |Alber verseczt, nachraÿsen |Uberlauff ~~hav~~ haẅ seczet |Durch wechsel zuck durch lauff abschnide, hende ~~dn~~ druck, heng vnd mitt blössen schlach vach streÿch ~~sch~~ stich mitt stossen ~·:·~ Glosa

|18r| |Mörck hie werden genampt die rechten haüptstucke der kunst deß langen schwerts wie ÿettlichs besunder haist mit dem namen, daß du die dester baß ver sten kündest |Der ist sibenzechen an der zal Und heben sich an den funff hewen an. ~

Long Sword

[24] The word *Text* is obviously missing here.

The text on the main pieces of the *Record*:
Wrathful cut, curved, crosswise
Has cockeye with parter.
The foolish one parries,
Pursue, overrun, set the cut,
Change through, pull,
Run through, cut off, press the hands.
Expose him with hanging, winding,
Strike, catch, sweep, stab with pushing.

Gloss: Remember, the main pieces of the *Record* and art of the long sword are correctly named here; each is individually called by its name and they are seventeen in number, and they begin at the five cuts.

Der text von den hawbt stucken der Zetl
z **Ornhaw**[25] **krump Twir hat schilher mit schaitler Alber versetzt, nachraisen vberlauff hew setzt durchwechsel zuck, durchlauf abschneid hendt druck heng wind mit plössen, Schlach vach streich stich mit stössenn** Glosa
Merck hye werden genant die rechten haubtstuck der Zetl vnd kunst des langen swertz wie ietlichs haisset besunder mit seinem namen Und send sibentzehen an der zal Und hebent sich an an den fünff hewen

[25] The small placeholder "z" remained because the initial, which should have extended over two lines, was not executed.

Likewise, now remember the first cut, called 'the wrathful cut'.
The second, the curved cut.
The third, the crosswise cut.
The fourth, the cockeyed cut.
The fifth, the part cut.

The sixth, they are the four guards.
The seventh, the four displacements.
The eighth, the pursuit.
The ninth, the overrunning.
The tenth, the setting aside.
The eleventh, the changing through.
The twelfth, the pulling.
The thirteenth, the running through.
The fourteenth, the slicing off.
The fifteenth, the hand pressing.
The sixteenth, the hangers.
The seventeenth, they are the winds.

And how you should expose him with the hangers and winds, and how you should drive all the pieces named previously, all this you will find written hereafter.

|Item ~~num~~ nun mörck der erst haẅ haist |der zorn haẅ |Der ander der krump haẅ |drit der zwer haẅ |Der viert der schill haẅ |Der funfft der schaittel/haẅ |der sechst daß sind die vier | 18ᵛ | hütten |Das sÿbent die vier verseczen |Das acht die nachraisen |Das neindt die überlauffen |Das ꝭ zeehend die abseczen |Daß aÿlfft daß durchwechßlen |Das zwülfft daß zucken |Daß drÿzehend die durchläuffen :/ |daß vierzehend die abschnÿden : |daß funffzehend die hend trucken |Daß sechzehend die hengen |Daß sibenzehend daß sind die winden |Vnnd wie du dich mitt den hengen und winden enplösen solt |Und wie du alle vorgenampte stuck trÿben solt daß vindestu alles her nach geschriben ~~~~~~~:·

The first cut is called 'the wrathful cut'.

The second, the curved cut.
The third, the crosswise cut.
The fourth, the cockeyed cut.
The fifth, the part cut.
Now the other twelve pieces begin
The first or sixth in number, they are called the four guards or lairs.
The seventh, the four displacements.
The eighth, the pursuit.
The ninth, the overrunning.
The tenth, the setting aside.
The eleventh, the changing through.
The twelfth, the pulling.
The thirteenth, the running through.
The fourteenth, the slicing off.
The fifteenth, the hand pressing.
The sixteenth, the hangers.
The seventeenth, the winds.

And how you should expose him with the hangers and winds and drive all the pieces named previously, all this you will find written and taught hereafter in the explanation and glosses of the *Record*, etc.

Der erst haw haist der zorenhaw der ander der krumphaw der dritt der twirhaw, der vierdt der schilhaw der funft der schaittlhaw Nun hebent sich die anderen zwelff stuck an Das erst oder an der zal das sechst haissent die vier huten oder legerr das sibent die versetzen das acht die nachraisenn das newndt die ẅberlauffen das .x. die absetzen das xj die durchwechseln das xij die zucken, das xiij die durchlauffen das xiiij die abschneiden das .xv. die hendt drucken das xvj die hengen das xvij die windenn |23v| Vnd wie dw dich mit den hengenn vnd winden emplossen sollt Und alle vorgenante stuck zetreiben findestu alles hyenach gelert vnd geschriben in der auslegung vnd glosen der zetl etc

10| This is the wrathful cut with its pieces:

27 Whoever hews over you,
The wrathful cut's point threatens him.

Gloss: Learn it like this: when someone hews in over you from his right side, then hew a wrathful cut in along with him, strongly with the long edge, also from your right shoulder. Then, if it is Soft on the sword, shoot the point in ahead of you, extended in toward the face, and threaten to stab him.

11| Again, a piece of the wrathful cut:

28 If he becomes aware of it,
Take it away above without danger.

Gloss: When you shoot the point in with the wrathful cut, if he then becomes aware of the point and displacements the thrust with strength, then pull your sword upward, off from his sword above, and hew back in above on his sword toward the other side, in toward his head.

19r| Das ist der zorn haw mitt sinen stucken ~

|Wer dir ober hawet |Zor haẅ ort im dröwet ·:
Glosa |Daß vernÿm also Wann dir ainer von siner rechten sÿtten oben oben ein hawet |so haẅ ainen zorn haẅ mitt der langen schniden och von diner rechten achslen mitt im starck ein |Ist der dann waich am schwert, so schüß Im den ort für sich lang ein zu dem gesicht |Vnnd träw im zu stechen ·~·:·~·

Aber ain stuck vß dem zorn haw

|Wirt er es gewar |So nÿms |19v| oben ab an far ·:
Glosa |Wann du mitt dem zorn haẅ den ort ein schüst wirt er dann deß orts gewar vnd verseczt den stich mit störcke |So ruck dein schwert übersich oben ab von dem sinen |Und haẅ im zu der andren sÿtten an sinem schwert wider oben ein zuo dem kopffe ~~

This is the wrathful cut with its pieces:
Whoever hews over it,
The wrathful cut's point threatens it.

Gloss: Learn it like this: when someone hews in over you from his right side, then hew a wrathful cut in along with him, strongly with the long edge, also from the right shoulder. Then, if he is Soft on the sword, shoot the point in ahead of you, extended toward his face, and threaten to stab him (as is done nearby hereafter).

Again, a piece of the wrathful cut:
If it becomes aware,
Take it away above without danger.

Gloss: This is when you shoot the point in at him with the wrathful cut (as is done previously nearby); if he then becomes aware of the point and displacements the thrust with strength, then pull your sword upward, off from his sword above, and hew back in above on his sword toward the other side, in toward his head (as is done here).

Das ist der Zornhaw mitt seinenn stuckenn.
Wer die oberhawet Zornhaw ort dem trauwet.
Glosa. Das vernim also, wenn dir einer vonn seiner Rechtenn seiten obin inhawet, So haw einenn Zorenhaw mit der lanngenn schneyden auch vonn der rechten achselnn mit Im starck, Ist er dann waich am schwert, So scheus im den ort fursich, lanng im zu dem gesicht, vnd dro Im zu stechenn, als hernach am nechsten gemachtt stehet.

Aber ein stuck aus dem Zornhaw.
|8r| Wirtt es gewar, so nim obenn abe onefar.
Glosa. das ist, wenn du Im mit dem Zornhaw denn ortt inschust, als vor am nechstenn gemachtt stet, Wirt er dan des orttes gewar, vnnd versetzt denn stich mit sterck So ruck dein schwert vbersich oben ab von dem seinen vnnd haw im zu der Anndernn seitenn ann seinem schwert wider oben in zu dem kopff als hie gemachtt stett.

12 | **Again, a piece of the wrathful cut:**

29 Be stronger back
And thrust. If he sees it, take it back.

Gloss: When you hew in with the wrathful cut, if he displaces you and remains strongly on the sword, then be strong against him on the sword, and go up with the Strength of your sword into the Weakness of his sword, and wind on the sword with your hilt in front of your head, and thus stab in above toward his face.

13 | **Again, a piece of the wrathful cut:**

When you stab in at him above with winding (as is previously), if he then goes high up with the hands and displaces the high thrust with the hilt, then remain standing in the winding like this and set the point down in between his arms and the chest.

Aber ein stuck vß dem zornhaw

|Biß störcker wider, vnd ~~sch~~ stich sicht erß so nÿm es wider ·:

Glosa |Wenn du im mitt dem zornhaw Inhawst verseczt er dir daß vnd pleibt dir damitt | 20r | starck am schwert |So bÿß gen im wider starck am schwert |Und far uff mit der störck dines schwerts in die schwöchi sines schwerts vnd wind am schwert ~~de~~ din gehülcz ~~f~~ vornen für dein ~~hopt~~ haupt vnd so stich in oben zu dem gesichte ~

Aber ain stuck vß dem zornhaw

|Wann du Im mitt dem winden oben ein stichst, alß vor stett, fört er den hoch vff mitt den henden vnd versetzt mitt dem gehülcze den obern ~~sttich~~ stich so plÿb ~~alst~~ also sten in dem winden vnd setz im den ort | 20v | niden zwischen sinen armen vnd der brust ~ ~

Again, a piece of the wrathful cut:
Be stronger back
And thrust. If he sees it, take it below.

Gloss: Remember, this is when you hew in with the wrathful cut (as is done previously nearby): if he displaces you and remains strongly on the sword, then be strong against him on the sword, and go up with the Strength of your sword into the Weakness of his sword, and wind on the sword with your hilt in front of your head, and then stab him above, in toward his face (as is pictured here).

Again, a piece of the wrathful cut:
When you stab in at him above with winding (as is pictured previously), if he then goes high up with the hands and displaces the high thrust with the hilt, then remain standing in the winding like this and set the point down on the chest in between his arms (as is pictured here below).

Aber ein stuck aus dem Zornhaw.
Bis stercker wider, vnnd stich, sihet ers so nim es nider.
Glosa. Merck das ist, wenn du Im mit dem Zorenn inhawest, als vorn am nechstenn gemacht stet, Versetzt er dir das, vnnd bleibt dir damit starck am schwert, So biß gen Im wider starck am schwert, Unnd far auf mit der sterck deines schwerts, Inn die schweche seines schwerts, vnnd winde am schwerte dein gehiltz, vorn vor deinem haupt, vnnd stich Im oben in zu dem gesicht als hie gemalt stet.

Aber ein stuck aus dem Zornhaw.
Wenn du Im mit dem winden obein instichst, als vor gemalt stet, fert er dann hoch auf mit den hendenn, Vnnd versetzt mit dem gehiltz den oberstich. So beleib also stehenn in dem winden, vnnd setz im den ortt niden zwischen seinen armen ann die brust, als hie vnden gemalt stet.

14| **A counter against the take-away:**
Remember, when you bind with someone strongly on the sword and he then pulls his sword upward, off from your sword above, and hews in again at you toward the other side of your head, onto the sword, then bind strongly with the long edge in above toward his head.

15| **Here remember a good lesson:**

30 Remember this precisely:
Cut, thrust, lair; Soft or Hard,[26]
31 Within, and drive after
Without running in. Do not rush your war.
32 Whoever fastens[!] the war
Above, they will become ashamed below.[27]

Gloss: This is that when you bind with someone on your sword (with a cut or with a thrust or otherwise), you should quite precisely notice if he is Hard or Soft on the sword, and when you have sensed that, Within it you should know which is best for you: if you should run in on him with the Before or with the After.

Ain bruch wider daß abnemen
|Mörck wenn du mitt ainem starck am schwert bindest |Ruckt er dan sein schwert übersich oben abe von dinem schwert vnd hawet dir zu der anderen sÿtten am schwert wider eÿn zuo dem kopffe |So bind starck mitt der langen schnÿden Im oben eÿn zu dem kopffe ~~

Hie mörck ain gutte lere
|Das öben mörck, |Haw stich leger waÿch oder hört |In des vnd fär nach, on hurt dein krieg sich |21r| nicht gäch |Weß der krieg ~~ri~~ riempt oben nÿder wirt er beschämpt ·:
Glosa |Daß ist dz du gar eben mörcken solt wann ~~du~~ dir ainer mitt ainem haẅ oder mit ainem stich oder sunst an din schwert ~~bintt~~ bindet ob er am schwert waich oder hört ist vnd wenn du das enpfunden hast |So solt du /In das/ wissen welchses dir am besten sÿ ob du mitt dem vor oder mitt dem nach an in hurten solt

[26] This line is the same as the second line of couplet 97, in the section on the hangers.
[27] This couplet uses similar phrasing to 80, in the section on overrunning, and dueling couplet 36.

A counter against the take-away:
Remember, when you bind with someone strongly on his sword and he then pulls his sword upward, off from your sword above, and hews in again at you toward the other side of your head, onto the sword, then wind strongly with the long edge in above toward his head.

A text of a good common lesson:
Remember this precisely:
Cut, thrust, lair; Soft or Hard,[26]
Within, drive after
Without running in. Do not rush your war.
Whoever aims at the war below,
They will become ashamed above in it.[27]

Gloss: This is that when you bind with someone on your sword (with a cut or thrust or otherwise), you should precisely notice if he is Hard or Soft on the sword; in this way, you sense or feel {and wind Within it according to the Hard or Soft to the nearest exposure with the war, since} you should also know Within it which is best for you: if you should work on him with the Before or with the After.

Here remember a good lesson:
Remember this precisely:
Cut, thrust, lair; Soft or Hard,[26]
Disciple, and drive after
Without running in. Do not rush your war.
Whoever aims at the war
Above, they will become ashamed below.[27]

Gloss: This is that when you bind with someone on your sword (with a cut or with a thrust or otherwise), you should quite precisely notice if he is Hard or Soft on the sword, and when you have sensed that, in it you should know which is best there: if you should work on him with the Before or with the After.

Ein bruch wider das abnemenn.
|8v| Merck du wenn du mit einem starck an sein schwert bindest, ruckt er dann sein schwert vbersich oben ab, vonn deinem schwert, vnd hawt dir zu der anndern seitenn am schwert wider in zu dem kopff, So wind starck mit der langen schneiden im oben in zu dem kopff.

24r| Ain text ainer guten gemainer lere
Das eben mercke hew stich leger waich oder herrte Indes var nach anhurt dein krieg sey nicht gach, Wes der krieg niden remet Indem wirt er oben beschemet Glosa
Das ist das dw eben mercken sollt wann dir ainer mit ainem haw oder stich oder sunst an dein swert pindt ob er am swert waich oder hert sey So dw das empfindest oder fülst [a]{So wind Indeß nach der waych oder hert in nach der nechsten plöß mit dem krieg} ~~so~~[28] [auch] soltu Indes wissen welches dir am pesten zimpt ob dw mit dem vor oder mit dem nach arbaiten solt

Hie merck ein gute lehre.
Das eben merck hew stich leger weich oder herte Junger, vnd far nach an hürt dein krieg, sey nit goch, wes der krieg remet oben nider, wirt er beschemet.

Glosa. Das ist, das du gar ebenn mercken solt, wen dir einer mit einem haw oder mit einem stich, oder sonst ann dein schwert bindet, ob er am schwert waich oder hert ist, vnnd wen du das empfunden hast, So solt in dan wissen, welches da am besten sey, ob du mit dem vor, oder mit dem nach arbeiten solt,

[28] This word is present in all witnesses, but for some reason was struck out in this one.

But you should not let yourself be rushed with the running in with the war, when the war is nothing other than the winding on the sword.

Likewise, drive the war like this: when you hew in with the wrathful cut, then as soon as he displaces, go up well with the arms and wind the point in on his sword toward his upper exposure. If he displaces the thrust, then remain in the winding and stab the lower exposure with the point. If he further follows after the sword with the displacement, then go through under his sword with the point and hang the point in above toward the other exposure on his right side. Thus he will be shamed above and below with the war, if you can otherwise correctly drive the threats.

|Aber du solt dir mitt dem an hurten nicht zu gauch laussen sÿn mitt dem krieg wenn der krieg ist nicht anders dann die winden am |21v| schwert

|Item den krieg trÿb also Wam du Im mitt dem zorn haẅ In haẅest |Alß bald er dann verseczt so far wol vff mitt den armen vnd wind im den ort am schwert ein zu der obern blöß verseczt er denn den stich |So blÿb sten in dem winden vnd stich mitt dem ort die vnder blöß folgt er dann fürbaß mitt der versaczunge dem ~~ss~~ schwert nach so far mitt dem ort vnden sÿn schwert durch vnd heng im den ort oben ein zu der anderen blöß siner rechten sÿtten |Also wirt er mitt dem krieg oben vnd vnden beschämpt |Ist daß du die ge |22r| fört andrest recht kanst trÿben ~

HANS MEDEL – HM

But you should not let yourself be rushed with the war, when the war is nothing other than the winding in the sword. These are to be driven awkwardly[29] at whoever cannot or does not understand.

ROSTOCK – JMR

But you should not let yourself be rushed with it with the war, when the war is nothing other when[!] the winding on the sword.

Likewise, drive the war like this: when you hew in with the wrathful cut, then as soon as he displaces, go up well with the arms and wind the point in on his sword toward his upper exposure. If he displaces the thrust, then remain in the winding and seek the lower exposure with the point. If he further follows after your sword with the displacement, then go through under his sword with the point and hang the point in above toward the other exposure on his right side. Thus he will be shamed above and below with the war, if you can otherwise correctly drive the threats.

Aber dw sollt mit dem anhurtẹn des kriegs nit zegach sein Wann der krieg ist nichtz anderst dann die winden in dem swert die sindt mißlich zetreibenn wer die nit wol kan oder versteet

aber du solt dir damit nit zu goch lassen sein, mit dem krieg, wen der krieg nit anderst ist, wen das windenn am Schwert.

Item denn krieg treib also, wenn du Im mit dem Zornhaw inhawest, als bald er denn versetzt, so far wol auf mit denn Armen, vnnd windt im den ort am schwert in zu der obern bloß, versetzt er dan den stich, so beleib sten in dem winden, vnnd such mit dem ort die vntern bloß. Volgt er dann furbas mit der versatzung deinem schwert nach, so far mit dem ortt vnter sein schwert durch, vnd heng Im denn ort oben |9r| in zu der anndern ploß seiner rechten seiten, also wirt er mit dem krieg oben vnd vnden beschemet, ist das du du[!] gefert anders recht kannst treybenn.

LONG SWORD

[29] '*mißlich*=awkward' is probably a misspelling of '*wißlich*=wisely'.

16 | How one should correctly find cuts and thrusts in all winding:[30]

33 In all winding,
Learn to correctly find cut, thrust.
34 You should also, with
Gauging, hew, stab, or slice
35 In all meetings
If you want to confound the masters.[31]

Gloss: This is that you should correctly find cuts, thrusts, and slices in all winding. When you wind, you should immediately gauge which of the three is the best to drive with it so you do not hew when you should stab, nor slice when you should hew, nor stab when you should slice.

And remember, when someone displaces one, that you meet them with another like this: if someone displaces your thrust, drive a cut; if someone runs in, then drive a slice under his arms. Remember this in all meetings and binds of the sword and you will confound the masters who set themselves against you.

Wie man In allen winden hew stich recht vinden sol ~

|In allen winden hew stich recht lern finden |Auch soltdu mit ~~brüch~~ brüffen, hew̃ stich oder schnitt In allen treffen den maistern wilt du sÿ effen ·:

Glosa |Daß ist daß du in allen winden hew̃ stich vnd schnitt recht finden solt |Also wenn du windest dz du da mitt zu handt solt brüffen weches dir vnder den ~~de~~ drÿen daß best sÿ zu triben ~~al~~ also dz du nicht haṽest wann du stechen solt vnd nit schnidest wann du hawen solt vnd nicht stechest |22[v]| wann du schniden solt |Und mörck wan man dir der aÿnes verseczt dz du in mitt dem andern treffest |Also, versecz man dir den stich so ~~trÿb~~ treÿb den haw |Laufft man dir eÿnn so treÿb den vndern schnitt |In sin arm des morck in allen treffen vnd anbinden der schwert wilt du anderst die maister effen die sich wider dich seczen ~~:~:·

[30] In the combined translation in Chapter 3, I render *winden* as "to turn", following the precedent in Ridgeway 2019, which I agree makes the text read more smoothly. In this chapter, I'm translating it "to wind" ⟶

How one should correctly find cuts and thrusts in the winding:[30]
In all winding,
Learn to correctly find cut, thrust, slice.
You should also, with
Gauging, hew, stab, or slice
In all meetings
If you want to confound their masters.[31]

Gloss: This is that you should correctly find cuts, thrusts, and slices in all winding. When you wind, you immediately gauge which of the three is the best to drive with it (the cut, or thrust, or slice) so you do not hew when you should stab, nor slice when you should hew.

And remember, when someone displaces one, that you meet them with another. Remember this in all meetings and binds of the sword and you will confound the masters who set themselves against you.

...Thus, do not hew when you should stab, nor stab when you should slice, nor slice when you should hew.

And when someone displaces one, meet them with another like this: if someone displaces your thrust, drive a cut; if someone runs in, then drive a slice under his arms. Remember this in all meetings and binds of the sword and you will confound the masters who set themselves against you and do not understand.

Wie mann in den winden hew, stich gerech[t] findenn sol.
Inn allenn windenn hewen stich, schnit recht lere find[en] auch soltu mit prufen hew, stich, oder schnit, in allen treffenn denn Meisten, wiltu sin effen.
Glosa. Das ist, das du in allen windenn, hew stich vnnd schnit recht findenn solt, Also wenn du windest, das du damit zu hand prufen, welches dir vnter den treyen das beste sey zu treybenn, der hew, oder stich, oder schnit, also das du nit hawest, wen du stechenn solt, Unnd nit schneidest, wenn du hawen solt. Unnd merck, wen man dir der eins versetzt, das du In mit dem andern triffest, das merck in allen treffen, vnnd anbinden der schwert, wiltu anders die meister effen die sich wider dich setzenn.

24v | ...Also das dw nicht hawest wenn dw stechen solt vnd nicht stechest wenn dw schneiden solt vnd wann man dir der ains versetzt das dw yn treffest mit dem anderen Also versetzt man dir den stich so treib den haw Laufft man dir ein so treib den vnderen schnidt in sein arem das merck in allen treffen vnd anpinden der swert wiltu anderst die maister effen die ich wider dich setzen vnd nit versteen

in order to preserve the distinction between it and the other two 'turning verbs', *wenden* and *kehren*; since *kehren* is rarely used, translating both of these as "to turn" seems less problematic.
[31] This couplet is very close to 90, in the section on pulling, and dueling couplet 63.

17 | On the four exposures:

36 Know four exposures;
Aim[32] so you strike wisely,
37 Without any danger,
Without regard for how he behaves.

Gloss: Here you should remember the four exposures on the man toward which you should always fence. The first exposure is the right side above the man's belt, the second is the left side. The other two exposures are also the right and left sides, below the man's belt.

In the approach, notice precisely which exposures he exposes to you, and artfully target the same without fear by shooting the extended point in, by pursuit, and otherwise with all threats, and care not for how he threatens and behaves against you. Thus you fence wisely and strike simply and excellently with this, and this does not allow him to come to his pieces.

On the four exposures:

Know four exposures;
Aim[32] at them so you strike wisely,
Without any danger,
Without regard for how he behaves.

Gloss: Here you should remember the four exposures on the man toward which you should always fence. The first exposure is the right side above the man's belt, the second the left side. The other two exposures are also the right and left sides, below the man's belt.

In the approach, notice precisely the exposures, which exposures he exposes to you, and boldly target the same without fear by shooting the extended point in, and by pursuit, and also by the winding on the sword, and otherwise with all threats, and care not for how he behaves against you. Thus you seek wisely and strike simply and excellently with this, and do not allow him to come to his pieces.

And always target the exposures and not the sword. If he will displace you, work further to the nearest exposure (with war or something else).

Von den vier blossen
Vier bloß wisse, rem so schlechstu gewisse an alle for, on zwifel wie er gebar ·:
Glosa |Hie soltu morcken die vier blossen an dem man da du all wegen zu fechten |23r| solt |Die erst bloß ist die recht ~~sÿtt~~ seÿtt die ander ist die link seÿtt oberhalben der girtel deß manß |Die ander zwuo sind och die recht vnd die linck seÿtten vnderhalben der girtel |Der blossen nÿm eben war in dem zufechten mitt welcher er sich gegen dir enblösse der selbigen reme künstlichen on far mitt einschiessen des langen orts mit nachraisen vnd sunst mit allen geförten vnd acht nitt wie er mit sÿnen geförten gegen dir ~~bar~~ gebar |So vichtest du gewisß vnnd schlechst schlege daruß die do treffenlich sind vnd ~~lau~~ laust in domitt zu seinen stucken nitt komen ~~:~

24v | Von den vier plössen
Vier plosse wisse, zw remen so schlechstu gewisse, An alle Var, on zweifell wie er gepar Glosa
Hye soltu mercken die vier plössen an dem manne da dw alweg zufechten solt Die erst plösse ist die recht seitten, die ander die linck seitten oberhalb der gürtl des mans Die ander zwo sind auch die recht vnd die linck seitten vnderhalb der gürtl Der plösse nym eben war in dem zufechten mit welcher plössen er sich gegen dir emplösse der selben reme künlich an far mit einschiessen des langen orts vnd mit nachraisen vnd auch mit dem winden am swerte vnd sunst mit allenn geferten vnd acht nit wie er gegen dir gepar so sichstu gewiß vnd schlechst schleg darauff die da trefflich seien vnd lest In zw sinen stucken nit kummen Und rem albeg der ploss vnd nit des swertz Wirt dir versetzt so arbait fürbas zu der nechsten plöss mit krieg oder anderem

[32] I generally read this verb as '*rähmen/rehmen*=aim, strive for, frame' in this translation, but it could also be '*raumen*=clear away, make space for' (which is how Andre Paurenfeyndt and his followers render it in their witnesses to the *Record*). Both verbs make sense in context here, and in Chapter 3 I use both at once.

On the four exposures:
Know four exposures;
Purify so you strike wisely,
Without any danger,
Without a doubt he will realize.

Gloss: Here you should remember the four exposures on the man toward which you should always fence. The first exposure is the right side above the man's belt, the second the left. The other two exposures are also the right and left sides, below the man's belt.

Likewise, you should know which exposures the man exposes to you, and boldly target the same exposures without fear by shooting the extended point in and by pursuit, and by winding on the sword, and also otherwise with all threats, and care not for how he threatens and behaves against you. Thus you strike in wisely.

In the approach, notice precisely which exposures he exposes to you, and boldly purify the same without fear by shooting the extended point in, by pursuit, and also by the winding on the sword, and otherwise with all threats, and care not for how he threatens and behaves against you. Thus you fence wisely and strike simply and excellently with this, and this does not allow him to come to his pieces.

Vonn denn vier blossenn.
Vier blosse wiß reine, so schlechstu gewiß ann aller var, one zweifel wirt er gewar.
Glosa. Hie soltu merckenn die vier blossenn ann dem man da du alleweg zu fechtenn solt, die erste bloß ist, die recht seytenn, die ander die linck oberhalb der gürtel des mans, die anndernn zwo sind auch die recht, vnnd die linck seyt vnterhalb der gürtel. Der blossenn |9^{v}| nim ebenn war in dem zufechtenn, mit welcher er sich gegenn dir emplöst, derselbenn reine kunlich an far mit inschiessenn des langenn orts mit nachreysenn, Und auch mit denn Winden am schwert, vnnd sonst mit allenn gefertenn, vnnd Achte nicht, wie er mit seinenn geferttenn gegenn dir gepar. So fichstu gewiß, vnnd schlechst schleg daraus die da treflich sindt, vnnd lest in damit zu seinenn stucken nicht kummenn.

17^{r}| |Item du solt wissen welche ploß der man gegen dir entplost der selben ploß reme künlichen an var mit ÿn schissen des langen ortz vnd mit noch ri^{e}ssen vnd mit winden am schwertt |17^{v}| vnd auch süns mit ~~namen~~ allen gefertenn vnd acht nicht wie er mit sinen gefertem gegen dir gebare so schlechstu In gewisß ~

18 | The text and the gloss on the redoubling and on the transmutation, how they exploit[33] the four exposures:

38 If you want to avenge,
Artfully exploit the four exposures:
39 Above, redouble;
Underneath, transmute correctly.
40 I truly say to you:
No master covers himself without danger.
41 If you have learned this correctly,
He can hardly come to blows.[34]

Gloss: This is when you want vengeance on someone such that you want to exploit his four exposures with art, then drive the redoubling toward the upper exposures against the Strength of his sword, and the transmutation toward the other exposures. Thus I say to you honestly that he cannot cover himself before it, and cannot come to strikes nor thrusts.

23v | Der text vnd die gloß von dem dupliern vnd von dem mutieren Wie die brechen die vier blossen ~

|Wilt du rechen |Die vier blossen künstlich brechen |Oben duplir, vnden recht mutier |Ich sag dir für war, sich ~~seu~~ schücz kain maister an far |Haustu es recht vernommen zu schlage mag er klain kommen ·:

Glosa |Daß ist Wann du dich an einem rechen wilt also, dz du im die vier blossen mitt kunst wilt brechen ~~d~~ |So trÿb dz dupliern zu der oberen ~~bo~~ blössen gen der störcki seines schwerts vnd daß mutiern zu der anderen blösß |24r| So sag ich dir für war daß er sich dar von nitt schüczen kan vnd mag weder zu schlachen noch zu stechen komen ~

[33] I read this as '*brauchen*=apply, use, abuse', but it could also be intended as '*brechen*=counter, break'; the latter doesn't seem to make sense here since elsewhere it describes countering an action or position of the opponent, not an inherent quality of their anatomy/geometry.

The text and the gloss on the redoubling and on the transmutation, how they exploit[33] the four exposures:

If you want to avenge yourself,
Artfully exploit the four exposures:
Above, redouble;
Below, transmute correctly.
I truly say to you:
No one covers himself without danger.
If you have learned,
He can hardly come to blows. [34]

Gloss: This is when you want vengeance on someone such that you want to soon exploit his four exposures with art, then drive the redoubling toward the upper exposures against the Strength of his sword, and the transmutation toward the other exposures. Thus I say to you honestly that he cannot guard himself before it, and cannot come to strikes nor thrusts.

Der text vnnd die gloss von dem dupliern vnnd von mutiern, wie die prechenn die vier blossenn.

Wiltu dich rechenn, die vier blossenn künstlichenn brechenn, obenn dupliern, niden recht mutiern, Ich sag dir furwar, sich schutzt kein man an far, hastu vernommenn, zu schlag mag er klein kommenn.
Glosa. das ist wenn du dich Rechen wilt, an einem, also das du Im die vier blossenn mit kunst balde wilt brechenn, So treib das dupliern zu der obernn bloß, gegenn der sterck seines schwerts, Vnnd das mutiern zu der Andernn bloß, So sag Ich dir fur war, das er sich darvor nicht behueten kann, vnd mag weder zu schlegenn noch zu stichen kummenn.

[34] This quatrain is very close to 100–101, in the section on the hangers.

19| **The redoubling:**[35]
Likewise, when you hew in above (with the wrathful cut or otherwise), if he displaces you with strength, then Within it, push the pommel of your sword under your right arm with the left hand and strike in on the sword with crossed hands, behind the blade of his sword (between the sword and the man), through the muzzle. Or strike him on the head with the piece.

20| **Remember the transmutation:**
Drive the transmutation like this: when you bind on his sword (by hewing over him or otherwise), then wind the short edge on his sword and go well up with the arms, and hang the blade of your sword over his sword, in from outside, and stab in toward the lower exposure.

Drive this to both sides.

Daß dupliern
|Item wann du in mitt dem zorn haw, oder sunst oben eÿn haẅst verseczt er dir mit stöck |So stos /|In des/ deines schwerts knopff vnder deinen rechten arm mitt der lincken hand vnd schlach in mitt gecruczten henden am schwert hinder sines schwerts ~~eli~~ klingen zwischen dem schwert |Und dem mann ~~vff~~ durch daß maul |Oder schlach im mit dem stück vff den kopff ~

Mörck daß mutiern
24v| |Daß mutiern treÿb also |Wenn du im mitt dem obern haw oder sunst an daß schwert bindest |So winde die ~~kun~~ kurczen schniden an sin schwert vnd far wol vff mit den armen vnd heng im dein schwerczs — clingen vssen über sein schwert vnd stich ~~v~~ im zu der vndern blösse vnd dz trÿb zu baÿden sÿtten ~

[35] This would literally just be “doubling”, but “redoubling” has the same meaning in English and “doubling” has taken on the very different meaning of “simultaneously hitting and being hit” in the parlance of contemporary *Kunst des Fechtens*.

Thc rcdoubling:[35]
Remember, when you hew in above (with the wrathful cut or otherwise), if he displaces you not strongly, then push...

Das dupliernn.
Merck wenn du Im mit dem Zornhaw, oder sonst oben in hawest, versetzt er dir nit starck, So stos...[36]

[36] Here the text at the bottom of the page not only cuts off in the middle of a piece, but JMR skips several paragraphs between the end of this folio and the beginning of the next one. Since ff. 5–9 are single leaves glued together rather than being a normal sewn booklet, it's likely they were cut out of the manuscript and then reinserted, and the folio including the missing text was lost in the process. This is confirmed by the original folio numbers, which skip one at this point (see Chapter 4).

21 | The curved cut[37] with its pieces:

42 Curve on nimbly,
Throw the point on the hands.

Gloss: This is how you should hew crookedly toward the hands. Drive the piece like this: when he hews at you with a cut over or under your[38] right side toward the exposure, leap away from the cut with your right foot, well toward his left side against him, and strike in with crossed arms, with the point on the hands.

And also drive this piece when he stands against you in the guard of the ox.[39]

22 | Again, a piece of the curved cut:

43 Whoever, curving, sets well,
Injures many cuts with stepping.

Gloss: This is how you should set aside cuts over you with the curved cut. Drive it like this: when he hews in toward your exposure from above his right side, step toward his left side with the right foot, [missing verb] over his sword, with the point on the ground in the lattice fence guard.[40] Drive this to both sides.

You can also strike him on the head from the setting aside.[41]

Der krumphaw mitt sinen ~~stüken~~ stucken

Krump vff behende, wirff den ort vff die hende ·:

Glosa |Daß ist wie du krump solt hawen zu den henden vnd daß stuck trÿb also wenn er dir von deiner rechten sÿtten mitt ainem obern oder vndern haẅ zu der blöss haẅet |So |25r| spring vsß dem haw mitt dinem rechten fuß gegen im wol vff sin |Lincke sÿtten vnd schlach in mit gecrüczenten armen mitt dem ort vff die hende vnd dz stuck trÿb och gem im wenn er gen dir ~~stätt~~ staut In der hüt deß ochsen ~

Aber ain stuck vß dem krumphaw

|Krump wer wol seczet mitt schrÿtten er vil heẅ leczet ~:·

Glosa |Daß ist wie du mitt dem ~~komp~~ krump haw die obern häw abseczen solt daß stuck trÿb also |Wann er dir von ~~sÿme~~ siner rechten sÿtten oben ein hawet zu der blosß so schrÿt mitt dem rechten fuß vff sÿn lincke sÿten ṽber sin schwert, mit dem ort |25v| vff die erden In die schranckhüte dz trÿb zu baÿden sÿtten Och magstu In vß dem abseczen vff dz haupt schlachen ~

[37] *Krumm* can mean both "curved" and "crooked"; it specifically describes a thing that is usually straight but has *unnatural* curvature. I use "curved" as the adjective form and "crookedly" as the adverb.

[38] This is often treated as a case of 'your/his' confusion by translators, but since it appears as "your" in both SR and HS and is a reasonable physical action, I see no reason to change it.

[39] Using this piece to counter the guard of the ox isn't described by Ainring, though Pseudo-Danzig, Lew, →

...This is how you should hew crookedly toward the hands. Drive the piece like this: when he hews at you with the cut over or cut under your[38] right side, leap away from the cut with the right foot, well toward his left side against him, and strike in with outstretched arms, with the point on his hands...

18ᵛ | ...daß ist wie du krumpt solt haulben zu den henden |Daß stuck tribe also wan er dir von diner rechten siten mit dem ober haulb oder vnter haulb zu haulbett so spring vß dem haulb mit dem rechten fus gegen ym woll vff sin lincke siten vnd schlag in mit vß gestreckten armen mit dem vff sin hende...

and Nicolaüs all cover it; it's described most clearly in the Lew gloss (JMR: 19ʳ.4–19ᵛ.1). In HS, the line about countering the ox is replaced by the entire text of Lew's piece.

[40] *Schränke* are the "barriers" or fences around a dueling or tournament field; *schranken* can also mean "to cross or form a lattice" and can describe the crisscrossed outward tilt of saw teeth. Mayr's Latin translator renders it '*custodia cancellata*=lattice guard' (PMM: 84ᵛ.1).

[41] This piece isn't described anywhere, but striking out of the lattice fence guard onto the opponent's hands is described in Pseudo-Danzig, Lew, and Nicolaüs; see the description in the Lew gloss (JMR: 19ᵛ.2–3)

23| Again, a piece of the curved cut:
44 Hew crookedly at the flats
Of the masters if you want to weaken them.

Gloss: This is when you want to weaken a master. Drive the piece like this: when he hews in at you from above his right side, then hew crookedly against his cut, onto his sword, with crossed hands.

24| Again, a piece of the curved cut:
45 When it sparks above,
Stand off, that I will praise.[42]

Gloss: This is when you hew in onto his sword with the curved cut: strike quickly from the sword up above in toward his head with the short edge, or wind the short edge on his sword with the curved cut and stab in toward his chest.

Aber ain stuck vsß dem krumhaw
Haw krump zu den flechen, den maistern wiltu sÿ schwechen :·
Glosa |Daß ist Wenn du ainen maister schwechen wilt |So trÿb dz stuck also wenn er dir oben einhaẅt von siner rechten sÿtten So haẅ krum mit ~~gekreucz-tem~~ gekrentzten henden gegen ~~sime~~ sinen haẅ vff sin schwert ~

Aber ain stuck vß dem krumhaw
|Wenn es kluczt oben, so stand ab dz will ich loben ·:
Glosa |Dast ist wenn du im mitt dem krumphaẅ vff sin schwert haẅst |So schlache vom schwert mitt der kurczen schniden |26r| bald wider vff, im oben ein zu dem kopff |Oder windt |Im mitt dem krumphaw die kurczen schnÿden an sin schwert vnd stich im zu der brust. ~ ~

[42] This couplet uses similar phrasing to 81, in the section on overrunning.

25 | **Again, a piece of the curved cut:**

46 Do not curve, hew short;
With it, see the changing through.

Gloss: This is when he wants to hew in at you from above his right side: do as if you want to bind on his sword with the curved cut, but shorten and go through under his sword with the point, and wind your hilt over your head on your right side, and stab toward his face.[43]

26 | **Remember how someone should counter the curved cut:**

47 Whoever leads you astray crookedly,
The noble[44] war confuses him.
48 That he truly
Does not know where he is without danger.

Gloss: Remember, this is when you hew over or under him from your right side: if he hews crookedly onto your sword from his right side with crossed arms, and with it he leads your strike astray, then remain with your sword strongly upon his and shoot the point in, extended under his sword toward his chest.

Aber ain stuck vß dem krumphaw

|Krum nicht kurcz haw, durch wechsel dar mitt schow :· **Glosa** |Das ist wenn er dir von siner rechten achseln oben ein will howen |So tu alß ob du mitt dem krumphaw an sin schwert wöllest binden Unnd kurcz vnd far mitt dem ort vnder ~~sn~~ sinem schwert durch vnd wind vff din rechte sÿttenn dein gehülcz über din höppt vnd stich im zu dem gesicht ~ ~

Mörck wie man den krumphaw brechen sol ~

| 26v | |Krump wer dich |Irret, der edel krieg in verwürret |Daß er fürwar nicht waÿst : wo er sÿ one far :· **Glosa** |Daß ist Wann du im von diner rechten sÿtten ober oder vnden zu haẅest |Haẅt er dann och von sÿner rechten sÿtten mit gekreutzen armen krump vff din schwert Und verirret dir do mitt dein heẅ So blÿb mitt dinem schwert starck an dem sinen |Unnd schüß im vnder dem schwert den ort lang ein zu der brust ~

[43] This is similar to the second half of paragraph 37 of the Pseudo-Danzig gloss; see PD: 18r.1.

[44] Apart from the obvious, '*edel*=noble' can also mean forthright, straightforward, simple, etc. This may be intended to establish a contrast between the 'crooked' cut and the 'straightforward' war that counters it.

...wants to hew in above: do as if you want to wind on his sword with the curved cut, but hew short and go through under his sword with the point, and wind your hilt over your head on your right side, and stab toward his face (as is pictured here).[43]

And this piece counters the...[45]

10ʳ | oben In wilt hawenn, so thu als du Im mit dem krumphaw ann sein schwert wollest winden, vnd haw kurtz vnnd far mit dem ort vnter seinem schwertt durch, vnnd wind auf dein recht seit deinen gehiltz vber dein haupt, vnnd stich im zu dem gesicht, als hie gemalt stet, vnd das stuck bricht die.

[45] This is probably intended to finish "guard of the ox". Using this piece to counter the guard of the ox isn't described by Ainring, though Pseudo-Danzig, Lew, and Nicolaüs all cover it; it's described most clearly in the Lew gloss (JMR: 20ᵛ.2).

27| **Another counter for the curved cut:**
Remember, when you hew in from above your right side, if he also hews {comes} onto your sword from his right side with crossed arms, and with it he presses your sword downward against the ground, then wind against your right side and go up well over your head with the arms, and set your point above onto his chest (Gloss).

If he displaces you, then remain standing like this with the hilt in front of your head and work nimbly with the point from one exposure to another.

This is called 'the noble war'. With it, you confuse him so much that he truly does not know where he should remain before you.

Ain andern brüch über den krumphaw
|Mörck wenn du im von diner rechten sÿtten oben ein haẅst |Haẅt er denn och von siner rechten sÿtten mit gekrenczten armen komp[46] |27^r| Vff dein schwert vnd drückt dir das da mit vnder sich gen der erden |So wind ge deiner rechten syten vnd far mit den armen wol vff ṽber dein haupt vnd secze Im dein ort oben an ~~dei~~ die brust Glosa |Versetzt er dir das so plÿb also sten mit dem gehũlcz vor dem haũpt vnd arbait behendtlich mit dem ort von ainer bloß zu der anderen |Das hayset der edel krig Da mit verwirstü In so gar Das er nit waysst wo er vor dir bliben sol fur war

[46] *Komp* added below the line by scribe II, whose section begins on f. 27^r.

Salzburg – HS

Likewise, a counter against the curved cut:
You shoot the point in under his sword toward his chest; if he presses your sword downward toward the ground with the curved cut, then wind against his right side and go up well over your head with the arms, and set the point above onto his chest.

If he displaces you, then remain standing thusly with the hilt in front of the head and work nimbly with the point from one opening to another.

This is called 'the war'. With that, you confuse him so much that he truly does not know where he should guard himself.

Rostock – JMR

Again, a counter against the curved cut:
Remember, when you shoot the point in under his sword toward his chest (as is written and pictured previously), if he presses your sword downward toward the ground with the curved cut, then wind against your right side and go up well over your head with the arms, and set the point above onto his chest (as is pictured hereafter).

If he displaces you, then remain standing thusly with the hilt in front of the head and work nimbly with the point from one opening to another.

This is called 'the noble war'. With that, you confuse him so much that he does not know where he should remain before you.

20^{v} | |Item ein bruch wider den krümpt haulb so du im den ortt vnter sin schwertt in schu^{e}st zu siner brust druck er dan mit dem krümpt haulb din schwertt vnter sich zu der ~~andern~~ erden so wind gegen siner rechten siten vnd far mit dem arm woll vff vber dem hauptt vnd setz im den ortt oben an sin brust versetz er dir dz so pleib also stett mit dem gehultz vor dem hauptt vnd arbeitt behendeglich mit dem ort von eÿner plos zu der andern dz | 21^{r} | heist der krieg do mit verirrest In so gar dz er nit weiß wo er sich hutten soll ~~

Aber ein bruch wider den krumphaw.
Merck wenn du Im den ort vnter seinem schwert ein scheust zu seiner brust, als vor geschriben vnd gemalt stet, druckt er dann mit dem krumphaw dein schwert vntersich zu der erdenn, So wind gen deiner rechtenn seittenn, vnnd far mit dem Arm wol auf vber dein haupt, vnd setz im den ort oben an die brust, Als hernach gemalt ist, versetzt er dir das, so beleib also stehenn mit dem gehiltz, vor dem haupt, vnd arbayt behentlich mit dem ort, vonn einer bloß zu der andernn, das heist der odel krieg, damit verwirrest du Inn so gar, das er nicht weis, wo er vor dir beleybenn sol.

28 | The crosswise cut[47] with its pieces:

49 The crosswise takes out
Whatever approaches from the day.

Gloss: Remember, the crosswise cut counters all cuts that are hewn down from above. Drive the cut like this: when he hews in at you above toward the head, then leap away from the cut with the right foot against him on his left side, and in the leap, turn your sword with the hilt high in front of your head so that your thumb comes underneath, and strike in with the short edge toward his left side so that you catch his cut in your hilt and you land in toward his head.

Der zwerhaw mit sinen stucken

|Zwerch benÿmp was vom tag her kümpt

Glosa |Merck der zwe^r h^a w bricht alle hew die von oben nÿde^r gehawen werden vnd den haw trÿb also Wen er dir oben In hawet zu dem kopf |So spring mit dem rechten fuß | 27^v | gen Im vß dem hawe Uff |Sin lincken sytten vnd im springen verwent din schwert mit dem gehültz houch vor deinem haupt das din doũm vnnden kome vmd[!] schlach In mit der kurtzen schniden zu sine^r lincken sytten |So vaschdü sinen haw In din gehültz vnd triffest In zu dem kopff ~

[47] *Twer* is translated as both "crosswise" and "across", depending on which fits the sentence better.

The crosswise cut:[47]
The crosswise takes out
Whatever approaches from the day.

Gloss: Remember, the crosswise cut counters all cuts that are hewn down from above. Drive the cut like this: stand with the left foot forward and hold your sword on your right shoulder, and when he hews in at you above (toward the head), then leap away from the cut with the right foot well against him on his left side, and in the leap, turn your sword with the hilt high in front of your head so that your thumb comes underneath, and strike in with the short edge toward his left side so that you catch his cut in the hilt and you press in toward his head (as is pictured here).

Der Twer haw.
+ Twer benimpt, was vom tag herkompt.
Glosa. Merck denn Twer haw pricht alle hew, die vonn obenn nider gehawenn werden, vnnd denn haw treib also, Stehe mit dem linckenn fus vor, vnnd halt dein schwert an deiner rechtenn Achsel, vnnd |10v| wann er dir obenn inhawet zu dem kopf, so spring mit dem rechtenn fus gen Im wol aus dem haw auf sein lincke seittenn, vnnd im springenn verwendt dein schwerd mit dem gehiltz, hoch vor deinem haupt, das dein dawme vnden komme, vnd schlag in mit der kurtzenn schneidenn, zu seiner linckenn seytten, fechstu seinen haw in dem gehiltz, vnnd truckst in zu dem kopf, als hie gemalt ist.

29 | A piece of the crosswise cut:

Cross with the Strength,
Remember your work with it.

Gloss: This is how you should work from the crosswise with the Strength, and do it like this: when you hew in with the crosswise, think of the Strength of your sword strongly on his with the crosswise. If he holds strongly against it, then strike in on his sword with crossed arms, behind his sword's blade on his head, or slice him through the muzzle with this piece.

30 | Again, a piece of the crosswise cut:

Remember, when you bind on his sword from the crosswise with the Strength of your sword, if he holds strongly against it, then push his sword from below with your hilt (on your right side), and strike around again swiftly with the crosswise, against his right side, in toward the head.

Ain stuck vß dem zwerhaw

|Zwer mit der sto^{e}rck den arbait do mit mo^{e}rck

Glosa Das ist wie dü mit der sto^{e}rck auss der zwer arbaiten solt vnd dem thun also, wen dü Im mit de^r zwer zu hauest |So gedenck das dü Im mit der zwer sterck deines schwerts starck In das Sin |Helt den er starck wyde^r |So schlach In am schwert mit ~~gekrütten~~ gekrüczten armen hinder seines schwertz klingen vff den kopff | 28^r | oder schnÿd In mit dem stuck dürch das maül ~

aber ain stuck vß dem zwerhaw

|Merck wen du Im vß der zwer mit der sto^{e}rck deines schwerts an sin Schwert bindest helt den er starck wÿde^r |So stoss mit deinem gehülcz sin schwert von dir vnderisch vff dein rechten sytten vnd schlach bald mit der zwe^r wyderum gen siner rechten syten Im zu dem kopffe ~

A piece of the crosswise cut:
Cross with the Strength,
Remember your work with it.

Gloss: This is how you should work from the crosswise with Strength, and do it like this: when you hew in with the crosswise, think that you wind strongly with the Strength of your sword on his. If he holds strongly against it, then strike in on his sword with crossed arms, behind his sword's blade on his head (as is done here) or slice him through the muzzle with this piece.

Again, a piece of the crosswise cut:
Remember, when you bind on his sword from the crosswise with the Strength of your sword, if he holds strongly by you, then push his sword from below with your hilt (on your right side, as is pictured here) and strike around again swiftly with the crosswise, against his right side, in toward the head.

Ein stuck aus dem Twerhaw.
Twer mit der sterck, den arbait domit merck.
Glosa. Das ist, wie du mit sterck aus der Twer arbaittenn solt, vnnd den thu also, wen du Im mit dem Twer zu hawest, So gedenck das du Im mit der sterck deines schwerts starck windest an das sein, helt er dann stark wider, so schlag in am schwert mit gecreutztenn Armen hin seines schwerts klingenn auf denn kopff, als hie ist gemacht, oder schneid im mit dem stuck durchs maul.

Aber ein stuck auf dem Twerhaw.
Merck wenn du im auß dem Twer mit der sterck deines schwerts ann sein schwert bindest, halt er dann starck bey dir, so stoß mit deinem gehiltz sein schwert vonn dir vntersich, auf dein rechte seyttenn, als hie gemalt stet, vnnd schlag balde mit |11r| der Twer wider vmb genn seiner rechtenn seyttenn im zu dem kopff.

31 | Again, a piece of the crosswise cut:
Likewise, when you bind him on his sword with the crosswise, if he is Soft on the sword, then lay the short edge in on the right side of his neck, and leap with the right foot behind his left, and pull him over it with the sword.

But likewise, when you bind on him with Strength and you sense he is Soft on the sword, then lay the short edge in over the right side of his neck.

32 | Another piece:
Likewise, when you bind on his sword with the crosswise, if he is Soft on the sword then press his sword down with the crosswise and lay the short edge forward, behind his arms on his neck.

33 |

Counter for the laying-on
Likewise, when someone lays the high work on your neck like this, release your sword from the left hand and push his sword from the neck with the right, and step against his left side with the left foot (in front of both of his feet), and go over both his arms close by his hilt with your left arm and make him dance.

Aber/ain stuck vß dem zwerhaw
|Item wan dü Im mit der zwer an sin schwert bindest Ist da er waich am schwert |So leg Im die kurtzen schnÿden zu seiner rechten sytten an den halß vnd spring mit dem rechten fuß hinde[r] seinen lincken vnd rücke In mit dem schwert darüber etc

27v | Item empfindestu aber wann dw mit sterck ym anbindest vnd er waich im swerte ist so leg Im die kurtz schneid zw seiner rechten seitten hinuber an den hals

Ain ander stuck
28v | |Item wen dü Im mit der zwer an sin schwert bindest Ist er dan waich am schwert |So truck mit der zwer sin schwert nÿder vnd leg Im die kurtzen schnÿden hinder sinen armen vornen an den halß

Bruch für das anlegen Wann dir ainer mit der oberen arbait also an den hals legt so laß dein swert aus der lincken handt vnd mit der rechten stoss sein swert von dem hals vnd schreit mit dem linckenn fues gegen seiner lincken seitten für sein baid fües vnd var mit dem lincken arem vber sein baid arem nahent bey seinem gehiltze vnd für In zw dem dantze

Likewise, if someone takes you by the neck toward your right side, then release your sword from your left hand and push his sword from your neck with the right, and step against his right side with your left foot (in front of both of his feet), and go over both his arms close by the hilt with your left arm and make him dance, or stab underneath between his legs toward his groin.

22v | Item nÿmpt dich eyner by dem hals zu diner rechten siten so loß din schwertt vß diner lincken hant vnd stos mit der rechten sin schwertt von dinem hals vnd schrit mit dinem lincken fus gegen siner rechten siten fur sin beide fus vnd far mit dinem lincken arm vber sin beide arm nohe by dem gehultz vnd fur in zu dem tantz oder stich im vnten zwuschen ~~dinem~~ beynen zu dem gemechtt ~~

Dresden – SR

Glasgow – GF

34 | A counter against the upper crosswise cut:

Likewise, when you bind on his sword from your right side (by hewing over him or otherwise), if he strikes around with the crosswise toward the other side, then come Before it, also with the crosswise, in under his sword onto his neck.

Here remember the counter against the upper crosswise cut:

Remember, when you bind on his sword from your right side (by hewing over him or otherwise), if he strikes around with the crosswise toward the other side, then come Before it, also with the crosswise, in under his sword onto his neck (as is pictured nearby hereafter), so that he strikes himself with your sword.

35 |

Here remember the counter against the lower crosswise strike:

Remember, when you bind on his sword from your right side, if he then strikes around from the sword with the crosswise toward the other exposure on your right side, remain with your hilt overhead, and turn the blade of your sword downward against his cut, and stab in toward the lower exposure (as is pictured hereafter nearby).

Long sword

ain bruch wider den obern zwerhaw

|Item wan dü Im von deine[r] rechten sytten mit ainem oberen haw, oder sunst an sin schwert bindest |Schlecht er dan mit der zwer vmb dir zu der anderen sytten |so kom vor aũch mit der zwerch vnder sin schwert Im an den halß ~

1[r] | Hie merck den pruch der wider den õbern twer haw

Merck wen du im von deiner reichtn seitten mit ainem ober haw oder sunst an sein schwert pindest, schlecht er dan mit der twer vmb, dir zu der andern seitten, so kum vor auch mit der twer vndter seinem schwert im an den hals, als am nächstn hernach gemalt stett, so schlecht er sich selber mit deinem schwert.

[Illustration]

1[v] | Hie merck den pruch wider die vnderen twer schleg.

Merck, wen du im von deiner rechten seitten an sein schwert pindest, schlecht er dan vom schwert vmb mit der twer dir zu der andern plõs deiner rechten seitten, so pleib mit deinem gehiltz vber deinem haubt, vnd verwendt dein schwertz kling vndter sich gegen seinem haw, vnd stich im zu der vndtern plõs, als am nachsten hernach gemalt stet.

[Illustration]

Likewise, a counter against the upper crosswise cut:

You bind on his sword from your right side by hewing over him; if he strikes around with the crosswise, then come Before it, also with the crosswise cut, in under his sword onto the neck.

A counter against the upper crosswise cut:

Remember, when you wind on his sword from your right side (by hewing over him or otherwise), if he strikes around with the with the [!] crosswise toward the other side, then come Before it, also with the crosswise, in under his sword onto his neck (~~as is pictured here~~).

A counter against the lower crosswise cut:

Remember, when you wind on his sword from your right side (with hewing over him), if he then strikes around from the sword with the crosswise toward the lower exposure on your right side, remain with your hilt overhead, and turn the blade of your sword downward against his cut, and stab in toward the lower exposure (as is pictured here).

21ᵛ | |Item ein bruch wider den zwerch ober haulb

so du Im von diner rechten siten mit eynen ober haulb an sin schwertt bindest schlecht er mit der zwierch vmb sich so küm auch vor mit dem zwerch haulb vnter sin schwertt an den hals ~~

Ein bruch wider den ober Twer haw.

Merck wenn du im von deiner rechtenn seytten, mit einem oberhaw, oder sonnst ann sein schwert windest, schlecht er denn mit dem mit der Twer vmb dir zu der anndernn seittenn, so kum vor auch mit der Twer vnter seinem schwert im an den hals ~~als hie stet gemalt~~.[48]

Ein stuck wider ein vnter Twerhaw.

Merck wenn du Im vonn deiner rechtenn seyttenn ann sein schwert windest mit einem oberhaw, schlecht er dann vom schwert vmb mit der Twer dir zu der vnternn bloß deiner rechtenn seittenn, So beleib mit deinem gehiltz ober deinem haupt, vnd verwendt dein schwert clingenn vntersich gen seinem haw, vnnd stich im zu der vntern bloß als hie gemalt stet.

[48] This phrase is also present in GF, but for some reason was struck out in JMR.

Dresden – SR

36 | **How one should strike with the crosswise toward the four exposures:**

51 Cross to the plow,
Yoke it hard to the ox.

Gloss: This is how you should strike with the crosswise at the four exposures in one round.[49] **Learn it like this:** when you approach to fence with him, notice when it seems right to you, then leap toward him and strike in with the crosswise toward the lower exposure of his left side. This is called 'striking toward the plow'.

37 | **Again, a piece of the crosswise cut:**
When you have struck in with the crosswise toward the lower exposure, strike quickly up above toward the other side with the crosswise, in toward the head. This is called 'striking toward the ox'. And then nimbly strike a further crosswise strike toward the ox and another toward the plow, crosswise from one side to the other, and hew in over him with it, above toward the head, and with that, pull yourself away.

Glasgow – GF

This is the text and the gloss on how one should strike with the crosswise toward the four exposures:
Cross to the plow,
Has yoked to the ox.

Gloss: This is how you should strike with the crosswise at all four exposures in one round.[49] Learn it like this: when you approach to fence with him, notice when it seems right to you, then leap toward him and strike in with the crosswise toward the lower exposure of his left side (as is pictured hereafter nearby). This is called 'striking toward the plow'.

Again, a piece of the crosswise cut:
Remember, when you have struck in with the crosswise toward the lower exposure (as is pictured previously next to this), strike quickly up above toward the other side with the crosswise, in toward the head. This is called 'striking toward the ox'. And then nimbly strike a further crosswise strike toward the ox and another toward the plow, always crosswise from one side to the other, and hew in over him, above toward the head, and with that, pull yourself away.

Long sword

Wie man zu den vier blossen mitt der zwer schlachen soll
|Zwer zu dem pflug, zu dem ochsen hart gefuge
Glosa |das ist wie dü In ainem zu gang | 29^{r} | mit der zwe^{r} zu den vier blossen schlagen solt **Das vernÿm also** |Wan dü mit dem zufechten zu Im kompst |So merck wan es dir eben ist |So spring zu Im vnd schlag In mit de^{r} zwer zu der vnderen bloß sine^{r} lincken sytten |Das hayst zu dem pflug geschlagen

2^{r} | **Das ist der text vnd die glõß wie man mit der twer zu den vier plõssn schlachen sol.**
Twer zu dem pflueg, zu dem ochsen hat gefueg.
Glosa. das ist wie du in ainem zu gang mit der twer zu allen vier plõsn schlachen solt, das vernym also, wen du mit dem zu fechtn zu im kumbst, So merck wen es dir eben ist so spring zu im, vnd schlag im mit der twer zu der vndteren plõs seiner lincken seitten, als am nechstn gemalt stet, das haist zu dem pflug geschlagen.
[Illustration]

Aber ain stuck vß dem zwerhaw
|Wen dü Im mit der zwe^{r} zu der vnderen bloß geschlagen hast so schlag bald vff mit der zwe^{r} |Im zu der anderen sytten oben In zu dem kopff das haisst zu dem ochsen geschlagen vnd schlach denn fürbas behendtlich ainen zwerch schlag zu dem ochsen vnd den anderen zu dem pflug crũczwÿß von aine^{r} sytten zu de^{r} anderen vnd haw Im do mit ainem oberen haw oben ein zu dem kopffe vnd zũch dich damit ab ~

2^{v} | **Aber ein stuck auß dem twer haw.**
Merck, wen du im mit der twer zu der vndteren plõss geschlagen hast, als vor am nächstn gemalt stet, so schlag pald auff mit der twer im zu der anderen seitten oben ein zu dem kopf, das haist zu dem ochssen geschlagen, vnd schlag den fũrbas behentlich albeg ainen twer schlag zu dem ochsn, vnd den andern zu dem pflũeg, creũtz weis von ainer seitten zu der andern, vnd haw im ein ober haw oben ein zu dem kopf, vnd zeuch dich damit ab.
[Illustration]

[49] This is the only time *zugangen* is used in the Liechtenauer glosses and may be a reference to the structure of *Gänge* (or 'rounds' of counted blows) used in *Fechtschule* competitions. Unlike other glossators, Ainring →

adds a strike toward the head at the end of the sequence followed by a very rare instruction to withdraw (*abzucken*). This would line up well with fencing a '*Gang* of five' blows at the *Fechtschule*.

38 |
52 Whatever is well crosswise
Threatens the head with leaping.

Gloss: This is that you should leap out on the side well with each crosswise strike when you want to strike in toward him; thus, you may well meet his head. And see that you are well-protected in the leap with your hilt above, in front of your head.

39 | Again, remember a piece of the crosswise hereafter, and it is called 'the misser':[50]
53 The misser misleads well,
Harries from below as he wishes.

Gloss: This is that all fencers that like to displace become misled and struck with the misser. Drive the piece like this: when you approach to fence with him, do as if you will freely a cut over him toward his left side, thus he is harried and struck underneath as you wish.

This is the text and a lesson from it:
Whatever is well crosswise
Threatens the head with leaping.

Gloss: This is that you should leap out on the side well with each crosswise strike when you want to strike in toward him; thus, you may well meet the head. And see that you are well-protected above, with the hilt well in front of your head.

Again, this is the text and the gloss on a piece of the crosswise cut, and called 'the misser':[50]
The misser misleads well,
Harries from below as he wishes.

Gloss: Remember, all fencers that like to displace become misled and struck with the misser. Drive the piece like this: when you approach to fence with him, do as if you will freely strike a crosswise cut toward the left side of his head, but with the cut, snatch your sword away and strike in with the crosswise toward the lower exposure of his right side (as is pictured hereafter nearby). Thus is he harried and struck underneath as you wish.

29v | Waß sich wol zwerch mit springen dem haupt geferet
Glosa |das ist das dü mit ainem yden zwerschlage wol vß solt springen Im vff die sytten, do |dü Im zu schlagen wylt so mag stü In wol treffen zu sienem haupt vnd wart das dü In dem sprũng oben vor dinem haupt mit dienem gehũltzen vol bedeck sÿest

Hie nach mörck aber ain stuck vß der zwer vnd das haÿsset der feler
|Feler wer wol furet von vnden nach wunsch er ruret
Glosa |Das ist mit dem feler werden alle fechter die da gern fersetzen ver fyrt vnd geschlachen das stu^{c}k trib also |Wan du mit |dem zu fechten zu Im kompst |So thu alß ob dü In mit ainem fryen ober haw zu sine^{r} lincken sÿtten |So ist er vnnden nach | 30r | wunsch gerũret vnd geschlagen

3r | Das ist der text vnd ain ler daũon
Was sich wol twer, mit springen dem haubt gefert,

Glosa das ist, das du mit einem yeden twer schlag wol auß solst springen, Im auff die seitten do du im zu schlagen wild, So magstu in wol zu dem haubt treffen, vnd wart das du in dem oben, vor deinem haupt mit dem gehiltz wol bedeckest seiest.

3v | Das ist der text vnd die gloß aber ains stucks auß dem twerhaw vnd haist der feler,
Feler wer wol fũret, von vndten nach wunsch er rũret.
Glosa Merck mit dem feler weren alle vechter die da geren versetzn verfũrt vnd geschlagen, das stuck treib also, wen du mit ^{dem} zu vechtn zu im kumbst, So thue als du mit einem freien twerhaw zu seiner lincken seitten zu dem kopf wellest schlagen vnd verzugk mit dem haw dein schwert vnd schlag im mit der twer zu der vndteren plõs, seiner rechten seitten, als am nachsten da hernach gemalt stet, So ist er vndten nach wunsch gerũrt vnd geschlagenn.
[Illustration]

[50] A *Fehler* is the opposite of a *Treffer* ("hitter") and signifies something that *doesn't hit* whatever target it was aimed at. Here, it's clearly an intentional miss.

Again, this is a piece of the crosswise; it is called 'the misser':[50]

The misser misleads,
Harries from below as he wishes.

Gloss: Remember, all fencers that like to displace become misled and struck with the misser. Drive the piece like this: when you approach to fence with him, do as if you will freely strike a cut over him toward the left side of his head, but with the cut, turn your sword and strike in with the crosswise toward the lower exposure of his right or left side (as is pictured here). Thus is he harried and struck underneath as you wish.

+ **Das ist aber ein stuck aus der Twer das heist der Feler.**

Feler wer furet, von vnden nach wunsch er ruret.
Glosa. Merck mit dem feler werenn al fechter die do gernn versetzenn verfurt vnnd geschlagen, das stuck treib also, wen du mit dem zu fechtenn zu Im kombst, So thu als du in mit einem |11v| freyenn oberhaw zu seiner linckenn seitten zu dem kopff wollest schlagenn, vnnd wend mit dem haw dein schwert, vnnd schlag in mit der Twer, zu der vndern bloß seiner rechtenn seyttenn oder linckenn als hie gemalet stet, so ist er vnten nach wunsch gerurt vnnd geschlagenn.

40 | **Again, a piece of the crosswise, and it is called 'the inverter':**

54 The inverter constrains,
Also run through with wrestling.
55 Knowingly take the elbow,
Leap into his balance.

Gloss: Remember, you should also drive this piece: when you bind on his sword with a cut in over or under him, invert your sword so that your thumb comes underneath, and stab toward his face above such that you constrain him so that he must displace you, and in the displacement, grab his right elbow with the left hand, and leap with the left foot in front of his right and push him over it like this. Or run through him with the inverter and wrestle, as you will find hereafter in the running through.

Again, this is the text and the gloss on a piece of the crosswise; it is called 'the inverter':

The inverter constrains,
Also run through with wrestling.
Knowingly take the elbow,
Leap into his balance.

Gloss: Remember you should also drive this piece: when you bind on his sword with a cut in over or under him, invert your sword so that your thumb comes underneath, and stab toward his face above such that you constrain him so that he must displace you, and in the displacement, grab his right elbow with the left hand, and leap with the left foot in front of his right and push him over it like this (as is pictured hereafter nearby). Or run through him with the inverter and wrestle, as you will find written hereafter in the running through.

aber ain stuck vsß dem zwerhaw vnd daß haÿsst der verkerer ~

|Verkerer zwinget |durch la^{e}ffer aũch mit ringet |den elenbogen gewisse nÿm spring Im In die wage
Glosa |Merck das stũck soltü also trÿb~~er~~en wen dü In mit ainem vnden oder oben haw an sin schwert bindest |So verker dein schwert das din doũme vnde kome vnd stich Im oben In zu dem gesichte |So zwingstu In das er dir versetzen muß Vnd In der verseczung begriff mit der lincken hand sin rechten elebogen vnd spring mit dem lincken fusse für sinen rechten vnd stosß In also daribe^{r}, |Oder lauff In mit dem verkerre^{r} durch vnd ringe, alß dü In dem du^{r}ch lauffen her nach wirst finden

4^{r} | **Das ist der text vnd die glos aber ains stucks aus dem twerhaw das haist der verkerer,**

Verkerere zwing^{e}t durch lauffer auch mit ringet, den elpogen, gewis nÿm spring im in die wagen.
Glosa. Merck das stuck solt du treiben also, Wen du im mit ainem vndter oder ober haw an sein schwert pindest, so verker dein schwert das dein dawm vndten kum vnd stich im oben ein zu dem gesicht, So zwingst du in das er dir versetzn muß, vnd in der versatzung Begreiff mit der lincken hand sein rechten elpogen, vnd spring mit dem lincken, fues fũr sein rechten, vnd stos in also darũber, als am nächsten da hernach gemalt stet, oder lauff im mit dem verkerer durch vnd ring, als du in dem durchlauffen hernach geschriben wirst vinden.
[Illustration]

51

[51] The text on HM: 28v is similar, but not enough to be considered a direct match.

Dresden – SR

41 | Again, a piece of the misser:

56 Miss twice,
If it meets, make a slice with it.[52]

Gloss: Remember, this is called 'the double misser' because one should drive a double misleading in one approach. Drive the first like this: when you approach to fence with him, leap with the foot against him and do as if you will strike in well with a crosswise strike toward the left side of his head, but snatch the strike away, in toward the right side of his head.

Glasgow – GF

Again, this is the text and the gloss on a misser:

Miss twice,
If it meets, make a slice with it.[52]

Gloss: Remember, this is called 'the double misser' because one should drive a double misleading in one approach. Drive the first like this: when you approach to fence with him, leap with the right foot against him and do as if you will strike in well with a crosswise strike toward the left [side] of his head, but snatch the strike away, and strike in toward the right side of his head (as is pictured hereafter nearby).

Long sword

30v | Aber ain stuck von ainem feler

|Fele[r] zwÿfach trifft man den schnit mit macht

Glosa |Merck das haysst der zwÿfach fele[r] |darümbe das man In ainem zu fechten zwaÿerlaÿ ferfÿrung darũß trÿben sol die ersten tryb also wen dü mit dem zu fechten zu Im kompst |So spring mit dem fuß gen Im vnd thun alß dü Im mit ainem zwer schlagen zu sine[r] lincken sÿtten zu dem kopff schlagen welest vnd ferzũck den schlag Im zu sine[r] rechten sÿtten an den kopff ~~

4v | Das ist der text vnd die glos aber von ainem feler

Veler zwifach, trifft man den schnit mit macht

Glosa Merck das haist der zwifach veler, darumb das man in ainem zu vechten zwayerlay verfũrung darauß treiben sol, die erst treib also, wen du mit dem zu vechten zu im kumbst, so spring mit dem rechten fũes gegen im vnd thue als du im mit einem zwerschlag zu seiner lincken zu dem kopf schlagen welst, vnd verzuck den schlag vnd schlag in zu seiner rechtn seittn an den kopf, als am nächstn da hernach gemalt stett. [Illustration]

[52] This couplet uses similar phrasing to 79, in the section on pursuit. The version in HM is an exactly match (apart from the insertion in another hand).

Again, a piece:
...**Miss twice,**
If it meets, make the ancient slice with skill.[52]

...Others[53] vary, and like this: this is called 'the double misser' because one should drive a double misleading in one approach. Drive the first like this: when you approach to fence with him, step or leap with the right foot so the left foot is in front of you against him and do as if you will strike in well with a crosswise toward the left side of his head, but snatch the strike away and leap forward on his right with the left foot, and strike in toward the right side of his head (if it is just like in the first piece, war).

Aber ain stuck

28v | **Fäler zwifach trifft man den altenn schnidt mit ^krafft macht**
...Etlich anderst vnd also das haist der zwifach feler darumb das man in ainem zwvechten zwaierlai verfurung darauß treiben sol Die ersten treib also wann dw mit dem zufechten zw ym kumbst so schreit oder spring mit dem rechten fues So dir der linck fues vornen stet gegen ym vnd thue als dw In mit ainer twir zw seiner lincken seitten zw dem kopf schlagen wöllest vnd vertzuck den schlag vnd mit dem lincken fues hinfur auf sein rechte gesprungen vnd schlag ym zw seiner rechten seitten an den kopf ist schier wie in dem ersten stuck krieg

[53] HM uses the label '*etlich*=others' to refer variously to Ainring or Nicolaüs when there is a separate gloss attributed to Hans Medel for the same text.

42| Again, a piece of the misser:

57 Twice further,
Step in left and do not be lazy.

Gloss: This is when you have struck in toward the right side of his head with the first misleading (as is pictured nearby):[54] strike in again quickly, around toward the right side of the head, and go over his sword with the short edge with crossed arms and leap in left (that is, on your left side), and slice in through his muzzle with the long edge.

Again, this is the text and gloss on a piece of the misser:
Twice further,
Step in left and do not be lazy.

Gloss: This is when you have struck in toward the right side of his head with the first misleading (as is pictured previously nearby): take the slice in under his arms, and strike in again quickly, around toward the left side of the head, and go over his sword with the short edge with crossed arms and leap in left (that is, on his left side), and slice in through his muzzle with the long edge (as is pictured).

Aber ain stuck vß dem feler

|Zwyfach es fyrbas schryt In lunck vnd biß nit laß

Glosa |Das ist wan dü Im mit der e^{r}sten verfÿrunge zu sine^{r} rechten sytten zu dem kepff geschlagen |31^{r}| haüst, alß am nesten gemelt ist |So schlach bald wyderümb Im zu der rechten sytten zu dem kopff vnd far mit der kurtzen schnÿden mit auß gecrütztcn armen ṽber sin schwe^{r}t vnd spring Imlincke |das ist auff dein lincken sÿtten vnd schnyd In mit der langen schnÿ-den durch das ma^{u}l ~

5^{r}| Das ist der text vnd glos aber ains stucks aus dem feler.
Zwifach es fũrbaß, schreit ein linck vnd pis nit laß

Glosa. Das ist wen du in mit der ersten verfũrung zu seiner rechten seitten zu dem kopf geschlagen hast, als vor am nachstn gemalt stet so nym damit den schnit vndten in sein armm, vnd schlag pald widerumb im zu der lincken seitten zu dem kopf, vnd far mit der kurtzn schneid auß gekreutznt armen vber sein schwert vnd spring im linck das ist auf sein lincke seitt, vnd schneid in mit der langen schneid durch das maul, als da gemalt stet.
[Illustration]

[54] This is the only instance in which the scribes for SR failed to remove an entire reference to an illustration.

Again, a piece:

Twice further,
Step in left and do not be lazy.

...But others say this is when you have struck in toward the right side of his head with the first misleading: strike in again quickly, around toward the right side of his head, and go over his sword with the short edge with crossed arms and leap in left (that is, on your left side), and slice in through his muzzle with the long edge and dig yourself out directly.

Aber ain stuck

28v | **Zwifach es fürbas schreit yn linck vnd bis nit lass**

...Etlich aber sagen das sey wenn dw yn mit der ersten verfürung zw seiner rechten seitten zw dem kopf geschlagen hast So schlach pald widerumb Im zw der rechten seitten ym zw dem kopf vnd var mit der kurtzen schneid aus gekreutzten armen v̈ber sein swert vnd spring ym linck das ist auf dein lincke seitten vnd schneid yn mit der langen |29r| Langenn[!] schneid durch das maul vnd heb dich resch ne [a]uß[55]

[55] Due to a fold in the upper area, some words in the digital copy are difficult or impossible to recognize.

43| The cockeyed cut[56] with its pieces:

58 The cockeye breaks in
Whatever the buffalo strikes or stabs.
59 Whoever threatens to change,
The cockeye robs him of it.

Gloss: Here remember, the cockeye is a cut that counters the cuts and thrusts of the buffalo (that takes mastery with violence). Drive the cut like this: when he hews in from above his right side, then also hew from your right side with the short edge against his cut, into the Weakness of his sword with outstretched arms, and strike him on his right shoulder.

If he changes through, then shoot in with the cut (extended toward the chest).

And also hew like this when he stands against you in the guard of the plow, or when he wants to stab toward you underneath.

This is the text and the gloss on the cockeyed cut[56] and of the pieces that one should drive from it:
The cockeye breaks in
Whatever the buffalo strikes or stabs.
Whoever threatens to change,
The cockeye robs him of it.

Gloss: Remember, the cockeye is a cut that counters the cuts and thrusts of the buffalo (that takes mastery with violence). Drive the cut like this: when he hews in from above his right side, then also hew from your right with the short edge against his cut, into the Weakness of his sword with outstretched arms, and strike him on his right shoulder (as is pictured hereafter nearby).

If he changes through, then shoot in with the cut (extended toward the chest).

Also hew like this when he stands against you in the guard of the plow, or when he wants to stab toward you underneath.

Der schilhaw mitt sinen stucken

|Schille^{r} ein bricht waß bufle^{r} schlecht ode^{r} stÿcht, We^{r} wech sel trawet schille^{r} In dar vß beraubet
Glosa Hie merck Der schille^{r} ist ain haw der dem ~~bufflen~~ buffeln die sich mayste^{r}schafft an nemmen mit gwalt In bricht |In hawen vnd stechen vnd ~~da~~ den haw trÿb also Wan er dir oben ein hawet |31^{v}| von sine^{r} rechten sytten |So haw och von dene^{r} rechten sytten mit der kurtzen schnÿden mit vff gerechten armen ge sinem hawe In die schwech sinenes schwerts vnd schlag In vff sinen rechten achsel Wechselt er dũrch |So schyß In mit dem hawe lang In zu der brũst vnd also haw a^{o}ch wan er gen dir stat In der hũtte de pflũgs Oder wen er dir vnden ~~s~~ zu wyll stechen

5^{v}| Das ist der text vnd die glõs von dem Schilhaw vnd von den stucken die man darauß treiben sol.
Schiler ein pricht was pũffl schlecht oder sticht, wer wechsell dra[ut] Schiller daraus in beraubt
Glosa. ~~Wechssel~~ Merck der schiler ist ain haw der dem pũff einpricht dy sich maisterschafft an nemen mit gewalt ein pricht in hew vnd in stich vnd den haw treib also, Wen er dir oben ein haut von seiner rechten seitten, So haw auch von deiner[57] rechten mit der kurtzn schneid mit auff gerackten armen gegen seinen haw in die schwech seins schwertz vnd schlag in auff sein rechte achsl, als am nãchstn da hernach gemalt stet, wechsselt er durch so scheũs im mit dem haw langk ein zu der prust, Also haw auch wen er gegen dir stet in der hutt des pflũgs, oder wen er dir vndten zu wil stechen.
[Illustration]

[56] *Schielen* is a tricky verb because it has a double meaning of "askew" and "cross-eyed" or "having a lazy eye" (both were once upon a time called "squint-eyed"), and the text relies on both meanings. "Cockeyed" is the closest we can get to both meanings in one English word. (Thanks to CHRISTIAN TROSCLAIR for suggesting this translation.) When *shielen* is used to tell you to look at rather than strike something, it's rendered "cock [i.e, turn] your eye".
[57] Corrected from *seiner*.

This is the cockeyed cut[56] with its pieces:

The cockeye breaks in
Whatever the buffalo strikes or stabs.
It threatens to change,
The cockeye robs him of it.

Gloss: Remember, the cockeye is a cut that counters the cuts and thrusts of the buffalo (that takes mastery with violence). Drive the cut like this: when he hews in from above his right side, then also hew from your right with the short edge against his cut. Thus you strike and displace together, and meet him with the cut (as is pictured here).

Das ist der Schilhaw mit seinen stucken.

+ Schilherein[!] pricht, was puffel schlecht oder sticht, Wechsel drewt, Schilher daraus in beraubt.
Glosa. Merck der Schilher ist ein haw, der denn Pufflen die sich meisterschafft an nemen, mit gewalt ein pricht in hawenn vnnd in stechenn, den haw treib also, wenn er dir obenn ein hawt von seiner rechtenn seyttenn, so haw auch von deiner rechtenn, mit der kurtzen schneid, gegenn seinem haw, also schlechstu vnnd versetzt miteinander, vnnd trifst in mit dem hew, als hie gemalt stet.

DRESDEN – SR

44 | **Again, a piece of the cockeye:**

60 Cock your eye; if he shortens on you,
See that you change through on him.

Gloss: Remember, this is a lesson that you should cock an eye (with the face) and quite precisely see whether he fences short against you; you should recognize it when he hews toward you and does not outstretch and extend his arms. So hew also and in the cut, go through in under his sword with the point, and stab in toward his face.

GLASGOW – GF

Again, this is the text and the gloss on a piece of the cockeye:

Cock your eye; if he shortens on you,
Changing through defeats it.

Gloss: Remember, this is a lesson that you should cock an eye (with the face) and quite precisely see whether he fences short against you; you should recognize it when he hews toward you and his arms are not outstretched and extended from him. So hew also and in the cut, go through in under his sword with the point, and wind your hilt overhead on your right side, and stab in toward his face (as is pictured nearby).

Likewise, all fencers who fence short, from the ox, from the plow, and with all windings in front of the man: joyfully change through from cuts and from thrusts with the extended point. With that, you retain them at the sword so that they must let you come to bind on them and be struck.[58]

LONG SWORD

Aber ain stuck vß dem schiller

|Schill kurßt er dich an durch wechsel er sigt Im an
Glosa |Merck das ist ain lerre |Das schillern solt mit dem gesichte vnd gar eben sechen ober kurtz gen dir vicht |Das solt bÿ dem erkenen wan er dir zu hawet vnd sin arm mit dem haw nicht lanck streckt |So haw |32^{r}| och Und far In dem haw mit dem ort vnder seinem schwert dürch vnd stiche In zu dem gesicht ~

6^{r}| **Das ist der text vnd die glos aber ains stucks aus dem schiler.**

Schil kurtzt er dich an, durch wegssel gesigt im an.
Glosa Merck das ist ain ler daz du schilen solt mit dem gesicht, vnd gar eben sehen ob er kurtz gegen dir ficht, das solt du pey dem erkennen, wen er dir zu haut vnd mit dem haw sein armm nicht lanck von im reckt, so haw auch vnd far in dem haw mit dem ort vndter seinem schwert durch vnd windt auff dein rechte seytt dein gehiltz vber dein haübt vnd stich im zu dem gesicht, als am nägsten gemalt stett.
Item allen vechternn die da kurtz fechten, auß dem ochsn, aus dem pflueg, vnd mit allen winden vor dem man, den wechsl frölich durch, auß hawen vnd aüs stechen mit dem langen ort, damit bestetigistũ sy an dem schwert das sy dich mũessen zu dem abent[!] lassen kumen, vnd sy schlagen
[Illustration]

[58] This combines phrases from paragraphs 51–52 of the Lew gloss (and specifically HS and JMR); see JMR: 24^{v}.3–25^{r}.2. *Abent* is a misspelling of *anbind*, evidenced by JMR which has *anpundt* (while HS has *an binden*). L, the witness most commonly studied in isolation, instead has *an winden* which obscures this parallel.

45 | Again, a piece of the cockeyed cut:

61 Cock your eye at the point,
And take the neck without fear.

Gloss: Remember, the cockeye counters the extended point. Drive it like this: when he stands against you and holds the point with outstretched arms against the face or the chest, stand with the left foot forward and cock an eye toward the point (with the face), and do as if you want to hew in toward the point, but hew strongly onto his sword with the short edge, and with that, shoot the point in extended toward the neck with a step in of the right foot, etc.

Again, this is the text and the gloss on a piece of the cockeyed cut:
Cock your eye at the point,
And take the neck without fear.

Gloss: Remember, the cockeye counters the extended point {with a deceit of the face}. Drive it like this: when he stands against you and holds the point with outstretched arms against the face or chest, stand with the left foot forward and cock an eye toward the point (with the face), and do as if you want to hew in toward it, but hew strongly onto his sword with the short edge, and with that, shoot the point in extended toward the neck with a step in of the right foot (as is pictured nearby hereafter).

Abere ain stuck uß dem schillhaw

|Schill zu dem ort Und nÿ den halß on forcht
Glosa |Merck der schiller bricht den langen ort vnd den tryb also wen er ge dir stat vnd helt dir den ort usß gerachten armen gen dem gesÿchte oder der brüst |So stand mit dem lincken fusß für, Und schill mit dem gesicht zu dem ort vnd thun alß dü Im zu dem ort hawen welest Und haw starck mit der kurtzen schnÿden vff sin schwert vnd schüß Im den ort, darmit lang In zu dem halß mit ainem zutrytt des rechten fuß ~: etc~

6v | Das ist der text vnd die glos, Aber ains stucks aus dem schilhaw
Schil zu dem ort, vnd nÿm den hals ane forcht.
Glosa. Merck der schiler pricht den langen ort ‡{mit ainer betrugnus des gesichtz} vnd den treib also, wen er gegen dir stett vnd helt dir den ort aus geracktn armen gegen dem gesicht oder prust, So ste mit dem lincken fües vor vnd schil mit dem gesicht zu dem ort, vnd thue als dü Im darzu hawen welst, vnd haw starck mit der kurtzn schneid auff sein schwert vnd schews im den ort da mit lanck ein zu dem ~~lanck~~ hals, mit ainem zu tritt des rechten füeß, als am nächsten da hernach gemalt stett
[Illustration]

59

Again, a piece of the cockeye:

Cock your eye at the point,
And take the neck without fear.

Gloss: Remember, the cockeye counters the extended point. Drive it like this: when he stands against you and holds the point with outstretched arms against the face or against the chest, stand with the left foot forward and cock an eye (with the face), and do as if you want to hew in toward the point, but hew strongly onto his sword with the short edge, and with that, shoot the point extended toward the neck with a step in of the right foot.

Aber ein stuck aus dem Schilher.

Schilh zu dem ort, vnd nim den hals one forcht.
Glosa. Merck der Schilher pricht denn langenn ort, vnnd denn treib also. Wenn er gegenn dir stehet, vnnd helt dir denn ort aus gerechten armenn gegenn dem gesicht, oder gegen der brust, |12ʳ| so stehe mit dem linckenn fus vor, vnnd schilh, mit dem gesicht, vnnd thu als du im zu dem ort wollest hawenn Unnd haw starck, mit der kurtzenn schneid auff sein schwert, vnnd schewß im denn ort, damit langkem zu dem hals, mit einem zu trit des rechten fuß.

[59] The text on HM: 30ʳ is similar, but not enough to be considered a direct match.

Dresden – SR

46 | Again, a piece of the cockeyed cut:

62 Cock your eye at the top
Of the head, if you want to harm the hands.

Gloss: Remember, when he wants to hew in above you, cock your eye (with your face) like you want to strike him on his head, but hew against his cut with your short edge, and strike in onto his sword blade with the point on the hands.

Glasgow – GF

Again, this is the text and the gloss on a piece of the cockeyed cut:
Cock your eye at the top
Of the head, if you want to harm the hands.

Gloss: Remember, when he wants to hew in above you, cock your eye (with your face) like you want to strike him on his head, but hew against his cut with your short edge, and strike in onto his sword blade with the point on the hands (as is pictured nearby hereafter).

Long sword

32v | Aber ain stuck vß dem schillhaw

|Schill zu dem oberen haupt hende wilt dü bedebren
Glosa |Merck wen er dir oben will In hawen |So schill mit dem gesicht alß dü In vff das haupt wylt schlagen Und ~~n~~ haw mit der kurtzen schnÿden gen sinen haw Und schlag In an siner schwertz klingen mit dem ort vff die hend ~

7r | Das ist der text vnd die glos aber ains stucks aus dem schiler.
Schil zu dem obern, haubt hendt wildu bedoberen
Glosa Merck Wen er dir oben ein wil hawen, So schil mit dem gesicht, als du in auff das haubt wellest schlagen, vnd haw mit der kurtzen schneid gegen seinem haw, vnd schlag in an seiner schwertzs klingen mit dem ort auff die hendt, als am nächsten da hernach gemalt stett.
[Illustration]

You can also do this: when he wants to hew a free cut in over you from above, cock your eye toward his head like you want to strike him on his head with it, but hew against his cut with your short edge, and strike in onto his sword blade with the point down on the hands.

Others speak like this: when he wants to hew in over you or stands against you in the extended point, cock your eye (with your face) like you want to strike him on his head. Hew against his cut with your short edge, and strike in onto his sword blade with the point on the hands.

27ʳ | ...dz magstu auch thün wen er dir von oben eyn haulbet eynen fryhen ober haulb so schilch zu sinem hauptt als du wollest yn dar vff sclagen vnd haulb mit der kurtzen schniden gegen sinem haulb vnd schlag in an sines schwertz klingen nider mit dem ortt vff die hende ~~[60]

30ʳ | Etlich sagen also wann er dir oben einhawenn will oder in dem langen ort gegen dir steet So schilh mit dem gesicht als dw yn auff das hawbt schlahenn wöllest haw mit der kurtzen schneid gegen seinemm haw vnd schlag yn an seines schwertz klingenn mit dem ort auf die hende

[60] This fragment is appended to Lew's description of countering the extended point.

DRESDEN – SR

47 | The parter[61] with its pieces:

63 The parter
Is dangerous to the face.

GLASGOW – GF

This is the text and the gloss on the parter[61] and of its pieces:
The parter
Is dangerous to the face.

Gloss: Here remember that the parter is dangerous to the face and the chest. Drive it like this: when he stands against you in the foolish guard, hew from the long part (down from above) with the long edge, and with the cut, remain high with the arms, and hang in toward his face with the point, etc.

Gloss: Remember that the parter is dangerous to the face and the chest. Drive it like this: when he stands against you in the guard of the foolish, hew from your part (down from above) with the long edge, and with the cut, remain high with the arms, and hang in toward his face with the point (as is pictured nearby hereafter).

48 | A piece of the parter:
64 With its turn,
Very dangerous to the chest.

Gloss: This is when you hang the point in toward his face with the parter: if he pushes your point firmly upward with his hilt in the displacement of the parter, then invert your sword with the hilt high in front of your head and set the point on his chest underneath, etc.

Again, a piece of the parter:

Likewise, if he displaces your parter firmly upward with his hilt, then invert your sword with the hilt high in front of your head and set your point on his chest underneath (as is then pictured).

LONG SWORD

~~Ain anderß stuck vß dem schillhaw~~
Der schayteler mitt sinen stucken

|Der schaytle^{r} dem antlytz ist gefer

Glosa |Hie merck der schaÿtle^{r} ist dem antlÿtz vnd der brüst gefa^{e}rlich |den tryb also |Wen er gen dir stat In der hut a^{u}lber |So haw mit der langen schnÿde von der lange schaÿttlen oben nÿder vnd belÿb |33^{r}| mit dem haw hoch mit den armen vnd heng Im mit den ort ein zu dem gesÿchte

7^{v} | Das ist der text vnd die glos von dem schaittler, vnd von den stucken daraus
Der schaittler, dem antlitz ist gefãr.

Glosa Merck der schaittler ist dem antlitz vnd der prust gar gefãrlich, den treib also, wen er gegen dir stett in der huet des albers, So haw mit der langen schneid von deiner schaittel oben nyder, vnd pleib mit dem haw hoch mit den armen, vnd heng im das ort ein zu dem gesicht, als am nächsten hernach gemalt stett.
[Illustration]

Ain stuck vß dem schaiteler
|Mit sine^{r} ker, der brust fast gefer
Glosa das ist wen dü Im den ort mit dem schaitle^{r} oben ein hengst zu dem gesicht |Stost er ~~dich~~ dir denn den ort In der versatzüng mit dem gehültz vascht ybe^{r} sich |So verker dein schwert mit dem gehültz hoch fyr din haupt vnd setz Im den ort vnden an die brust etc

8^{r} | Aber ain stuck aüs dem schaittler

Item versetzt er dir mit dem gehültz vast übersich den schaittler, so verker dein schwert mit dem gehültz hoch für dein haubt vnd setz im dein ort vndten an die prüst, als dan da gemalt stet.
[No illustration]

[61] *Scheitel* refers to the top of the front of the head, the hairline, and the part of the hair. The verb *scheiden* means to divide or separate, and the noun *Schaiden* refers to damage or harm. Thus, while Ainring is →

This is the parter[61] with its pieces:

The parter
Is dangerous to the face.
With its turn,
Very dangerous to the chest.

Gloss: Remember, drive the parter like this: hew from your part (down from above) with the long edge, in toward his head. If he displaces, hang in the point with the long edge over his hilt and stab toward his face (as is pictured here).

Likewise, remember: if you hew in above with the parter[53] and hang toward his face, and he pushes your point firmly upward with his hilt, then invert your sword with the hilt firmly upward in front of your head and set the point on his chest underneath.

Likewise, if he pushes your point firmly upward with his hilt, then turn your sword with the hilt high in front of your head so your thumb comes underneath, and set the point on his chest under his hands (as is pictured here).

Das ist der Schatler mit seinen stucken.

Der Schatler dem antlitz ist geuer, mit seiner ker, der brust vast geuer.
Glosa. Merck denn Schaytler treyb also, Haw mit der langenn schneydt, oben nider vonn deiner Schaytel, Im zu dem kopff, versetzt er, so heng im denn ort, mit der langenn schneid, vber sein gehiltz, vnnd stich im zu dem gesicht, als hie gemalt stet.

28v | Item merck so du im mit dem schittler oben yn haulbest vnd gehengt zu dem gesicht stost er dan mit dem gehultz den ort vast vbersich so verkere din schwertt mit dem gehultz fast vber sich fur din hauptt vnd setz im den ortt vnten an sin brust ~~

Item stost er dir denn ortt, mit seinem gehiltz fast vber sich, so verwendt dein schwert mit dem gehiltz, hoch für dein hauptt, das dein dawmen vnden kum, vnnd setz im den ort vnder seinen hendenn ann die prust, als hie gemaltt stet.

clear that the name refers to starting from the part of *your* hair (i.e., over your head) rather than targeting *their* hairline, the other potential meaning of "to part" (divide the opponent in half) may also be intended.

49 | **How the crown counters the parter:**

65 Whatever comes from it,
The crown takes that away.

Gloss: Remember, when you hew in above with the parter, if he displaces with the hilt high above his head, the displacement is called 'the crown', and it runs in at you with it.

Another piece:

50 | **How the slice counters the crown:**

66 Slice through the crown,
Counter the hard beautifully.
67 Press the sweep,
Pull yourself away with slicing.

Gloss: Remember, when he displaces the parter (or other such cut) with the crown and runs in with it: take the slice into his arm under his hands, and press firmly upward; thus, the crown is countered again. And turn your sword from underneath into the slice over, and pull yourself away with it.

This is the text and the gloss on how the slice counters the crown:

Slice through the crown,
Counter the hard beautifully.
Press the sweep,
Pull yourself away with slicing.

Gloss: Remember, when he displaces the parter (or other such cut) with the crown and runs in with it: take the slice into his arm under his hands, and press firmly upward (as is pictured nearby); thus, the crown is countered again. And turn your sword from underneath into the slice over, and pull yourself away with it.

Wie die kron den schaÿtler bricht

|Waß von Im komp, die kron das abnÿmpt

Glosa |Merck wan dü Im mit dem schai^{tt}le^{r} oben ein hawest, versetzt er mit dem gehulcze hoch ob |33^{v}| ob sinem haupt Die versatzũng hayst die kron vnd laufft di^{r} do mit eÿm

8^{v}| **Ain ander Stück**

Wie der schnitt die kron bricht

|Schnid dũrch die krone, |So brichest dü sÿ hart schon
Die strich drucke mit schnitten sÿ ab zu^{c}ke

Glosa |Merck wen er dir den schailtle^{r} oder sũnst ainen haw vesetzt mit der kron, vnd dir da mit ein laũfft, |So nÿm den schnit vnder sin henden In sin arm Vnd trũck vast ṽber sich so ist die kron wyde^{r} gebrochen Vnd wende din schwert vß dem vnderen schnit In den oberen vnd zũch dich da mit abe

9^{r}| **Das ist der text vnd die glos wie der schnidt pricht die kron**

Schneid durch die kron, So prichst du sÿ hart schon,
die strich druck mit schnitten sy ab zugk

Glosa. Merck Wen er dir den schaittler oder sunst ain haw versetzt mit der kron, vnd dir damit einlaufft So nym den schnidt vnder seinen henden an sein armm vnd truck vast ṽbersich, als am nachsten gemalt statt, So ist die kron wider geprochen vnd wendt dein schwert auß dem vndteren schnitt in den obern vnd zeuch dich damit ab.

[Illustration]

Another piece:
Whatever comes from it,
The crown takes that away.
Slice through the crown,
Counter the hard beautifully.
Press the thrust,
Pull yourself away with slicing.

Gloss: Remember, when you hew in above with the parter, if he displaces with the hilt high above his head, the displacement is called 'the crown', and it runs in at you with it.

Likewise remember, when he displaces the parter (or other such cut) with the crown and runs in with it: take the slice into his arms under his hands, and press firmly upward; thus, the crown is countered well. And wind your sword from underneath into the slice over, and pull yourself away with it.

Take the slice underneath, into his arms under his hands, and press firmly upward; thus, the crown is countered against.

Ein annder Stuck.
Was vonn Im kumpt, die kran das abnymptt, Schneydt durch die kran, so prichstu Sie hart schön, die Stich truck, mit schniten sie abzuck.
Glosa. Merck wenn du Im mit dem Schaytler, obenn einhawest, versetzt er mit dem gehiltz hoch ob seinem haupt, die versatzung heist die kran, vnd laufft dir damit ein,

28v | Item merck wan er dir den schittler oder sünst ein haulb versetzt mit der kron vnd dir do mitt yn laufft so nym den schnitt vnder sin henden in sin arme vnd druck vast vber sich so ist die kron wol gebrochen vnd wind din schwertt in den obern schnidt vnd zuch do mit abe ~~

so nim denn vnternn |12v| schnit, vnder sein hendenn in sein Arm, vnnd druck fast vbersich, so ist die kron wider gebrochenn.

51 | These are the four lairs:[62]

68 Four lairs alone:
Hold from them, and flee the common.
69 Ox, plow, foolish,
From the day are not unknown to you.

Gloss: This is that you should not hold any lairs other than the four that will be named here.

52 | The first guard

The ox.[63] Arrange yourself in it like this: stand with the left foot forward and hold your sword by your right side in front of your head, and let the point hang against his face.

53 | The second guard

The plow. Arrange yourself in it like this: stand with the left foot forward and hold your sword by your right side, hands crossed, over your knee, so that the point is against his face.

Das sind die ~~vier~~ vier leger

|Vier leger allain |Da von halt vnd flűch die gemaim
|34r| Ochs pflug, alber vom tag, sÿ dir nit vnmer

Glosa |Ist das dü von kainem leger nicht halten solt, denn alain von den vieren die hie genant worden sind ~

Die erst hutt

|Der ochß, do schick dich also mit stand mit dem lincken fuß vor vnd halt din schwert neben diner rechten sytten vir din haupt Vnd laß Im den ort gegen dem gesicht hangen

Die ander hutt

|Der pflug do schick dich also mit stand mit dem lincken fuß vor vnd halt din schwert gecrűtzten henden neben diner rechten sytten, ober dinem knÿe |Also das Im der ort gegen dem gesicht ste

This is the text and the gloss on the four lairs:[62]

Four lairs alone:
Hold from them, and flee the common.
Ox, plow, foolish,
From the day are not unknown to you.

Gloss: Remember, this is that you otherwise should not hold any lairs other than the four named here for you.

Likewise, the first guard is called 'the ox'.[63] Arrange yourself in it like this: stand with the left foot forward and hold your sword by your right side with the hilt in front of the head, and let the point against his face (as is pictured nearby hereafter).

Likewise, the second guard is called 'plow'. Arrange yourself in it like this: stand with the left foot forward and hold your sword by your right side with crossed hands, over the knee, so that the point is against his face (as is pictured nearby hereafter).

9v | Daz ist der text vnd die glos von den vier legeren.

Vier leger allein, daűon halt vnd fleűch die gemayn,
Ochs pflu[g] Alber, vom tag seÿ dir nit vnmăr.

Glosa. Merck daz ist, das [du] sunst von chainem leger nichtz halten solt, dan allein von den firen die dir hie genant werden

Item. Dye erst huet haist der ochs, da schick dich also mit sten mit dem lincken fues vor, vnd halt d[ein] swert neben deiner rechten seitten mit dem gehiltz vor dem ha[ubt] vnd las im den ort gegen dem gesicht als am nächstn gemalt stett

Item. die ander hűt haist der pflueg, da schick dich also mit sten mit dem lincken fues vor, vnd halt dein swert mit gekreutzten hendten neben deiner rechten seÿtten ob dem knye also das ym d[er] ort gegen dem gesicht stet, als am nächstn hernach gemalt stet
[Illustration]

[62] A *Leger* is a lair or place to lie down and also a military encampment (especially a siege camp). A *Hut* is a defensive position or guard. It has been proposed based on these meanings that *Leger* originally indicated a position used offensively and *Hut* a position used defensively, but there's no evidence of this distinction in any treatises from Liechtenauer's tradition and the terms are used interchangeably.

On the four lairs:[62]
Four lairs alone:
Hold from them, and flee the common.
Ox, plow, foolish,
From the day; the three things are not unknown to you (says master Hans Medel)

Gloss: This is that you should otherwise hold no lairs other than the four lairs that are named here; they are the oft-named four guards, that is the ox,[63] plow, foolish one, from the roof.

31v | Von den vier legernn
Vier leger allain da von haltman, vnd fleuchtz die gemain Ochs pflueg alber vom tag die drei ding sind dir nit vnmer Sagt Maister hans medl Glosa Das ist das dw sunst von kainem leger halten solt dann allain von den vier legernn die hye genant sein Sy werden offt genennt die vier huten das ist der ochs pflueg alber vom tag

[63] While 'oxen' refers to bovines trained as draft animals—usually castrated males, but sometimes bulls or cows—the term *Ochs* also encompassed the aurochs, the wild ancestor of domestic cows that was prized for its leather and hunted to extinction by the 17th cen. Unlike domestic cattle, the aurochs had large horns that pointed inward.

54 | The third guard
The foolish one.[64] Arrange yourself in it like this: stand with your right foot forward and hold your sword with arms outstretched away from you, with the point on the ground.

Likewise, the third guard is called 'the foolish one'.[64] Arrange yourself in it like this: stand with the right foot forward and hold your sword with arms outstretched away from you, with the point on the ground (as is pictured nearby hereafter).

55 | The fourth guard
From the day.[65] Arrange yourself in it like this: stand with left foot forward and hold your sword on your right shoulder, or hold it overhead with arms outstretched.

And how you should fence from the guards, you will find it written in this book.

Likewise, the fourth guard is called 'from the day'.[65] Arrange yourself in it like this: stand with the left foot forward and hold your sword on your right shoulder (as is pictured nearby hereafter), or hold it overhead with arms outstretched.

And how you should fence from the guards, you will find it all written.

Die dritt hutt
34v | |Der alber do schick dich also mit stand mit deinem rechten fuß vor vnd halt din schwert mit vßgerechten armen vor dir mit dem ort vff die erden

Die vierd hutt
|Vom tag do schick dich also, mit stand mit lincken fuß vor vnd halt din schwert an diner rechten achseln |Oder halt es mit vßgerechten armen ṽber din haup vnd wie dü vß der hutten fechten solt, das findest dü In dissem buch geschrÿben ~ ~

10r | Item. dye dritt hutt haist der Alber, da schick dich also mit sten, mit dem rechten fues vor, vnd halt dein schwert mit ausgerachkten armen vor dir mit dem ort auff der erden, als am nächsten hernach gemalt stett.

Item. die viert hũt haist vom tag, da schick dich also mit sten, mit dem lincken fues vor, vnd halt dein schwert an deiner rechten achssel Als am nachsten hernach gemalt stet, oder halt es mit auß gestrackten armen vber dein hãubt, vnd wie du auß den hũtten vechten solt das vindest alles geschriben.
[Illustration]

[64] *Albern* is an adjective meaning "foolish" whereas *Alber* is a noun meaning "poplar tree"; Meyer interprets it as the former, which he justifies with the explanation that only a fool would use such a worthless guard; Mayr reads it as the latter, and his Latin translator renders it '*populus arbor*=poplar tree' without any explanation (PMM: 89r.2–90r.3). There's no way to be sure which one Ainring intended, but PD: 2r renders it →

Alwer, a MHG word meaning "simple or worthless", and Hans Medel uses '*alberlich*=foolishly' in relation to it (HM: 32r.5), suggesting a shared understanding early in the tradition.

65 Could also be '*von Dach*=from the roof'. I went with "day" to match the pastoral theme of the others, and for the similarity to HF and J, which uniquely have '*von himmel*=from the sky'.

56| These are the four displacements[66] that injure or counter the four lairs:

70 Four are the displacements
That also greatly injure the lairs.
71 Guard yourself from displacement.
If that happens, it greatly rewards you.

Gloss: Remember, you have heard previously that you should fence solely from the four lairs, and you should also know that the four displacements, they are four cuts.

The first cut is the curved cut; it counters the guard of the oxen.[67]

The second cut is the crosswise cut; it counters the guard from the day.[67]

The third is the cockeyed cut; it counters the guard of the plow.

The fourth is the parter; it counters the foolish guard.

This is the text and the gloss on the four displacements[66] that counter the four lairs:

Four are the displacements
That also greatly injure the lairs.
Guard yourself before crosswise displacement.
If that happens, it troubles you more.

Gloss: Remember, you have heard previously that you should fence solely from the four lairs, and you should also know that the four displacements counter the same four lairs, and the four displacements, they are four cuts.

The first is the curved cut; it counters the guard which is called 'the ox'.[67]

Likewise, the second cut is the crosswise cut; it counters the guard from the day.[67]

Likewise, the third is the cockeye; it counters the guard of the plow.

Likewise, the fourth, it is the parter; it counters the guard which is called 'the foolish one'.

Das sind die vier verseczen die die vier leger leczen oder brechen ~

|Vie^{r} sind versetzen |Die die leger aüch ser letzen vor versetzen hiet dich, geschicht es se^{r} es |35^{r}| mÿt dich

Glosa Merck dü hast vo^{r} geho^{r}t das dü alain vß viern legern vechten solt |So solt dü och nün wyssen die vier versetzüng Das sind vier hew Der erst/haw |Ist der krumpfhaw der bricht die hüt des ochsen der ander |ist der zwe^{r}haw der bricht die hütt vom tage der dritt ~~dr~~ |ist de^{r} schÿllhaw der bricht die hüt deß pflugs der vierdt |ist der schaittle^{r} der bricht die hut albe^{r},

10^{v}| Daz ist der text vnd die glos von den vier versetzen, die die vier leger prechen.

Vier sind versetzen, dy die leger auch ser letzen, vor zwir versetzen hütt dich, geschicht es mer es müet dich.

Glosa. Merck du hast vor gehört das dü allain aus vier legeren vechten solt, So soltu auch nu wissen die vier versetzen die dy vier selbigen vier leger prechen, vnd die vier setzen das sind vier hew, der erst ist der krumbhaw, der pricht die hüet die da haist der ochs. Item der ander ist der twerhaw, der pricht die hutt vom tag. Item die dritt ist der schiler, der pricht die hautt des pfluegs. Item der viert daz ist der schaittler, der pricht die hutt die da haist der alber,

[66] In the 15th century, the word *Versetzen* encompassed both '*versetzen*=to move or shift location' and '*vorsetzen*=to place in front', and in this text it usually refers to a defensive action we might label a "parry", which is how it's translated in Chapter 3. In this segment, though, it refers instead to four attacks that were covered earlier in the five strikes. Many people have tried to find some obscure translation of *versetzen* that covers both attacking and defending, but I prefer to embrace the ambiguity and let people grapple with →

with the confusion. (I'll just note that nothing in SR's text, at least, states that countering guards is *why* they're called '*versetzen*', merely that there are certain things called *versetzen* which *also* each counter a guard.)

[67] Ainring doesn't describe how the curved cut counters the ox or how the crosswise counters the day. These pieces are described most clearly in the Lew gloss (JMR: 19r.4–19v.1 & 21^r.2–21^v.1).

57| And guard yourself from all displacements that the simple fencers drive, and remember, when he hews then also hew, and when he stabs then also stab; and how you should hew and stab: you will find this written in the five cuts and in the setting aside, etc.

And otherwise guard yourself before all displacements that the simple fencers drive, and remember, when he hews then also hew, and when he stabs then also stab; and how you should hew and stab: you will find this written in the five cuts and in the setting aside.

Vnd hÿt dich vor allen versetzen die die schlechten vechte[r] trÿben vnd merck wen er hawt, |So haw och, vnd wen er sticht so stych och, Vnd wie dü hawen vnd stechen solt, das findest dü In den fünff hewen vnd In den absetzen geschryben etc

vnd hütt dich sunst vor allem versetzen, die dy schlechten fechter treiben, vnd merck wen er hawt, So haw auch vnd wen er sticht so stich auch, vnd wie du haũen vnd stechen solt das vindestu in den fũnff heũen vnd in dem absetzen geschriben.

Guard yourself before displacement.
If you need it, it rewards you.

You should guard yourself before displacing as other fencers displace: they allow the point to go out before someone low or high to the side, and whoever does that cannot seek the exposure on someone. Therefore do not displace, but remember, when he hews then also hew, and when he stabs then also stab; and how you should hew against his cuts, you will find this described in the five cuts; and how you should fence against his thrusts, you will find that described in the setting aside.

30r | **Vor versetzen hut doch**
Geschicht dir not es mut dich

|Du solt dich hütten vor dem versetzen als yttlich vechter versetzen so lossen sie den ortt vor dem man nider oder hoch vff ein siten vß gan vnd wer dz thutt kan der bloß am man nit suchen dar vmb versetz nit vnd merck wan er haut so haulb auch vnd wan er sticht so stich auch vnd wie du gegen sinen haulben haulben solt dz vindest in den fünff heulben geschrebenn |30v| vnd wie du gegen sinen stichen vechten solt dz vindestu in den ab setzen geschriben ~

58 | A piece against displacement:

72 If you are displaced,
And however it comes to that,
73 Hear what I advise:
Wrench off, hew quickly with haste.

Gloss: This is that however it happens that you are displaced, remember: if someone displaces your upper cut, then in the displacement, go against him with the pommel in front of his front hand and wrench downward with it, and with the wrench, strike him on the head with the sword.

This is the text and the gloss, a piece against displacement:
If you are displaced,
And however it comes to that,
Hear what I advise:
Wrench off, hew quickly with haste.

Gloss: This is that however it happens that you are displaced, remember: if someone displaces your upper cut, then in the displacement, go against him with the pommel over his front hand and wrench downward with it, and with the wrench, strike him on the head with the sword (as is pictured nearby hereafter).

59 | Again, a piece against displacement:
Remember, when you hew a lower cut from the right side, if he then falls with the sword on yours so you cannot come up with it, then go nimbly over his sword with the pommel, and strike in with the snapping, with the long edge toward the head. Or, if he falls on your sword against your left side then strike in with the short edge.[68]

Again, a piece against displacement:
Remember, when you hew a lower cut from the right side, if he then falls with the sword on yours so you cannot come up with it, then go nimbly over his sword with the pommel, and strike in with the snapping, with the long edge toward the head (as is pictured nearby hereafter). Or, if he falls on the sword against your left side then strike in with the short edge.[68]

35^{v} | Ain stuck wider die versäczung

|Ob dier versetzt ist, vnd wie das dar komen ist, ho^{e}re was ich rate, rays abe, haw schnel mit dratte
Glosa |das ist wie es dar zu komen ist, das dir versetzt ist worden |So mercke versetzt man dir ainen ober haw So var Im in der versaczung mit dem knopff vir sin vo^{r} gesetzte hand vnd rayse da mit vnder sich vnd mit dem raysse schlage In mit dem schwert vff den kopff ~

11^{r} | Daz ist der text vnd die glos ains stucks wider die versatzung
Ob dir versetzet ist, vnd wie das darkumen ist · hõr was ich ratt, raiß ab haw schnell mit drat.
Glosa. daz ist wie es darzu kumen ist, das dir versetzt ist worden, So merck, versetzt man dir ^{ain} ober haw, So far im in der versatzung mit dem knopf ṽber sein versetzte handt, vnd raiss damit vntersich, vnd mit dem reissen schlag in mit dem swert auff den kopf, als am nächsten da hernach gemalt stett.
[Illustration]

Aber ain stuck wider die versäczung ~
|Merck wen dü hawest ainen vnder haw von der rechten sytten fellet er dir dan da mit dem schwert vff das dine, das dü do mit nicht vff komen magst |So far behendlich | 36^{r} | mit dem knopff ṽber sin schwert vnd schlag jn mit dem schnappen mit der langen schnÿden zu dem kopffe |Oder velt er dir gen diner lincken sytten vff das schwe^{r}t so schlach In mit der kũrtzen schnÿden

11^{v} | Aber ain stuck wider die versatzung
Merck, wen ^{du} hawst ain vndterhaw von der rechten seitten, velt er dir dan mit dem schwert auff das dein, daz du da mit nit auff magst kumen, So far behentlich mit dem knopf vber sein swert, vnd schlag in mit dem schnappen mit der langen schneid zu dem kopf, als am nächsten hernach gemalt stet, oder velt er dir gegen deiner lincken seitten auff das schwert so schlag in mit der kürtzen schneid.
[Illustration]

[68] This piece is instead associated with verse 93 (see piece 82, "Slicing off", below) in Pseudo-Danzig, Lew, and Nicolaüs. Ainring offers no separate gloss for that verse, and instead lumps it together with 94.

This is a piece against displacement:

If you are displaced,
And however it comes to that,
Hear what I advise:
Wrench off, hew quickly with haste.

Likewise, if your upper cut is displaced and he comes in nearby, then go with the pommel over his front hand and wrench downward with it, and with the wrench, strike him on the head with the sword.

Gloss: This is when your upper cut is displaced: in the displacement, go with the pommel over his front hand and wrench downward with it, and with the wrench, strike him on the head with the sword.

Another:
Likewise remember, if you hew a lower cut from the right side and he then falls with his sword on yours so you cannot come up with it, then go nimbly over his sword with the pommel, and wrench your blade from under his, and strike in with the snapping, with the long edge toward him. But if he falls on your sword against the left side, then strike in with the short edge.[68, 69]

Das ist ein stuck wider das versetzenn.

Ob dir versatzt ist, vnd wie das darkommen Ist, hör was ich rath, reys ab, haw schnel mit drat.

31r | |Item ob dir der ober haulb versetzt ist vnd nohent an in chümpt so far mit dem knoff vber sin versetzte hantt vnd rieß do mit vndersich vnd mit dem reissen schlag in do mit dem schwertt vff din kopff ~~

Glosa. Das ist, wenn dir einer versetzt ein oberhaw, so far im in der versatzung mit dem knopf, vber sein vorgesatzte handt, vnd reys damit vndersich, vnnd mit dem reyssenn, schlag in mit dem schwert auf denn kopff.

alliud ~
|Item merck so du haulbst ein vnter haulb von diner rechten siten felt er dir dan mit sinem schwert vff dz din dz du dar mit nit vff kümen magst so far behendlich mit dem knopff vber sin schwert vnd rieß din clingen vnten von der siner vnnd schlag in mit den schnappen mit der langen schniden züm felt er dir aber gegen ~~dir~~ der lincken siten vff dz schwert so schlag in mit der kurtzen schniden ~

[69] These two paragraphs also appear on HM: 32v–33r, but there's only space for four witnesses here so the Medel text is included in an appendix.

60 | **Again, a piece against displacement:**

74 Set upon four extents.
Learn to remain on them if you want to end.

Gloss: When you hew in from above your right shoulder, if you want to quickly end with him, then when he displaces, remember to strike around quickly with the crosswise and with the sword, grab with the left hand in the middle of the blade and set the point in his face, etc.

Or set upon him (toward whichever of the four exposures you can approach best), etc.[70]

This is the text and the gloss, again a piece against displacement:
Set upon four extents.
Learn to remain on them if you want to end.

Gloss: This is when you hew in from above your right shoulder: if you want to quickly end with the sword, then when he displaces, remember to strike around quickly with the crosswise and with the strike, grab your sword with the left hand in the middle of the blade and set the point in his face (as is pictured hereafter).

Or set upon him (toward whichever of the four exposures you may approach best).[70]

61 | **Again, a piece against displacement:**

Likewise, when you set the point in his face with the half sword, if he displaces that, bash in toward the other side of his head with the pommel. Or leap with the right foot behind his left and go in with the pommel over his right shoulder, around the neck in front, and pull him over your right leg with it.

This is the text and the gloss, again a piece against displacement:
Remember, when you set the point in the face with the half sword (as is pictured nearby previously), if he displaces that, bash in toward the other side of his head with the pommel. Or leap with your right foot behind his left and go in with the pommel over his right shoulder, around the neck in front, and pull him over your right leg with it (as is pictured nearby hereafter).

aber ain stuck wider die versaczung

|Setz an vier enden, blieb dar vff lere wiltu enden
Glosa Wen dü Im von dener rechten achselm oben ein ha^{u}wst, |wilt dü denn bald mit Im enden So mo^{e}rck wen er versetzt |So schlach bald vmb mit der zwer Und begrüff mit dem schwe^{r}t mit der lincken hand miten In de^{r} klingen vnd setz Im den ort in das gesycht |Oder setz Im an, zu den vier blössen zu ~~weche~~ welcher dü am basten komen kanst etc~

12^{r} | **Das ist der text vnd die glos aber ains stucks wider das versetzen.**
Setz an vier ennden, pleib darauff ler wildü enden
Glosa. daz ist wen du im von deiner rechten achssel oben ein hawst, wil du dan pald mit ~~in~~ dem schwert[71] enden, So merck wen er versetzt, so schlag pald vmb mit der twer vnd begreiff mit dem schlag dein schwert mit der linken handt mitten in der klingen, vnd setz im das ort in das gesicht, als am nãchstn gemalt stett, oder setz im an zu den vier plössen zu welcher du am pesten kümen magst.
[Illustration]

36^{v} | **Aber ain stuck wider die versaczung ~**

|Item wan dü Im mit dem halben schwert den ort In sin gesicht setzest, versetz e^{r} dir das |So stoß In mit dem knopff zu der anderen sytten zu sienem haupt |Ode^{r} spring mit dem rechten fuosß hinder sin lincken vnd far Im mit dem knopff ṽber sin rechten achseln vornen vmb den halß vnd ruck In da mit ṽber din rechte bain ~

12^{v} | **Daz ist der text vnd die glos aber ains stucks wider daz versetzen**
Merck, wen du im dem halben schwert das ort in das gesichst setzest, als am nãchsten vor gemalt stet, versetzt er dir das, so stõs in mit dem knopf zu der andern seytten zu seinem haubt, Oder spring mit deinem rechten fües hinter seinen lincken vnd far im mit dem knopf ṽber sein rechte achssel vorn vmb den hals, vnd ruck in da mit vber dein rechts pain als am nãchsten hernach gemalt stett.
[Illustration]

[70] This piece isn't described Ainring, though Pseudo-Danzig, Lew, and Nicolaüs all cover it; it's described most clearly in the Lew gloss (JMR: 28^{r}.2–28^{r}.5). '*Ansetzen*=setting upon' is common in the armored fencing gloss.
[71] *Dem schwert* is inserted above *in*, which is struck out but which matches *Im* in SR.

Another piece against displacement.
Text:
Set upon upon[!] four extents.
Learn to remain on them if you want to end.

...According to the common gloss, others speak like this: when you hew in from above your right shoulder: if you want to quickly end with it, then when he displaces, remember to strike around quickly with the crosswise and with the strike, grab your sword with the left hand in the middle of the blade and set the point in his face.

Or set upon him (whichever of the four exposures you can or may approach best).[70]

And if someone displaces the setting upon, strike in toward the other (his left) side with the pommel. Or go in with the pommel over his right shoulder, on the neck in front, leap in with your right foot behind his left, and pull and throw him over it.

Or with the pommel on the other side around the neck in front, step to the side, and pull over the foot.

Counter: take the elbow.

Step behind his foot and push.[72]

33r | Ain ander stuck wider die versatzung text

Setz an an vier enden bleib darauf lere wiltu enden
...Etlich nach gemainer glos sagen also wenn dw ym von deiner rechten achsel obenn einhawest wiltu dann pald domit enden so merck wann er versetzt so schlag pald vmb mit der twir vnd begreiff mit dem schlag dein swert mit der lincken handt mitten in der klingen vnd setz ym den ort in das gesicht oder setz ym an die vier plössen zw welcher dw am pasten kummen magst oder kanst

33v | Und so er dir der ansetzen ains versetzt so Schlag In mit dem knopf zw der anderen seiner lincken seitten Oder var ym mit dem knopf über sein rechte achsel vornen an den hals doch spring ym mit deinem rechten fues hinder seinen lincken vnd darüber geruckt vnd geworffen Bruch nym den elenpogen

33r | oder mit dem knopff uff die andern seyten vmb den halß vornen vnd geruckt uber den fuß tritt auff dise seyten
33v | tritt hinach fur seinen fuß vnd stoß

[72] This is the only time P is quoted in this chapter. While P is generally textless or only has single-letter captions, it sometimes has brief text such as fragments of the *Record*. In this case, it has a paraphrase of Ainring as well as an additional, unique instruction for the counter named in HM. No highlighting will be attempted since this doesn't align with the witnesses of Ainring's text very well.

62 | **On the pursuit:**

Learn the pursuits
Twofold, or slice into the weapon.

Gloss: This is that you should learn the pursuits well, and they are dual. Drive the first when he wants to hew in above you; remember: while he pulls his sword upward to strike, pursue him with the cut or with a thrust, and land into the upper exposures ere he comes against you with the cut. Or fall in above into his arm with the long edge, and with that, press him from you.

63 | **Again, a pursuit:**

Likewise, when he hews toward you from above, let his sword go down toward the ground with the cut, and pursue him with a cut in above toward the head ere he comes up with the sword.

Or, if he wants to stab you, remember: while he pulls the sword toward himself for the thrust, pursue him and stab him ere when he completes his thrust.[73]

Von Nachraÿßen

|Nach raÿsse lere zwÿfach oder schnÿd In die were

Glosa Das ist das dü die nach raysen wol lernen solt, wan sÿ sind zwayerlay Die ersten trÿb, wen er dir oben In will hawen |So me^{r}ck |37^{r}| die wil er dz schwert vff zücht zu dem schlag so raÿse im nach mitt dem hawe oder mitt ainem stich vnd triff in zu den[74] obern blöß ee er mit dem haw wider kump oder fall im mitt der langen schniden oben in sin arm vnd truck in domitt von dir ~

Aber ain nachraisen

|**Item** wann er dir von oben zu hawst laß er dam sein schwert mitt dem haẅ nider gen zu der erden |So raÿß im nach mitt ainem haẅ oben ein zu dem kopfe ee er mitt dem schwert vff kumpt

|Oder will er dich stechen so mörck die wil er dz schwert zu im zücht zu dem stich so raÿß im nach vnd stich in ee wan er sinen stich valbringt ~~

[73] This paragraph seems like a better conceptual match for the first pursuit than for the second, and is a candidate for being a scribal omission that was added after they noticed the error.
[74] Corrected from *dem*.

On the pursuit, a teaching:
Learn the pursuits
Twofold, or slice into the weapon.

...And according to the text: This is that you should learn the pursuits well, and they are dual. Drive the first like this: if he wants to hew in above you, remember that while he pulls his sword upward to strike, pursue him with the cut or with a thrust, and land into the upper exposures ere he comes down with the cut. And if he binds on you and wants to work thereafter with the sword, then follow after [him] and afterward, Within, take the slice in above toward the arm with the long edge, and with that, press him from you strongly, etc.

This is on the pursuit:
Learn the pursuits
Twofold, or slice into the weapon.

Gloss: Remember, this is that you should learn the pursuits very well. Drive it like this: when he wants to hew in above you, remember that while he pulls his sword upward to strike, pursue him with a cut or with a thrust toward the exposure ere when he comes down with the cut.

Another piece:
The other pursuit is: when he hews toward you from above, let his sword toward the ground with the cut, and pursue him with a cut in above toward the head ere he comes up with the sword.

Or, if he wants to stab you, remember: while he pulls the sword toward himself for the thrust, pursue him and stab ere he then completes his thrust, etc.[73]

Or if he hews down from above and you let his sword go down toward the ground with the cut, pursue him with a cut over in toward the head ere when he comes up with the sword. Thus he is struck.

33v | Von dem nachraisenn *ain lere*[75]
Nachraisen lere zwifach oder schneid in die were
...Und nach dem text ist das dw die nachraisenn wol lernen sollt Wann sy sindt zwaierlai Die ersten treib also So er dir oben ein will hawen So merck weil er das swert auff zeucht zw dem schlag so raiß ym nach mit dem haw oder mit ainem stich Und triff yn zw der oberen plös ee er mit dem haw nider kumpt Und ob er dir anpünde vnd darnach von dem swert arbaiten wolt so volg nach vnd nym Indeß, hinach den schnit mit der langen schneiden oben ein zw dem arem vnd druck starck yn domit von dir etc So hat er kain macht Also raist ym alltzeit starck nach oben zw dem kopf

Das ist vonn Nachreysenn.
Nachreyßen lehre, Zwifach oder schneid in die wehre.
Glosa. Merck das ist das du die nachraysen gar wol lernenn solt, vnnd die treyb also, wen er dir obenn ein wil hawenn, so merck dieweil er das schwert aufzeucht zu dem schlag, so reys im nach, mit einem ~~schlag~~ haw oder mit einem stich, zu der bloß, ee wen er mit dem haw niderkumbt.

Ain ander stuck
Das ander nachraisen ist Wann er dir von oben Zuhawt lest er dann sein swert mit dem haw zw der erdenn So raiß ym nach mit ainem haw oben ein zw dem kopf Ee er mit dem swert aufkumbt
Oder will er dich stechen So merck weil er das swert zw ym zeucht zw dem stich so raiß ym nach vnd stich ee dann er sein stich verbringt etc

Oder hawt er von obenn nider, vnnd lest sein schwert mit dem haw nider gehenn zu der erden, so rays im nach mit einem oberhaw zu dem kopf, ee wenn er mit dem schwert aufkumbt, so ist er geschlagenn.

[75] *ain lere* inserted by another hand.

64 | On the outer take:[76]

76 Two outer takes;
The work begins after that.
77 And gauge the threats,
Whether they are Soft or Hard.[77]

Gloss: Remember, the two outer forms are two pursuits upon the sword. Drive them like this: when he mis-hews[78] in front of you and you pursue him, if he displaces you, then remain with your sword on his and gauge whether he is Soft or Hard in the threat. If he rises up with the sword with strength, then extend your sword outside over his and stab in toward the lower exposure, etc.

65 | The second outer take:

Likewise, when you fence cautiously against him from hewing underneath (or otherwise from threats underneath), if he lays over you and winds on the sword ere when you come up with it, then remain strong with the sword against his underneath. If he works toward your upper exposure, follow after with the sword and take the Weakness of his sword with the long edge, and press down and stab toward his face.

37v | Von der eüßern nÿm

|Zwai eußren nÿnn dein arbat dar nach beginn Und brieffe[!] die gefert, Ob sÿ sind waych oder ho^r^t

Glosa |Merck die zwi eüssere mÿm das sind zwai nachraisen am schwert das trib also wan er sich vo^r^ dir verhawet |So rayß Im nach Ve^r^setz er dir den das So plÿb mit dem schwert an dem sinen vnd brieff, ob er mit denn gefo^e^rt waich oder ho^r^t ist |Hept dan er mit dem schwert das dim mit sterck ṽbersich sicht |So lang din schwert vssen yber das sin vnd stich In zu dem vnderen blöß etc

Die ander eüßer nÿm

|Item wen du geim fichtest vß vnder hawen oder sũnst vß |38r| den vnderen geferten ṽberylet er dich den vnd windt dir vff da schwert /ee/ wan dü do mit vff komest |So blieb mit dem schwert starck vnden an dem sinen wend er den vnd arbait dir zu dem obern blöß |So volg mit dem schwert nach vnd mit der langen schnÿden nÿm Im die ~~stich~~ schwo^e^ch sines schwerts vnd truck nyder vnd stich Im zu dem gesicht

[76] The second word in the name of this piece is sometimes spelled '*myn*=form' and sometimes '*nym*=take'. *Myn* tends to appear more commonly in the 15th cen. and *nym* in the 16th, but it's hard to say which one is the original term and both are used in this passage. Either term would capture the essence of the piece, which is actions from a bind on the "outside" (left) instead of the "inside" (right).

[77] This couplet is very close to 109, in the section on the winds.

78 The prefix *ver-* can indicate a thing done preemptively or in front, but also a thing done poorly. *Verhawen* seems to refer to the second pursuit (described above) in which the opponent hews to the ground in front of you; this could be rendered 'fore-hewing' (whether meaning hewing before you or hewing in front of you), but I choose 'mis-hewing' to emphasize that it's an error. Mayr's Latin translator has '*ictus longius*=long strike' (PMM: 91^{v}.3–92^{r}.1, 92^{v}.4), describing the nature of the error.

Dresden – SR

66 | **On the feeling and on the word 'Within':**

78 Learn the feeling;
'Within', that word slices sharply.

Gloss: This is that you should learn and understand well the feeling and the word 'Within', since these two things belong together and are the greatest art of fencing.

And learn it like this: when one of you binds the other on the sword, then meanwhile, just as the swords flash together, you should immediately feel boar[!] he has bound you Soft or Hard. And as soon as you have sensed that, think on the word 'Within'; that is, that in that same sensing, you should work nimbly with the sword toward the nearest exposure, according to the Soft and according to the Hard, so that he will be struck before he knows it.[79]

Glasgow – GF

This is the text and the gloss on the feeling and on the word 'Within':

Learn the feeling;
'Within', that word slices sharply.

Gloss: This is that you should learn and understand well the feeling and the word 'Within', since these two things belong together and are the greatest art in fencing.

Likewise, learn it like this: when one of you binds the other on the sword, then meanwhile, as the swords slide together, you should immediately feel (that is, sense) whether he has bound you Soft or Hard. And as soon as you have sensed that, think on the word 'Within'; that is, that in that same sensing, you should work nimbly with the sword toward the nearest exposure, according to the Soft and according to the Hard, so that he will be struck before he knows it.[79]

Long sword

Von dem fülen vnd von dem wort Indes

|Das fulen lere In des das wõrt schnÿdet sere

Glosa |das ist das dü das fülen, vnd das wo^{r}t In das wol lernen, vnd verstan solt wan die zwÿ ding geho^{e}-ren zu samen vnd sind die gro^{e}ßte künste zu dem fechten

Vnd das vernÿm also |38^{v}| wen aine^{r} dem anderen an das schwert bindt, |So soltü In dem alß de schwert zu samen glütz schon zu hand fullen, ebe^{r} waych oder ho^{r}t angebunden hab, Und alß bald dü das empfindest haüst |So gedenck an das wort /In des/ das ist das dü In den selben enpfinden behendtlichen nach der waich vnd nach der ho^{r}t mit dem ~~mit~~ schwert solt arbaiten zu der nechsten bloß |So wirt er/geschlagen ee wann er sin gewar wirt ~

13^{r}| **Das ist der text vnd die glos von dem fũlen vnd von dem wort Inndes.**

Das fũlen lere, inndes das wort schneidet sere.

Glosa. daz ist, das du das fũlen vnd das wort indes wol leren vnd versten solt, wen die zway ding gehõrn zu samen, vnd sind die grõstn kunst in dem vechten

Item. das vernÿm also, wen ainer dem anderen an das schwert pindt so solt du in dem als die swert zesamen glischent ze hant fũlen das ist enpfinden ob er waich oder hert angepunden hat, vnd als pald du das enp-funden hast, so gedenck an das wort inndes, das ist daz du in dem selbigen enpfinden behentlich nach der waich vnd nach der hert mit dem schwert solst ar-baitten zu der nãchsten plõs, so wirt er geschlagen ee wen er sein gewar wirt

[79] This is similar to paragraphs 92 of the Pseudo-Danzig gloss, 85 of the Lew gloss, and 75 of the Nicolaüs gloss; see PD: 29^{r}.2, JMR: 30^{r}.2, & AR: 10^{r}.3.

A lesson on the feeling and on the word 'Within':

Learn the feeling;
'Within', that word slices sharply.

Gloss: This is that you should learn and understand well the word 'Within', since these two things belong together and one may not exist without the other, and they are the greatest art of fencing.

Learn it like this: when one of you binds the other on the sword, then meanwhile, just as the swords flash together, you should immediately feel (or sense) whether he has bound you Soft or the Hard. And once you have sensed that, think on the word 'Within'; that is, that in that same sensing, you should work nimbly toward the nearest exposure, according to the Soft or Hard, so that he will be easily struck before he knows it.[79]

35r | Von dem fülen Und dem Wort indes ain lere

Das fulen lere Indes das wort schneidet sere Glosa Das ist das dw das wort Indes wol lernen vnd versteen solt Wann die zway ding gehören zesamen vnd mag ains an das andert nit gsein vnd send die grosse kunst zw dem vechten

Das vernym also wann ainer dem anderen an das swert pindet so soltu in dem als die swert zwsamen klitzen zwhandt fülen oder empfinden ob er waich oder hert angebunden hab Und als dw das empfunden hast so gedenck an das wort Indes das ist das dw in dem selbigen empfinden behentlich nach der waich oder der herte arbaiten solt zw der nechsten plös so wirt er geren geschlagen ee er sein gewar wirt

Likewise, you should think of the word 'Within' in all binds of the sword, since

'Within' redoubles
and 'Within' transmutes,
'Within' runs through
and
'Within' takes the slice,
'Within' co-wrestles,
'Within' takes the sword.
In the art, 'Within' does whatever your heart desires.[80]

'Within' is a sharp word, and all fencers by[!] do not know the word will be cut down with it. And the word 'Within' is also the key by which all of the art of fencing will be unlocked.[81]

Likewise, remember: you should think of the word 'Within' in all binds of the sword, since

'Within' redoubles,
'Within' transmutes,
'Within' runs through,
'Within' changes through,
'Within' takes the slice,
'Within' co-wrestles,
'Within' takes the sword.
In the art, 'Within' does whatever your heart desires.[80]

'Within' is a sharp word, and all fencers who do not know the word will be cut down with it. And the word 'Within' is also the key by which all of the art of fencing will be locked out.[81]

|Item du solt an das wort /Indes/ In allen an binden des schwertz gedencken wann, ~~wan~~ /|Indes/ dupliert vnd /Indes/ muttiert /Indes/ laufft durch vnd /Indes/ nimpt den schnitt /|Indes/ ringt mitt, ~~vn~~ /|Indes/ nimpt im dz schwert /|Indes/ tut in der kunst |39r| waß dein hercz beger /Indes ist/ ein scharpffes wort damitt alle fechter verschnitten werden bÿ des ~~wortz~~ worts nichten wissen Und dz wort Indes ist och der schlissele damit alle kunst de fechtens vffgeschlossen wirt ~ ~ ~ ~ ~:·

Item Merck, du solt an das wort inndes in allen anpinden des schwertzs gedencken, wen indes dupliert indes mütiert, indes laufft durch, indes wechsselt durch, indes nÿmbt den schnidt, indes ringet mit, indes nymbt im das schwert, indes thuet in der kunst wes dein hertz begert, inndes ist ein scharffs wort damit all vechter verschnytten werdent, die des worts nit wissen, vnd das wort indes ist der schlissl, da mit alle kunst des vechtns aus geslossen wirt.

[80] These eight lines are rhyming couplets in German and appear at this point in every gloss of the 'long' sword (though in other versions, the changing through and running through are reversed), but they aren't part of Liechtenauer's *Record* so they may be a remnant of another teaching or a separate mnemonic. →

Likewise, you should think of the word 'Within' in all binds of the sword, since

'Within' redoubles
'Within' transmutes,
'Within' runs through,
'Within' changes through,
'Within' takes the slice,
'Within' co-wrestles,
'Within' takes the sword.
In the art, 'Within' does whatever your heart desires.[80]

'Within' is a sharp word, and all fencers who do not know the word will be cut down with it. And the word 'Within' is also the key by which all of the art of fencing will be unlocked.[81]

Item dw solt an das wort Indes in allen anbinden des swertes gedenckenn wann Indes duplirt Indes mutiert Indes laufft Indes wechselt durch Indes nymbt den schnidt, Indes ringet mit, Indss |35v| nymbt ym das swert Indes thut in der kunst was dein hertz begert Indes ist ain scharpf wort domit alle vechter verschniten werden die des wortes nit wissen Und das wort Indes ist der schlüssell damit alle kunst des vechtens aufgeschlossen werden

Note that '*mitringen*=co-wrestling' appears as a surrogate for Within in the wrestling treatise of Ott Jud.
[81] This is a close match to paragraphs 94 of the Pseudo-Danzig gloss, 87 of the Lew gloss, and 77 of the Nicolaüs gloss; see PD: 29r.4, JMR: 30r.4, & AR: 10r.5.

67 | Pursuit:

79 Pursue twice,
If it hits, make with the same old slice.[82]

Gloss: This is when he mis-hews in front of you, so pursue him with a hew toward the upper exposure. If he then goes up and winds under you on the sword, then yet remember: as soon as one sword flashes on the other, fall in over his arms from the sword with the long edge and press him from you like this. Or, slice in from the sword through his muzzle. Drive this to both sides.

This is another piece of the text and the gloss on pursuit:

Pursue twice,
If it hits, make with the same old slice.[82]

Gloss: Remember, this is when he mis-hews in front of you, so pursue him with a hew toward the upper exposure. If he then goes up and binds under you on the sword, then remember: as soon as one sword flashes on the other, fall in over his arms from the sword with the long edge and press him from you like this (as is pictured nearby). Or, slice in from the sword through his muzzle. Drive this to both sides.

Nachraÿsen ~

|Nachraisen zwifach, trifft man den alten schnitt mitt macht ·:
Glosa |Das ist wann er sich vor dir verhawet So raÿse im nach mitt ainem hawe zu der obern blöß fört er dann vff vnd windet dir vnden an dz schwert |So merck aber so bald ain schwert vff dz ander blitzscht |So val im vom schwert mitt der langen ~~schw~~ schnÿden über sein arm[83] vnd truck in also von dir |**Oder** | 39v | schnÿd in vom schwert durchs maul dz trÿb zu beden sÿtten ~~~

13v | Das ist aber ain stuck text vnd glos von nach raÿssen.

Nach rayssen zwyfach, trifft man den alten schnidt mit macht.
Glosa Merck, das ist wen er sich vor dir verhawt, So rays im nach mit einem haw zu der obern ploß, vert er den auff vnd pindt dir vndten an das schwert, So merck als pald ain swert auff das ander glischt so vall im vom schwert mit der langen schneid vber sein armm vnd druck in also von dir, als am nachstn gemalt stett, oder schneid im vom schwert durchs maul das treib zu paiden seytten.
[Illustration]

[82] This couplet uses similar phrasing to 56, in the section on the misser. The "old slice" is probably meant to convey "the previous slice", referring to that verse and its gloss (piece 41). (HM confuses this issue by adding an 'old slice' to that verse as well.)
[83] Corrected from *arnn*.

84

84 The text on HM: 35v is similar, but not enough to be considered a direct match.

68 | On overrunning:[85]

80 Whoever aims below,
Overrun, then he becomes ashamed.[86]
81 When it sparks above,
Then strengthen, that I will praise.[87]
82 Make your work,
Or press hard twofold.

Gloss: This is when he aims at your lower exposure in the approach with a cut or with a thrust: you should not displace him, but see that you run over him with a cut in above toward the head, or set upon him above with the point. Thus he becomes ashamed by you, since all upper cuts and all upper setting upon overreaches the lower.

Von überlauffen ~
|Wer vnden remet ~~überlan~~ vberlauffenden, der wirt beschemet |Wenn es klutzscht oben, so störck dz will ich loben |Dein arbait mach, oder hert druck zwifach ·:

Glosa |Das ist wann er dir in dem zufechten mitt ainem haẅ oder mitt ain stich remet der vndern bloß dz soltu im nicht verseczen |Sonder wart dz du in überlauffest mitt ainem haw obe ein zu dem kopffe oder im den ort oben anseczest so wirt er von dir beschemet wann alle ober haw vnd alle ober anseczen überlangen die vndern ~ ~

This is the text and the gloss on overrunning:[85]

The aiming (winding) below
Overrun, then he becomes ashamed.[86]
When it sparks above,
Then strengthen, that I will praise.[87]
Make your work,
Or press hard twofold.

Gloss: Remember, this is when he aims at your lower exposure in the approach with a cut or with a thrust: you should not displace him, but see that you run over him with a cut in above toward the head, or set in at him above with the point (as is pictured nearby). Thus he becomes ashamed by you, since all upper cuts and all upper setting upon overreaches the lower.

14r | Das ist der text vnd die glos von vberlaüffen.
Der vndten rembt, ÷{windet}[88] ṽber lauff den der wirt peschemt, Wen ^es^ klischt oben, So sterck das wil ich loben, dein arbait mach, oder herte drück zwifach.

Glosa. Merck das ist, wen er dir in dem zu fechten mit einem haw oder mit ainem stich rent der vndern plõß, das sol du im nicht versetzen Sunder wart das du in vber lauffest mit ainem haw oben ein zu dem kopf oder im das ort oben einsetzen, als am nächstn gemalt stett So wirt er von dir beschemt, wen all ober hew, vnd all ober an setzen ṽber lengen die vnderen.
[Illustration]

[85] This is 'overrunning' in the sense of a river that overruns its banks or a cup that runneth over.
[86] This couplet uses similar phrasing to 32, in the section on the wrathful cut, and dueling couplet 36.
[87] This couplet uses similar phrasing to 45, in the section on the curved cut.
[88] Written above the word *rembt*. This odd insertion diverges from the standard *Record*, but matches the way the verse is written in the Lew gloss in JMR: 30v.1, which has *Wer vndenn windet*.

Dresden – SR

69 | Likewise, how one should set aside[89] cut and thrust:

83 Learn setting aside;
Artfully injure cut, thrust
84 Whoever stabs at you,
So that your point hits and counters his.
85 From both sides,
You'll meet every time if you step.[90]

Gloss: This is that you should learn to set aside cuts and thrusts with art like this, so that your point meets him and his will be countered. Learn it like this: When someone stands against you and holds his sword like he wants to stab you underneath, stand against him in the guard of the plow from your right side and give yourself an exposure with the left. If he stabs toward your same exposure underneath, then wind with the sword on your left side against his thrust, and step toward him with the right foot so your point meets and his misses.

Glasgow – GF

This is the text and the gloss, how one should set aside[90] cut and thrust:

Learn setting aside;
Artfully injure cut, thrust
Whoever stabs at you,
And your point hits and counters his.
From both sides,
You'll meet every time if you step.[90]

Gloss: This is that you should learn to set aside cuts and thrusts with art like this, so that your point meets him and his will be countered to him. Remember it like this: When someone stands against you and holds his sword like he wants to stab you underneath, stand against him in the guard of the plow from your right side and give yourself an exposure with the left. If he stabs toward your same exposure underneath, then wind with sword on your left side against his thrust, and step toward him with the right foot so your point meets and his one misses (as is pictured hereafter).

Long sword

40r | Item wie man hew und stich abseczen sol ~

|Lern abseczen, haw̃ stich kunstlich leczen Wer vff dich sticht, dz din ort trifft, vnd sinen prücht |Von baiden sÿtten, triff alle mal wilt du schrÿtten ·:

Glosa |Das ist dz du lernen solt hew vnd stich also mitt kunst abseczen |Dz dein ort In treffe vnd im der sin gebrochen werde |**Und Das vernÿm also** |Wenn ainer gegen dir ~~sticht~~ steet vnd helt sein schwert alßer dir vnden zu stechen wölle |So stee wider gegen im in/ der hut des pflugs von diner rechten sÿtten vnd gib dich blöß mitt der | 40v | ~~lingen~~ lincken |Stich er denn dir vnden zu der selbigen blöß |So wind mitt dem schwert gegen sÿnen stich vff din lincke sÿtten vnd schrÿt zu im mitt dem rechten fuß |So trifft din ort vnd der sin fält ~ ~ ~ ~ :·

14v | Das ist der text vnd die glos wie man haw vnd stich sol absetzen

Lern absetzen, hew stich kunstlich letzen, wer auff dich sticht, vnd dein ort trifft, vnd seinen pricht von paÿden seitten, trifft allemal wildu schreitten.

Glosa, das ist, das du leren solt hew vnd stich, also mit kunst absetzen das dein ort in treff, vnd der sein im geprochen werd, das merck also, Wen ainer gegen dir stett vnd helt sein swert als er dir vndten zu well stechen, so ste wider gegen im in der hutt des pfluegs von deiner rechten seÿtten, vnd gib dich plos mit der lincken, Sticht er dir den vndten zu der selben plõs, So wint mit swert gegen seinen stich auff dein lincke seitten, vnd schreitt zu mit dem ~~r lincken~~ rechten fues, So trifft dein ort vnd das sein verfelt, als hernach gemalt stett.

[Illustration]

[89] *Absetzen* is generally translated "setting aside" or "setting off", but another sense of the word is "to lure someone away", and Ainring (and other gloss writers) rely heavily on this meaning in the pieces they present for it.

[90] This couplet uses similar phrasing to 108, in the section on the winds.

Dresden – SR

70| Again, a piece of setting aside:
Likewise, when you stand against him in the guard of the plow on your left side, if he hews toward the upper exposure of your left side, then go up with the sword and against his cut on your left side, the hilt in front of your head, and step toward him with the right foot and stab in toward his face.

Glasgow – GF

Again, a piece of setting aside:
Remember, when you stand against him in the guard of the plow on your left side, if he hews toward the upper exposure of your left side, then go up with the sword and wind against his cut on your left side, the hilt in front of your head, and step toward him with your right foot and stab in toward his face (as is pictured nearby hereafter).

Aber ain stuck von abseczen
Item wann du gen im stäst In der hut de pflugs von diner lincken sÿtten |Hawt er dir dann zu der obern blöß diner lincken sÿtten so far vff mitt dem schwert vnnd vff die lincken sÿtten gegen sinen haw, dz gehülcz für din haupt vnd schrÿt zu im mitt dem rechten fuß vnd stich im in dz gesicht ~ ~ ~~~

15r| Aber ain stuck von absetzen :~
Merck Wen du gegen im stest in der hutt des plugs von deiner lincken seitten, haut er dir dan zu der oberen plõs deiner lincken seitten, So far auff mit dem schwert, vnd wind auff dein lincke seitt gegen seinem haw das gehiltz fũr dein haubt, vnd schreitt zu im mit deinem rechten fues, vnd stich in in das gesicht, als am nãchstn da hernach gemalt stett.
[Illustration]

Long Sword

71 | **On changing through:**

86 Learn changing through
From both sides with great thrusts.
87 Whoever binds on you,
Changing through quickly finds him.

Gloss: This is that you should learn the changing through well. Drive it like this: In the approach, when you hew in or stab toward him, if he wants to bind on your sword (with a cut or with displacement), then let your point sneak through under his sword and stab in greatly toward the other side with it. Thus you find his exposure quickly.

This is the text and the gloss on changing through:

Learn changing through
From both sides with great thrusts.
Whoever binds on you,
Changing through quickly finds him.

Gloss: This is that you should learn the changing through well. Drive it like this: In the approach, when you hew in or stab toward him, if he binds on your sword (with a cut or with displacement), then let your point sneak through under his sword and stab in greatly toward the other side with it. Thus you find his exposure quickly (as is pictured).

Likewise, if he becomes aware of the thrust and goes after it with the displacement, change through toward the other side again.

Likewise, another:

As you approach him, set your left foot forward and hold it[91] extended against his face. If he hews toward the sword above or below and wants to strike it away, let your point sink underneath and stab toward the other exposure of his other side, and do this against all cuts.[92]

Von durchwechslen

41r | |Durchwechslen lere, von bayden sÿtten stich mitt sere |Wer vff dich bindet durchwechsel in schier ~~bin~~ findet ·:

Glosa |Daß ist daß du dz durchwechslen wol lernen solt vnd dz trÿb als Wenn du im in dem zufechten —[93] ein hawest oder zu stichst will er dir denn mitt ainem hawe oder/ mitt/ veseczen an daß schwert binden so lauß den ort vnder sinem schwert ~~vn~~ durch wischen |Unnd stich im do mitt sere ~~ain~~ eÿn zu der anderen sÿtten so fündest du an im schier de blöß ~

15v | **Das ist der text vnd die glos von durch wechsseln.**

Durch wechssel lere, von paiden saitten stich mit sere, wer auff dich pindt, durch wechssel in schier vindet

Glosa. das ist, das du das durchwechssel wol leren solt, vnd das treib also, Wen du im in dem zu fechten ein haust oder zu stichst, pindt er dir den mit einem haw oder mit versetzen an das schwert, So laß das ort vndter seinem schwert durch wischen vnd stich im da mit ser ein zu der anderen seitten, So vinst du an im schier die plõs, als da gemalt stett.

Item wirt er dan des stichs war vnd fert im mit der versatzung nach, so wechsl aber durch zu der anderen seÿtten.

Item ein anders, so du zu im kumbst, so setz dein lincken fueß vor vnd halt im dz lang gegen dem gesicht haut er dan oben oder vndten zu dem schwert, vnd wil das hin weg schlagen, so laß dein ort vntersich sincken vnd stich in zu der andern plõs der anderen seytten vnd thue das gegen allen hewen.

[Illustration]

[91] Should say "hold *the point* extended" but '*ort*=point' is missing so it's just "hold *it* extended".

[92] This is similar to paragraphs 95 of the Lew gloss and 84 of the Nicolaüs gloss; see JMR: 32r.2 & AR: 11r.5–11v.1.

[93] Illegible deletion.

72| On the pulling:

88 Step close in binding.
The pulling gives good findings.
89 Pull; if he meets, pull more;
Work, devise, and do woe.
90 Pull in all meetings
If you want to confound the masters.[94]

Gloss: This is when you approach to fence with him: hew in strongly from above the right shoulder toward the head. If he binds you on the sword (with displacement or otherwise), then step close toward him in the bind, and pull your sword off from his above, and again hew in above toward the other side, in toward his head. If he displaced you on the other side, then again strike in above toward the other side, and work nimbly after the upper exposure however you can, with the redoubling and with other pieces.

This is the text and the gloss on pulling:

Step close in binding.
The pulling gives good findings.
Pull; if he meets, pull more;
If he works, slice and do him woe.
Pull in all meetings
If you want to confound the masters.[94]

Gloss: This is when you approach to fence with him: hew in strongly from above the right shoulder toward his head. If he binds you on the sword (with displacement or otherwise), then step close toward him in the bind (on the sword), and pull your sword off from his above, and again hew in above toward the other side, in toward his head (as is pictured nearby), and work nimbly with the redoubling (or otherwise with other pieces) in toward the nearest exposure.

Or, do as if you want to pull, but remain on the sword and quickly stab in again on the sword toward the face. If you do not correctly meet with the thrust, then work with the redoubling (or otherwise with other pieces).

Von dem zucken

|Tritt nahend In binden daß zucken git gut fünden
Zuck trifft er |41v| zuck mer, arbait erfinde dz tut we
|Zuck in allen treffen, den maistern wiltu sÿ effen ~ ·:

Glosa |Das ist wenn du mitt dem zufechten zu im kumst so haẅ von der rechten achslen starck oben ein zu dem kopff bindt er dir dann mitt verseczen oder sunst an dz schwert |So tritt im ~~bamd~~ bande nahent zu im vnd zuck dein schwert oben abe von den sinen vnd haẅ im zu der andern sÿtten wider oben ein zu dem kopff |Verseczt er dir dz zu dem andern male |So schlach wider zu der andern sÿtten oben ein vnd arbait behentlich nach der obern blöß die dir werden mag mitt ~~den~~ dem dupliernn vnd mitt andern stucken ~~~:·

16r| Das ist der text vnd die glos von Zücken :~

Tritt nachentt Im pinden, daz zucken gibt gutt finde,
zuck trifft er zück mer, arbaitt er schnitte das thuet Im
wee, zuck in allen treffen den maister wildu sy äffen.

Glosa. daz ist wen du mit dem zu fechten zu im kumbst, So haw im von der rechten achssel starck oben ein zu dem kopf, pindt er dir den mit versetzen oder sunst an das schwert, So tritt im pand nachent zu im an das schwert, vnd zuck dein schwert oben ab von dem seinen vnd haw im zu der anderen seitten wider oben ein zu dem kopf, als am nächstn da gemalt stet, vnd arbait behentlich mit dem dupliern oder sunst mit anndern stucken im zu der nächstn plõs,

Oder thue als ob du wellest zucken, vnd pleib am schwert vnd stich pald am schwert wider ein zu dem gesicht, triffst du in dan nicht recht mit dem stich, so arbait mit dem duplieren oder sunst mit anderen stucken.
[Illustration]

[94] This couplet is very close to 35, in the section on the wrathful cut, and dueling couplet 63.

73 | On running through:
91 Run through, let hang
With the pommel; grab if you want to wrestle.
92 Whoever strengthens against you,
Remember to run through with it.

Gloss: Remember, this is when someone runs in on another: if he goes up with the arms and wants to overwhelm you above with strength, then go up with the arms as well and hold your sword overhead by the pommel with the left hand, and let the blade hang behind over your back, and run through with your head under his right arm, and leap with the right foot behind his right, and with the leap, go in with the right arm well around in front of the body, and in this way hold him on the right hip and throw him in front of you.

This is the text and the gloss on running through:
Run through, let hang
With the pommel ~~grab~~ if you want to wrestle.
Whoever strengthens against you,
Remember to run through with it.

Gloss: Remember, this is when someone runs in on another: if he goes up with the arms and wants to overwhelm you with strength, then go up as well and hold your sword overhead by the pommel with the left hand, and let your blade hang behind over your back, and run through with the head under his right arm, and leap with the right foot behind his right, and with the leap, go in with the right arm well around in front of the body, and in this way hold him on your right hip and throw him in front of you (as is pictured nearby).

42r | Von durchlauffen ~
|Durchlauff lauß hangen : mitt dem knopffe grÿff wilt du rangen Wer gegen dir störcke, durchlauff damitt mörcke ·:
Glosa |Mörck dz ist Wann ainer dem andern einlauffe: fört er dann vff mitt den armen vnd will dich oben mitt störck überdringen |So far och vff mitt den armen, vnd halt din schwert mitt der lincken hand bÿ dem knopff ober dinem ~~havpt~~ haupt |Und lauß die clingen hinden ṽber deinen ~~rucken~~ Ruck hangen |Und lauff mitt dinem haupt durch vnder sinem rechten arm vnd spring mitt dem rechten fuß hinder sinen rechten, vnd mitt dem sprünge far im mitt dem |42v| rechten arm vornen wol vm den lÿb vnd fasse in also vff dem rechten hüffe vnd wirff in für dich ~ ~ ~ :·

16v | Das ist der text vnd die glos von dürch laüffen.
Durch lauff las hangen, mit dem knopff ~~greÿff~~[95] wildü rÿngen, wer gegen dir sterck, durch lauff da mit merck.
Glosa. Merck das ist, wen einer dem anderen einlaufft fert er dan auff mit den armen, vnd wil dich mit sterck ṽber dringen, So far auch auff vnd ~~far~~ halt dein schwert mit der lincken hant pey dem knopf, vber deinem haubt, vnd las dein kling hinden vber deinen ruck hangen, vnd lauff mit dem haubt durch vndter seinem rechten armm, vnd spring mit dem rechten füeß hintter seinen rechten, vnd mit dem sprung far im mit dem rechten armm voren wol vmb den leib, vnd faß in also auff dein rechte hüff vnd wirff in für dich als am nächstn gemalt stett
[Illustration]

[95] This word is present in all witnesses, but for some reason was struck out in this one.

This is on wrestling:
Run through, let hang
With the pommel; grab if you want to wrestle.
Whoever strengthens against you,
Remember to run through with it.

Gloss: Remember, this is when someone runs in on another: if he goes up with the arms and wants to overwhelm you above with strength, then go up with the arms as well; hold your sword overhead by the pommel with the left hand, and let the blade hang down behind you, and run through him with the head under his right arm, and leap with the right foot behind his right, and with the leap, go in with the right arm well around in front of the body, and in this way hold him on your right hip and throw him in front of you (as is pictured here).

Das ist vonn Ringenn.
|13r| Durchlauff las hanngenn, mit dem knopf greif wiltu rangenn, wer gegenn dir sterck, durch lauf damit merck.
Glosa. Merck das ist, wen einer dem anndernn einlaufft, fert er dann auf mit den armen vnnd wil dich obenn mit sterck vberdringen, so far auch auf mit denn Armen, so halt dein schwert mit der linckenn handt, bey dem knopf, vber deinem haupt, vnnd las die kling hinter dir nider hanngenn, vnd lauf im mit dem haupt durch, vnder seinen rechtenn Arm, vnnd spring mit dem rechtenn fus hinter sein rechtenn, vnnd mit dem spring vor im, mit dem rechten arm, vornn wol vmb den leyb, vnd fas in also auf dein rechte hüf, vnnd wirf in fur dich als hie gemalt stet.

74 | Again, a run through:
Likewise, when he wants to overwhelm you above with strength in the running in with the sword: hold your sword by the pommel with the left hand, and let the blade hang over your back, and run through with your head under his right arm, and remain with the right foot in front of his right, and go with your right arm well behind him around his body, and hold him on your right hip, and throw him behind you.

Again, this is a piece of run through:
Remember, when he wants to overwhelm you above with strength in the running in with the sword: hold your sword by the pommel with the left hand, and let the blade hang over your back, and run through with your head under his right arm, and remain with the right foot in front of his right, and go with your right arm well behind him around his body, and hold him on your right hip, and throw him behind you (as is pictured).

75 | A wrestle at the sword:
Likewise, when someone runs in on another: release your sword from the left hand and hold it in the right, and push his sword away from you on your right side with your hilt, and leap with your left foot in front of his right, and go in with the left arm well around behind his body, and hold him on your left hip, and throw him in front of you.

{But see that you do not miss.}

...in with the left arm well around behind his body, and hold him on your left hip, and throw him in front of you (as is pictured nearby hereafter).

Aber ain durchlauffen
|Item wann er dich in dem einlauffen mitt dem schwert oben mitt störck über dringen will |So halt dein schwert mitt der lincken hannd bÿ dem knopff vnd lauß die clingen über deinen |Ruck hangen vnd lauff mitt dem haüpt durch vnder sinem rechten arm vnd blib mitt dem rechten fuß vor sinem rechten, vnd far mitt dem rechten arme hinden wol im vmb den lÿb vnd fasß in vff din rechte hüff vnd wirff in hinder dich ~ ~ ~ ~ ~ ~ ~ :·

17r | Das ist ein stuck aber von durchlauffen.
Merck, wen er dich in dem ein lauffen mit dem schwert oben mit sterck wil v̈ber dringen, So halt dein schwert mit der lincken handt pey dem knopf, vnd las dein kling v̈ber deinen rücken hangen, vnd ~~h~~ laufft mit dem haubt durch vnder seinem rechten armm, vnd pleib mit dem rechten fües vor seinem rechten, vnd far Im mit dem rechten armm hindten wol vmb den leib, vnd fas in auff dein rechte hüff, vnd wũrff in hintter dich, als gemalt stett ~
[Illustration]

Ain Ringen im schwert
43r | |Item wann ainer dem andern einlaufft so lauß dein schwert vß der ~~k~~ lincken hand vnd halt es in der rechten vnd stoß sein schwert mitt dinem gehültze vff din rechte sÿtten von dir, vnd spring mitt dem lincken fuß für sinen rechten vnd far im mitt dem lincken arm hinden wol vm den lÿb vnd fasse in vff din lincke hüffe vnd wirff in für dich ~

☞ Doch lug dz es dir nit fel ~

17v* | *...im mit dem lincken arm hinden wol vmb den leib, vnd fas in auff dein lincke hufft vnd wirff in fu dich Als am nachsten hernach gemalt stett* [96]

[96] This passage is filled in based on an attempt at revealing the text under the paper patch. It's the only instance in which text could be read that corresponds to a piece that isn't present elsewhere in GF.

Again, a wrestle:
Remember, when he also goes up with the arms in the running in, hold your sword by the pommel with the left hand: and let the blade hang down behind you, and run through with your head under the right arm, and remain with the right foot forward in front of his right, and go in with your right arm well behind him around his body, and hold him on your right hip, and throw him behind you (as is pictured here).

Aber ein Ringenn.
Merck wenn er in dem einlauffenn auch auffert mit denn armenn, so halt dein schwert mit der linckenn handt, bey dem knopf, vnd las die kling hinter dir nider hangenn, vnd lauf mit dem haupt durch vnter dem rechten armenn, vnnd bleyb mit dem rechtenn fus vorn fur seinem rechtenn, vnnd far mit dem Rechten arm Im hindenn wol vmb denn leib, vnnd fas in auf dein rechte huf vnnd wirf in hinter dich als hie gemalt stet.

76 | Again, a wrestle at the sword:
Likewise, when someone runs in on another: release your sword from your left hand and hold it in the right, and push his sword away from you on your right side with the hilt, and leap with the left foot behind his right, and go in with the left arm well around his body in front under his breast, and throw him backward over your foot.

77 | Again, a wrestle at the sword:
Likewise, when you run in with another: release your sword from the left hand and hold it in the right, and go in with the pommel outside over his right arm and pull downward with it, and grab his right elbow with the left hand, and leap with the left foot in front of his right, and in this way pull him over your foot on your right side.[97]

Again, a wrestle at the sword:
Remember, when you run in on another: release your sword from the left hand and hold it in the right, and go in with the pommel above over his right arm and pull downward with it, and grab his right elbow with the left, and leap with the left foot in front of his right, and in this way pull him over your foot on your right side (as is pictured nearby).[97]

Aber ain ringen im schwert
|**Item** wann ainer dem andern ein laufft so lauß din schwert vsß der lincken hand vnd halt es in der rechten vnd stoß mitt dem ghültz sein schwert vff din[98] rechte sÿtten von dir, vnd spring mitt dem lincken fuß hinder |43ᵛ| seiner rechten vnd far im mitt dem lincken arme vornen vnder sin brüste wol vm den lÿb vnd wirff vnd vff dein fuß hindersich ~

Aber ain ringen im schwert
|**Item** wann du mitt ainem eÿnlauffest so lauß dein schwert vß der lincken hannd vnd halt es in der rechten |Und far im mitt dem knopff vssen über seinen rechten arm vnnd züch domitt vndersich vnd mitt der lincken hannd begrÿff sein rechte elnbogen vnd spring mitt dem lincken fuß für seinen rechten vnd ruck in also über den fuß vff din rechte sÿtten ~~~~

17ᵛ | **Aber ain ringen Im schwert**
Merck, wan du ainem ein laufst, So laß dein schwert auß der lincken hant, vnd halt es in der rechten, Und far Im mit dem knopf oben vber sein rechten armm, vnd zeuch damit vndtersich, vnd mit der lincken begreiff seinen rechten Eelpogen, vnd spring mit dem lincken füeß für sein rechten, vnd ruck in alzo ṽber den füeß, auff dein rechte seytten als am nächstn da gemalt stet.
[Illustration]

[97] This is similar to paragraphs 112 of the Pseudo-Danzig gloss, 105 of the Lew gloss, and 93 of the Nicolaüs gloss; see PD: 33ᵛ.3, JMR: 34ʳ.4, & AR: 12ᵛ.2.
[98] Corrected from *dim*.

78 | **Again, a wrestle at the sword:**

Likewise, when someone runs in on another, go over his right arm with your inverted left hand and grab your right arm with it, and press his right arm over your left with your right arm, and leap with your right foot behind his right, and turn yourself from him on your left side, so you throw him over your right hip.

Remember, when someone runs in on another, go with the left arm over his right and grab his arm with hand inverted, and press his left arm over your left with the right arm, and leap with the right foot behind his right, and turn yourself from him on his left side, and throw him over his right hip (as is pictured nearby).

79 |

Another wrestle:

When someone runs in on you at the sword, etc., let your sword fall, and invert your right hand and grab his right outside with it, and with the left, hold him by the right elbow, and leap with the left foot in front of his right, and push his right arm over your left with the right hand, and rise upward with it; in this way, he is locked, and you can break his arm like this or throw him in front of you over the leg.[99]

Aber ain ringen im schwert

|Item wenn ainer dem andern |44r| eÿnlaufft |So far mitt lincker verkörter hand über sinen rechten arm Und begrÿff da mitt dinem rechten[100] arm vnd ~~d~~ druck mitt dinem rechten arm sinen rechten über dinen lincken vnd spring mitt dinem rechten fuß hinder sinen[101] rechten Und wende dich von im vff dine lincke sÿtten |So wirffestu in über din rechten hüffe ~

18r | Merck, Wan ainer dem andern einlaufft, So far mit dem lincken arm v̈ber sein rechten, vnd begreiff damit sein rechten arm mit verkerter hant, vnd druck mit dem rechten armm sein lincken v̈ber dein lincken, vnd spring mit dem rechten füeß hindter sein rechten, vnd went dich von Im auff sein lincke seyten, vnd würff In vber sein rechte hiff, als am nechsten da gemalt stet.

Aber ain ander ringen Item wan dir ainer ein laufft im schwert, etc so laß dein schwert fallen, vnd verker dein rechte hant, vnd begreiff damit sein rechte außwendig, vnd mit der lincken fas in peÿ dem rechtn Elpogenn, vnd sprin[g] mit dem lincken füeß für sein rechten, vnd stos mit der rechten han[t] sein rechtn armm, vber dein lincken, vnd heb in damit vbersich, also is[t] er uerriegl t, vnd magst Im also den armm prechen, oder für dich vbe[r] das pain werffen

[No illustration]

[99] This is a close match to paragraphs 113 of the Pseudo-Danzig gloss, 108 of the Lew gloss, and 95–96 of the Nicolaüs gloss; see PD: 33v.4, JMR: 34v.3, & AR: 12v.3–4.

[100] Corrected from *rechtem*.

[101] Corrected from *sinem*.

DRESDEN – SR

GLASGOW – GF

80 | Another wrestle at the sword:
Likewise, when someone runs in on another, invert your left hand and go over his right arm with it, and grab his sword by the handle with it between both his hands and pull it on your left side. In this way, you take his sword.[102]

A disarm:
Remember, when someone runs in on another, invert your left hand and go over his right arm with it, and grab his sword by the handle with it between both his hands and pull it on his left side (as is pictured nearby).[102]

{This will be bad for him.}

81 | A disarm:
Likewise, when he binds on your sword (with displacement or otherwise), grab both swords in the blades with your inverted left hand, and hold them fast together, and go through underneath him with the pommel, with the right hand against your left side, over both his hands, and pull it upward over you on your right side. In this way, you remain with both swords.[103]

Again, this is a disarm:
Remember, when he binds on your sword (with displacement or otherwise), grab both swords by the middle of the blades with your left hand, and hold them fast together, and go through underneath him with the pommel, with the right hand against your left side, over both his hands, and pull it upward over you on your right side. In this way, you remain with both swords (as is pictured nearby).[103]

LONG SWORD

Ain ander ringen Im schwert
|**Item** wann ainer dem andern eÿnlaufft |So verkör dein lincke hannd vnnd far domitt über sinen rechten arm vnd begriffe domitt sÿn schwert bÿ der handhäbe zwischen sinen baiden henden vnd rucke vff dein lincke sÿtten |So nimest du im dz schwert ~

18ᵛ | Ein schwert nemen
Merck, wan ainer dem andern ein laufft, so verker dein lincke hant, vnd far da mit vber sein rechten armm, vnd begreifft damit sein schwert peÿ der hant heb, zwischen seinen baÿden hendten, vnd ruck auff sein lincke seyten, als am nechsten da gemalt stet
[No illustration]

Das wurd in übel kÿen ~

44ᵛ | Ain schwert nemen
|**Item** wenn er mitt verseczten oder sunst an din schwert bindet So ~~begrüff~~ begrÿff mitt lincker verkörter hand baÿde schwert In den clingen vnd halt sÿ veste zu samen, vnd far mitt der rechten hand gegen diner lincken sÿtten mitt dem knopff vnden durch im über sein baÿde hende vnd ruck do mitt übersich vff din rechte sÿtten |So belÿben dir baÿde schwert ~:·

19ʳ | Das ist aber ain schwert nemen
Merck, wan Er mit versetzen, oder sunst an dein swert pindts so greyff mit der linckn hant peyde swert mitten in den klingen, vnd halt sÿ vest zu samen, vnd far mit der rechten hant gegen deiner lincken seyten, vndten durch mit dem knopf, vber sein pede hent, vnd ruck da mit ṽber sich auff dein rechte seytt, so peleiben dir bayde swert, als am nechsten da gemalt stet.
[Illustration]

[102] This is a close match to paragraphs 114 of the Pseudo-Danzig gloss, 108 of the Lew gloss, and 97 of the Nicolaüs gloss; see PD: 33ᵛ.5–34ʳ.1, JMR: 34ᵛ.2, & AR: 12ᵛ.5.
[103] This is a close match to paragraphs 115 of the Pseudo-Danzig gloss and 98 of the Nicolaüs gloss; see PD: 34ʳ.2 & AR: 12ᵛ.6.

82 | On slicing off:

93 Slice off the hard ones
From below in both paths.
94 The slices are four:
Two below, with two above.

Gloss: Remember, there are four slices. Drive the first like this: when he runs in on you and goes up high with his arms and wants to overwhelm you above with strength against your left side, turn your sword and fall in underneath his hilt with crossed hands, with the long edge on his arms, and press upward with the slice. Or, if he runs in against your right side, then fall in with the short edge on his arm and press upward as before.

83 | Again, a slice:

Likewise, when you bind strongly on his sword (with a cut or otherwise), if he allows his sword to snap off of yours and strikes above at your head, turn your sword with the hilt in front of your head and slice through his arm underneath, and with the slice, set the point on his chest underneath.

Von abschniden ~

Schÿde ab die hörten, von vnden in baÿden geferten
Vier sind der schnitt, zwen vnden zwen oben mit ·:
Glosa |Mörck |45ʳ| es sind vier schnitt |Den ersten trÿb also |Wann er dir einlaufft vnd hauch vff fört mitt den armen vnd dich oben gegen diner lincken sÿten überdringen will mitt störck |So verwend din schwert vnd fall im mitt der langen schniden mitt ge crützen henden vnnder sinem gehülcze In sin arm vnd truck mitt dem schnitt übersich |Oder laufft er dir ein gegen diner rechten sÿtten |So fall Im mitt der kurczen schniden in sirn arm vnd truck übersich alß vor ~~:·

Aber ain schnitt

|Item wann du im mitt ainem haw oder sunst starck vff sin schwert |45ᵛ| bindest |laust er/dann sein schwert abschnappen von dem dinem vmd schlecht dir oben zu dem kopff |So verwende din schwert mitt dem gehültz für dein haupt vnd ~~schiy~~ schnÿd im vnden durch sin arm vnd secz im mitt dem schnitt ~~vnden~~ den ort vnden an sin brust ~

This is the text and gloss on slicing off:

Slice off the hard ones
From below in both paths.
The slices are four:
Two below, with two above.

Gloss: Remember, there are four slices. Drive the first like this: when he runs in on you and also goes high with his arms and wants to overwhelm you above against your left side, turn your sword and go in underneath his hilt with crossed hands, with the long edge on his arms, and press upward with the slice (as is pictured nearby). Or, if he runs in against your right side, then go in with the short edge on his arm and press upward as before.

19ᵛ | **Das ist der text vnd glos von abschneiden**

Schneid ab die hartn, von vndten in baÿden geferten, vier sind der schnit, zwen vndten, zwen oben mit. Glosa. Merck es sind vier schnit, den Ersten treib also Wan Er dir ein laufft, vnd hoch auch fart mit den armen, vnd wil dich oben gegen deiner lincken seiten vber dringen, So verwent dein schwert, vnd far Im mit der langen schneid mit gekreutztn hendten, vndter seinem gehũltz Im sein armm, vnd truck mit dem schnidt ṽbersich, als am nechstn da gemalt stet Oder laufft er dir entgegen deiner rechten seÿten, so far Im mit der kürtzen schneid in sein armm, vnd druck ṽbersich als vor.
[Illustration]

84 | Again, a slice:

Likewise, drive the slice like this: when someone binds against the sword on your left side and strikes around from the sword toward your right side (with the crosswise or otherwise), leap away from the cut with the left foot on his right side and fall in with the long edge above, over both arms. Drive this to both sides.

85 | On the transformation of the slice:

Turn the edge,
Press the hands to escape.

Gloss: This is when, in your running in, you can[!] with the under slice in his arm (so your point goes out against his right side): press firmly upward with the slice, and with the pressing, leap with the left foot on his right side and turn your sword with the long edge above over his arms (so your point goes out against his left side), and press his arm from you with it.

This is the text and gloss on the transformation of the slice:

Turn your edge,
Press the hands to escape.

Gloss: Remember, this is when, in the running in, you come with the under slice in his arm (so your point goes out against your right side): press firmly upward with it, and with the pressing, leap with the left foot on his right side and turn your sword with the long edge above over his arms (so your point goes out against his left side). In this way, you have transformed the under slice into the over. Drive this to both sides.

Aber ain schnitt

|**Item** den schnidt trÿb also |Wenn dir ainer an dz schwert bindt gegen diner lincken sÿtten vnd schlecht vmb vom schwert mitt der zwerche oder sunst dir zuo der rechten sÿtten |So spring mitt dem lincken fuß vß dem haw vff sin rechte sÿttenn vnd fall im mitt der langen schniden oben uber baÿde arm |Das ~~tr~~ trÿb zu |46^r| baiden sÿttenn ~

Von der verwandlung der schnitt

|Die schnÿde wende, zu flechen druck die hende :·

Glosa |Daß ist wann du im in dein einlauffen kanst mitt dem vndern schnitt vnden in sin arm daß dein ort gegen sine~~n~~r rechten sÿtten außgät |So tuck mitt dem schnitt vast übersich, vnd mitt dem trucken spring mitt~~e~~ dem lincken fuß vff sÿn rechte sÿtten vnd wende din schwert mitt der langen schniden oben über sein arme dz din ort gegen siner lincken sÿtten vß gee vnd truck sin arm domitt von dir ~

20^r| Das ist der text vnd glos von der verwandlung der schnÿdtt

Dein schneyd wendt, zu flachen druck die hendt

Glosa. Merck daz ist, wen du im in dem ein lauffen kümbst mit dem vnderen schnit vndten in sein armm daz dein ortt gegen deiner rechten seytten aus gett, So druck damit vast v̈bersich, vnd mit dem drucken spring mit dem lincken fueß auff sein rechte seitten, vnd wendt dein schwert mit der langen schneid oben ober sein armm das dein ort gegen seiner lincken seÿtten ausgee, also hastũ verwandelt den vndteren schnid in den oberen das treib zu paiden seÿtten

This is the slice over:
Remember, drive the slice like this: when someone binds against the sword on your left side, if he strikes around from the sword toward your right side, then leap away from the cut with the left foot on his right side and fall in with the long edge above, over both arms and press him away from you (as is pictured here).

Text:

Turn your edge,
Press the hands in toward the flat.

Gloss: Remember, this is when, in the running in, you come with the under slice in his arm (so your point goes out against your right side): press firmly upward with it, and with the pressing, leap with the left foot on his right side and turn your sword with the long edge over his arms (so your point goes out against his left side). In this way, you have transformed the under slice into the over. Drive this to both sides.

13v | **Das ist der oberschnit.**
Merck denn schnit treib also, wenn dir einer an des schwert pindt, gegenn deiner linckenn seitten, schlecht er dann vmb vom schwerdt, dir zu der rechtenn seittenn, so spring mit dem lincken fus aus dem haw, auf sein rechte seittenn, vnd fal im mit der langenn schneydt oben vber Pede Arm, vnd druck von dir, als hie gemalt stet.

13r | [104] Text.

Dein Schneid wend. zu der Flech druck in die hend.
Glosa Merck das ist, wenn du Im in dem einlauffen kumbst mit dem vndern schnit vnden in sein arm. das dein ort. gegen deiner rechten seitten außgeet. so druckh damit vast übersich, vnd mit dem drucken spring. mit dem linckhen fuß, auff sein rechte seiten. vnd wende dein Schwert mit der langen Schneid ob seinem arm, das dein ort gegen |13v| seiner Lincken seitten außgee. also hastu verwandlet. den vndern Schnit, in dem obern. Das treib zu baiden seiten. ~

[104] AR begins with the Nicolaüs gloss (ff. 1r–13r.2), but switches to Ainring at this point without any indication that the source has changed.

DRESDEN – SR

86| On two hangers:[105]

96 Two hangers come
From each hand from the ground.
97 In all paths,
Hew, stab, lairs; Soft or Hard.[106]

Gloss: Remember, there are two hangers from each hand and from each side from the ground. Drive them like this: when you bind on his sword with the lower setting aside on your left side, hang the pommel of your sword against the ground and stab upward out of the hanger, toward his face from below. If he pushes your point upward with a displacement, then remain on his sword like this and go up with him, and hang the point down from above toward his face. And you should nimbly drive all threats in the two hangers like this, and cut, thrust, and slice thereafter as you sense if he is Hard or Soft in the bind up of the swords.

GLASGOW – GF

This is the text and the gloss on two hangers:[105]

Two hangers come
From each hand from the ground.
In all paths,
Hew, stab, lairs; Soft or Hard.[106]

Gloss: Remember, there are two hangers from each hand and from each side from the ground. Drive them like this: when you bind on his sword with the lower setting aside on your left side, hang your pommel against the ground and stab upward out of the hanger, toward his face from below. If he pushes your point upward with a displacement, then remain on his sword like this and go up as well, and hang your point down from above toward his face. And you should nimbly drive all threats in the two hangers like this, and cut, thrust, and slice thereafter as you gauge or sense if he is soft or hard in the bind of the swords.

The hangers from the ground are the plow on both sides.

LONG SWORD

Von zwaÿen hengen

Zwaÿ hengen werden, vß ainer hand von der erden |In allem |46v| gefert, hew stich Leger ~~oder~~ waich oder hert ·:

Glosa |Mörck eß sind zwaÿ hengen vß ainer hand vnd von ainer sÿtten von der erden die trÿb also |Wann du im mit dem vndern abseczen gegen diner lincken sÿtten an sin schwert bindest |So heng dinen schwert knopff gegen der erden vnd stich im von vnden auff vß dem hengen zu dem gesichte stost er dann mitt verseczenn dinen ort übersich uff |So blÿb also am schwert vnd far mitt im vff vnd heng im den ort von oben nider zu dem gesicht Und in den zwaÿen hengen solt du alle gefert behentlich trÿben |Hew stich vnd schnitt |Darnach alß du in dem anbinden der schwert enpfindest ober damitt waich oder hört ~~sÿ~~ ist ~~

Das ist der text vnd dÿ glos von zwaÿen henngen

Zway hengen werden, aus einer hand von der erden, in allem gefertt, hew stich leger waich oder hertt.

Glosa. Merck es sind zway hengen aus ayner hendt vnd von ainer seitten von der erden, die treib also, wen du im mit dem vnden absetzen gegen deiner lincken seÿtten an sein schwert pindest, So heng dein knopf gegen der erden, vnd stich im von vnden auff aus dem hengen zu dem gesicht, Stost er denn mit versetzen dein ortt v̄ber sich auff, So pleib also am schwert vnd far auch auff vnd heng Im dein ort von oben nÿder zu dem gesicht, vnd in den zwaÿen hengen solt du alle gefert behentlich treiben, hew stich, vnd schnitt darnach, als du in dem an pinden der schwert enpfindest oder prüeffest, ob er damit waich oder hertt ist, die hengen von paiden seÿtten das ist der pflüg von paiden seÿtten

[105] "To hang" is the cognate for *hengen* and I use it in this chapter because I like cognates, but note that in the combined translation in Chapter 3, I instead use the translation "angle", which is the definition that will make the most sense and be the most actionable for most readers. (Mayr's Latin translator agrees →

107

Text:
Two hangers[105] come
From any point on the earth.
In all paths,
Hew, stab, lairs; Soft or Hard. [106]

Gloss: Remember, there are two hangers from each hand and from each side from the ground. Drive them like this: when you bind on his sword on your left side with the lower, hang the pommel of your sword against the ground and stab upward out of the hanger, toward his face from below. If he pushes your point upward with a displacement, then remain on his sword like this and go up as well, and hang the point down from above toward his face. And you should nimbly drive all threats in the two hangers like this, and cut, thrust, and slice thereafter as you gauge or sense if he is soft or hard in the bind of the swords.

Text.
Zway hengen werden auß Jedem ort auff die erden in allem gefert. haw. Stich. leger. Waich oder hert.

Glosa. Merckh es seind zway hengen auß ainer hend vnd von ainer seiten von der erden die treib also: Wenn du Im mit dem vndern gegen deiner Linckhen seitten an sein Schwert pindest. So heng deinen knopff gegen der Erd. vnd Stich Im von vnden auff. auß dem hengen zu dem gesicht. Stost Er dir denn mit versetzen deinen ort übersich auf. so bleib also amm Schwert. vnd var auch auff. vnd heng Im den ort von oben nider zu dem gesicht. vnd in den zwain hengen soltu alle geuert. behendigclich treiben, haw. Stich vnd Schnit, darnach als du dann in dem anpinden der Schwert empfindest. oder prufest. ob Er damit waich oder hert ist.

and used '*inclinatio*=inclination or angle'; see PMM: 82r.3, 93v.3, 96r.3, & 97r.1–2.)
106 This line is the same as the second line of couplet 30, in the section on the wrathful cut.
107 The text on HS: 40v is similar, but not enough to be considered a direct match.

87 | On the spreading window:[108]

98 Make the spreading window;
Stand freely,[109] examine his intention.
99 Strike in so that it snaps at
Whoever pull himself away in front of you.
100 I truly say to you,
No one covers himself without danger.
101 If you have learned correctly,
He can hardly come to blows.[110]

Gloss: Remember, this is called 'the spreading window': when he binds on your sword with a cut or a displacement, remain strong with the long edge on his sword, arms outstretched, with the point in front of his face, and stand freely[109] and examine his intentions (whatever he wants to drive against you).

Likewise, if he strikes around from the sword with a cut over your other side, then bind after his hew with strength, in above toward his head with the long edge. Or, if he strikes around with the crosswise, then fall into his arms with the over slice. Or, if he pulls his sword back on himself and wants to stab at you below, then pursue him on the sword and set upon him above.

This is the text and the gloss on the spreading window:[108]

Make the spreading window;
Stand freely,[109] examine his intention.
Strike in so that it snaps at
Whoever pulls himself away in front of you.
I truly say to you,
No one covers himself without danger.
If you have learned,
He can hardly come to blows.[110]

Gloss: Make the spreading window like this: when he binds on your sword with a cut or with a displacement, remain strong with the long edge on his sword, arms outstretched, with the point in front of his face, and stand joyfully[109] and examine his intentions (whatever he wants to drive against you).

Likewise, if he strikes around from the sword with a cut over your other side, then bind in against his hew with strength, in above toward his head. Or, if he strikes around from the sword with the crosswise, then fall into his arms with the over slice. Or, if he pulls his sword back on himself and wants to stab at you below, then pursue him on the sword with the point and set upon him above.

47r | Von dem sprechfenster

|Sprechfenster mach stand frÿlich besiche sin sach |Schlage in dz er schnappe wer sich vor dir zühet abe |Ich sag dir für ware, sich schüczt kain man one fare Haustu recht vernommen zu schlage mag er klain kummen ·:

Glosa |Mörck dz haisset dz sprechfenster |Wenn er dir mitt hewen oder verseczenn an dz schwert bindet |So belib starck auß gerechtenn armen mitt der langen schniden am schwert mitt dem ort im vor dem gesichte vnd stand freÿlich vnd besich sin sach was er gegen dir trÿben wöll

|Item schlecht er vom schwert vmb mitt ainem oberhaw dir zu der anderen sÿtten So bind mitt storck sinem haẅ ~~mitt~~ nach |47v| mitt der langen schniden im oben ein zu dem kopff |Oder schlecht er vmb mitt der zwer |So fall im mitt dem obern schnitt in die arme |Oder zuckt er sin schwert an sich vnd will dir vnnden zu stechen |So raise im nach an dem schwerte vnd secz im oben an

20v | Das ist der text vnd die glos von dem sprechvenster

Sprech venster mache stand freÿleich besich sein sache, slach In das er schnapp, wer sich vor dir zuechet ab, Ich sag dir fũrwar, sich schũtzt kain man anefar, ~~Glosa~~ hastu vernumen, zu schlag mag er klain kummen.

Glosa. das sprech venster mach also, wen er dir mit hawen oder mit versetzen an das schwert pindt So pleib starck mit ausgestrackten armen mit der langen schneid am schwert, mit dem ort Im vor dem gesicht, vnd stee frõlich vnd besich sein sach, was er gegen dir treiben wil.

Item Schlet er vom schwert vmb mit einem oberhaw dir zu der andern seÿtten, so pindt im gegen seinem haw mit sterck ein oben zu dem kopf, Oder schlecht er vom schwert vmb mit der twer so vall im mit dem obern schnidt in die armm, Oder zuckt er sein schwert an sich vnd wil dir vndten zu stechen so raÿß im nach mit dem ort am schwertt, vnd setz im oben an.

[108] *Sprechfenster* is often read as *Sprachfenster*, literally "window for speaking" and the term for the screen through which cloistered monastics could communicate with the outside world (but not, as far as I can tell, a wicket—the small window inside a door that can be used to talk without opening the door). However, Mayr's Latin translator renders it '*fenestra patula*=spread-open window' (PMM: 96v.1–4), which seems to be based on reading it as *Spreichfenster* from '*spreien*=spread'. This is an intriguing possibility, so I use →

Make the spreading window;[108]
Stand joyfully,[109] examine his intention.
Strike in so that it snaps at
Whoever pulls himself away in front of you.
I truly say to you,
No one covers himself without danger.
If you have learned this,
He can hardly come to blows.[110]

Gloss: Make the spreading window like this: when he binds on your sword with a cut and a displacement, remain strong with the long edge on his sword, arms outstretched, with the point in front of his face, and stand joyfully[109] and examine his intentions (whatever he wants to drive against you).

If he strikes around from the sword with a cut over your other side, then bind against his hew with strength, in above toward his head. If he strikes around from the sword with the crosswise, then fall into his arms with the over slice. If he pulls his sword back on himself and wants to stab at you below, then pursue him on the sword with the point and set upon him above.

Spirch venster mach, Stand froelich besich sein sach. Schlach in das Er schnabe, wer sich vor dir zeuchet abe. Ich sag dir fürwar. Sich Schutzet kain Man ane vare. hastu es vernummen. zu Schlag mag Er klain kummen.

Glosa. Das Spirchvenster mach also. Wenn Er dir mit hawen vnd versetzen. an das Schwert pindt. so bleib starckh auß geräckhten armen mit der langen Schneid am Schwert. mit dem ort Im vor dem gesicht. vnd stee frölichen vnd besich sein sach. was Er gegen dir treiben will.

Schlecht Er vom Schwert vmb mit einem Oberhaw dir zu, der andern seiten. So pind gegen seinem haw mit Sterckh ein Im oben zu dem kopff. Slecht Er vom Schwert vmb mit der twir, So vall Im mit dem obern Schnit in die arm, zuckht Er sein Schwert an sich. vnd will dir vnden zustechen. so raiß Im nach mit dem ort amm Schwert. vnd setz Im oben an.

it here. (It could, of course, be intended to encompass both meanings.)

[109] This could equally be meant as '*freilich*=freely' or 'safely (due to freedom)' or '*fröhlich*=joyfully, cheerfully, or funnily'; both spellings are attested here and in other early witnesses of Liechtenauer's *Record*. (Again, it could be intended to encompass both.)

[110] This quatrain is very close to 40–41, in the section on the four exposures.

Or likewise, if he wants neither to pull himself away from the sword nor to strike around, then work on the sword with the redoubling, and otherwise thereafter with other pieces as you sense the Hard and the Soft on the sword, etc.

Remember, if he wants neither to pull himself away from the sword nor to strike around, then work on the sword with the redoubling, and otherwise thereafter with other threats as you sense the Hard and the Soft on the sword.

88 | Hereafter remember that which is called 'the extended point':

Remember, ere you approach close to fence with him, set your left foot forward, arms outstretched, and hold the point extended against his face or his chest. If he hews down from above toward your head, then wind against his hew with the sword and stab toward his face. Or, if he hews toward your sword (down from above or up from below) and wants to strike your point way, then change through and stab in toward the exposure on the other side. Or, if he meets the sword with strength with the cut, then let your sword snap around so that you land toward his head. If he runs in on you, then drive the wrestling or the slice, etc.

{See that you do not fail.}

Here remember how you should stand in the extended point, and what pieces you should drive from it:

Remember, when you approach to fence with him, ere he binds you on the sword, set the left foot forward with outstretched arms and hold your point extended against his face or against his chest. If he hews in from above toward your head, then wind against his hew with the sword and stab toward his face. Or, if he hews toward your sword (down from above or up from below) and wants to strike your point way, then change through and stab in toward the other exposure or side. Or, if he meets the sword with strength with the cut, then let it snap around so that you land toward his head. Or, if he runs in on you, then drive the slice or await the wrestling.

|Item oder will er sich vom schwert nicht abziechen noch umbschlachen |So arbait du am schwert mitt dem duplieren vnd sunst mitt anderen stucken darnach alß du enpfindest die waich vnd die hört am schwert etc

Hie nach mörck waß do haÿsset der lang ort

|Mörck Ee wenn du mitt dem zufechten zu nahent an In kümst |So secze dinen lincken fuß vor vnd halt Im den ort auß gerächten |123ʳ|[111] armen lang gegen dem gesicht oder der brüst Hawt er denn dir oben nider zu dem kopffe |So wind mitt dem schwert gegen sinen haw vnd stich im zu dem gesicht |Oder hawt er von oben nider oder von vnden auff dir zu dem schwert vnd will dir den ort wegk schlachen |So wechsel durch vnd stich im zu der anderen sÿtten zu der blöß |Oder trifft er dir mitt dem haw dz schwert mitt störcke so lauß din schwert vmb schnappen |So triffest du in zu dem kopff Laufft er dir ein So trÿb die ringen oder den schnitt ~ ~ ~ ~ etc

Lug dz es dir nitt fel

Merck, wil er sich vom schwert nit abziechen noch vmb schlagen so arbeit dw am schwert mit dem duplieren vnd sunst mit anderen geferten, darnach als du enpindest die weich vnd die hert am sch[wert]

Hie merck wie du solst sten in dem langen ort vnd was stuck du daraus treiben solst.

Merck, wen du mit dem ~~schwert~~ zu fechten zu im kumbst, So setz den lincken fues vor ee wen er dir an das schwert pindt vnd halt im dein ort lanck mit ausgerackten[112] armen gegen dem gesicht, oder gegen der prust, hautt er dir dan denn oben ein zu dem ~~gesicht, oder haut er dir~~ kopf, So wint mit dem schwert gegen seinem haw, vnd stich im zu dem gesicht, oder hawt er von oben nÿder oder von vndten auff dir zu dem schwert vnd wil dir dein ortt weg schlagen, so wechsl durch vnd |21ʳ| stich im zu der anderen plõs oder seytten, oder trifft er dir mit dem haw das schwert mit sterck so las vmb schnappen so trifst du in zu dem kopf, Oder laufft er dir ein so treib den schnit oder wart der ringen.

[111] The sheet containing ff. 123 & 124 got mixed up when SR was bound, some time before the modern foliation was applied. This is their original location (see Chapter 4 for the full collation of SR).
[112] Corrected from *ausgeratten*.

If he wants neither to pull himself away from the sword nor to strike around, then work on the sword with the redoubling, and otherwise thereafter with other threats as you sense the Hard and the Soft on the sword.

Likewise remember, when you approach to fence with him, ere he binds you on the sword, set your left foot forward, arms outstretched, and hold the point extended against his face or his chest. If he hews in from above toward your head, then wind against his hew with the sword and stab above toward his face. Or, if he hews toward your sword (down from above or up from below) and wants to strike your point way, then change through and stab in toward the exposure on the other side. Or, if he meets the sword with strength with the cut, then let it snap around so that you land toward his head. Or, if he runs in on you, then drive the slice or await the wrestling.

Will Er sich vom Schwert nit abziechenn noch vmbschlachen, so arbait du amm Schwert mit dem tuplirn vnnd sonst mit andern geuerten. darnach als du empfindest die Waich vnd die hert amm Schwert. ~

14r | Item merckh wenn du mit dem zu vechten zu Im kumbst. So setz dein Lincken fuß vor, ee wann Er dir an das Schwert pindt. vnd halt Im den ort lanckh auß geräckten armen gegen dem gesicht. oder der prust. hawt Er dir denn oben ein zu dem kopff, so wind mit dem Schwert gegen seim haw, vnd Stich Im oben zu dem gesicht. oder hawt Er von oben nider, oder von vnden auff. dir zu dem Schwert, vnd will dir den ort wegkhschlachen. So wechsell durch, vnd Stich Im zu der andern seiten. zu der plöß. oder trifft Er dir mit dem haw das Schwert. mit Sterckh, So laß vmb schnappen, So triffstu in zu dem kopff, laufft Er dir ein, so treib den Schnit, oder wart der Ringen. ~

89 | The text of the conclusion of the *Record*:

102 Whoever leads well and counters correctly,
And finally indeed experiences,
103 And counters particularly,
Each in three wonders.
104 Whoever correctly hangs well,
And brings winding with it,
105 And eight winds
With correct weighing considers.
106 And yet each one,
(The winds, I mean) is a trio,
107 So they are twenty
And four, just count them.
108 From both sides
Learn eight winds with steps,[113]
109 And gauge the threats,
No more than Soft or Hard.[114]

Gloss: This is a lesson in which the *Record* is grasped in short. Learn it like this: that you should very well take account of and practice the art against whoever you fence with, so you know to drive your counters against his pieces, and you can work with three particular wounds from each counter.

This is the text of the closure of the whole *Record*:

Whoever leads well and counters correctly,
And finally indeed experiences,
And counters particularly,
Each in three wonders.
Whoever correctly hangs well,
And brings winding with it,
And eight winds
With correct weighing considers.
And you each one
(The winds, I mean) is a trio,
So they are twenty
And four, just count them.
From both sides
Learn eight winds with steps,[113]
And gauge the threats,
No more than Soft or Hard.[114]

Gloss: This is a lesson in which the art of the *Record* is grasped artfully. Learn it like this: that you should very well practice and take account of the art, and in this way you nimbly lead all threats and pieces against whoever you fence with, so you know to correctly drive your counters against his pieces, and you can work with three particular wonders[115] from each counter.

123ᵛ | Der text von der beschliessung der zedel ~

|Wer wol fürett, vnd recht bricht, vnd entlich garbericht Und bricht besunnder ÿgclichs im drew wunder Wer recht wol henget, vnd winden do mitt brenget |Und winden acht, mitt rechten wegen etracht, |Und Jo Ir aÿne, |der winden selb dritt ich maÿne ~~So~~ |So sind ir zwinczig, vnd fürzel sÿ einzigen |Von baÿden sÿttenn, acht winden ler mit schritten |Unnd brieff ~~da~~ die gefert, nicht mer dann waich oder hert ~~

Glosa |Das ist ain lere do die zedel kurcz Inen begriffen ist die lert also |Daß du in der kunst gar wol ~~gern~~ bericht vnd geüpbt |124ʳ| solt sin gegen den do du mitt ~~fech~~ fichtest daß du dine brüch gegen sinen stucken also wisset zü trÿben |Daß du vß ainem ÿeden bruch besunder magst gearbaitten mitt ~~drÿ~~ dreÿen wunden

Das ist der text von der verschliessüng der gantzen zettl :~

Wer wol füert vnd recht pricht, vnd entlich gar bericht, vnd pricht besunder, yedlichs in dreũ wunder, wer recht woll hengt, vnd winden damit pringt, vnd winden acht, mit rechten wegen betracht, vnd doch ir aine der winden selb tritt, Ich maÿn so sind ir zwaintzigk, vnd vier zell sÿ aÿntzick von paiden seitten, acht winden ler mit schreitten, vnd prüeff die gefert, nicht mer dan waich oder hert.

Glosa. Das ist ain ler, darin die kunst der zetl kurtzlich begriffen ist die lest, also das du in der kunst gar wol geübt vnd bericht solst sein, also das du alle gefertt vnd stuck behentlich daraus kundest füren gegen dem da du mit fichst, das du deine prüch gegen seinenn stucken also gerecht wist zu treiben, das dũ aus einem yeden prüch besünder mügst gearbeitten mit dreÿen wunderen

[113] This couplet uses similar phrasing to 85, in the section on the setting aside.
[114] This couplet is very close to 80, in the section on the pursuit.
[115] Sometimes translated as "wounder", based on '*Wunde*=wound', but this form of the word doesn't seem to be supported in any dictionary or other source. *Wunder* is pretty much always "miracle", "wonder", →

AUGSBURG – AR

Text:
Whoever leads well and counters correctly,
And finally indeed experiences,
And particularly,
Each in three wonders.
Whoever correctly hangs well,
And brings winding with it,
And eight winds
With correct weighing considers.
And yet each one
(The winds, I mean) is a trio,
So they are twenty
And four, just count them.
From both sides
Learn eight winds with steps,[113]
And gauge the threats,
No more than just Soft or Hard.[114]

Gloss: This is a lesson in which the art of the *Record* is grasped in short. Learn it like this: that you should practice and take account of the art, and in this way you nimbly lead all threats and pieces against whoever you fence with, so you know to correctly drive your counters against his pieces, and you can work with three particular wonders[115] from each counter.

ROSTOCK – JMR

This is the conclusion of the whole *Record*:
Whoever leads well and counters correctly,
And finally indeed experiences,
And counters particularly,
Each in three wonders.
Whoever correctly hangs well,
And brings winding with it,
And eight winds
With correct weighing considers.
And yet each one
(The winds, I mean) is a trio,
So they are twenty
And four, just count them.
From both sides
Learn eight winds with steps,[113]
And gauge the threats,
No more than if they are Soft or Hard.[114]

Text.

Wer wol furet vnd recht pricht. vnd endtlich gar bericht. vnd besunder, yedlichs in drey wunder, Wer recht wol hengt. vnd winden damit pringt. vnd winden acht. mit rechten wegen betracht. vnd Je Ir aine, der winden selb drit Ich maine, So sind Ir zwaintzig. vnd vier zel Sy aintzig, von baiden seitten, acht winden lern mit Schreiten. vnd prueff die geferrte, nicht mer nur waich oder herte.

Glosa. Das ist ein ler. da die kunst der zetell kurtz Inn begriffen ist, die lert also. Das du in der kunst geuebt. vnd bericht solt sein, also das du alle gefert vnd Stuckh behendigclich darauß kündest füeren. gegen dem da du mit vichst. das du deinen pruch gegen seinen stucken also gerecht wissest zetreiben. Das du auß ainem Jeden pruch besunder mügst gearbeiten mit drey wundern,

Das ist die beschliessung der gantzen Zedtl.

Wer wol fürt vnnd recht bricht, vnd entlich gar pericht, vnnd pricht besunder, iglichs in drey wunder, wer recht wol hengt, vnnd winden domit pringt, vnnd winden acht, mit rechten wegenn betracht, Vnnd yo ir eine der winden selb drit Ich meine, So sindt Ir zwantzig, vnd vier zel sie embzig. Von beiden Seiten, die acht windenn ler mit schreittenn, vnd pruf die geuert nicht mer, ob sie sind waich oder hert.

LONG SWORD

or similar. Thanks to REBECCA L. R. GARBER for pointing this out to me. It's possible, of course, that readers were expected to notice the parallelism between these two words and read "thing that wounds" into the term, but there's no way to be sure of that. Since this is the only part of the *Record* and gloss that uses the term, the specific meaning may not actually be all that important.

You should also hang on the sword well, and you should bring eight winds from these hangers, and you should consider the winds like this so that you know which of the same three to drive.

Likewise, you should also hang on the sword well, and you should bring eight winds from these hangers, and you should consider and correctly weigh the winds like this so that you know which of the same three to drive.

90 | Here remember how you should drive hangers and the winds:

Learn it like this: there are four binds of the sword, two over and two under. You should drive two particular winds well from each bind of the sword.

Learn it like this: if he binds above against your left side, then wind the short edge on his sword and go well up with the arms, and hang the point in above and stab toward his face. If he displaces the thrust, then let the point hang in above on the sword and wind on your right side. These are two winds on one side of the sword.

Here remember how you should drive the hangers and the winds:

Likewise, learn it like this: there are four binds of the sword, two over and two under. You should drive two particular winds well from each bind.

Likewise, do it like this: when you approach to fence with him, if he binds above against your left side, then wind the short edge on his sword and go well up with the arms, and hang your point in above and stab toward his face. If he displaces the thrust with strength, then let your point hang in above on the sword and wind on your right side and stab. These are two winds on one sword-side.

|Auch solt du wol hengen am schwert Und auß den hengen solt du bringen acht windenn vnd die windenn solt du also betrachten daß du der aines selbs dritt wissest zu treÿben ~

Hie mörck wie du hengen vnd die winden treÿben solt

|Das vernÿm also |Es sind vier anbinden der schwert |Zwaÿ obere vnd zwaÿ vndere |Nun solt du vß ÿedem anbinden der schwert ~~zwaÿ obere besund~~ besunder ~~tr~~ trÿbenn zwaÿ windenn

|Das vernÿm also |124v| |Bindt er dir oben an gegen deiner lincken sÿtten |So winde die ~~kucz~~ kurcz schnide an sin schwert Und far wol vff mitt den armen vnd heng im den ort oben ein Und stich im zu dem gesicht |Verseczt er den stich |So lauß den ort am schwert oben ein hangen vnd wind vff dein rechte sÿtten |Daß sind zwaÿ winden an ainer sÿtten des schwerts ~ ~

Item auch solt du woll hengen am schwert, vnd aus dem hengen solt du pringen acht winden, vnd die winden solt du also betrachten, vnd recht wegen, das du der ains selb dritt wist zetreiben.

21v | Hie merck, wie du die hengen vnd die winden treiben solst.

Item das vernym also, es sind vier an pinden der schwert zwaÿ obere vnd zwaÿ vndre, nu soltu aus yedem an pinden pesunder treiben zwaÿ winnden.

Item. dem thue also, wen du mit dem zu fechten zu im kumbst, pindt er dir denn oben an gegen deiner lincken seytten, So wint die kurtz schneid an seinem schwert, vnd far wol auff mit den armen, vnd heng im dein ort oben ein, vnd stich im zu dem gesicht, versetzt er den stich mit sterck, So las im dein ort am schwert oben ein hangen, vnd wint auff dein rechte seytt vnd stich das sind zwaÿ winden an einer seÿtten schwertz,

You should also hang on the sword well, and you should bring eight winds from these hangers, and you should consider and correctly weigh the winds like this so that you know which of the same three to drive.

Likewise, learn it like this: there are four binds of the sword, two over, two under. You should drive two particular winds well from each bind.

Do it like this: when you approach to fence with him, if he binds above against your left side, then wind the short edge on his sword and go well up with the arms, and hang the point in above and stab toward his face. If he displaces the thrust with strength, then let the point hang in above on the sword and wind on your right side and stab. These are two winds on one side of the sword.

auch soltu wol hengen am Schwert. vnd auß den hengen soltu pringen. acht winden. vnd die winden soltu betrachten also vnd recht wegen. Das du der ains selb drit wissest zutreibenn. ~

Item das vernymb also, es send vier anpinden der Schwert. zway oben, zway vnden. Nu soltu auß Jedem anpinden besunder treiben. zway winden.

dem thue also. Wenn du mit |**14**^v^| dem zuvechten zu Im kumbst, pindt Er dir den oben an gegen deiner linckhen seiten, So wind die kurtz Schneid an sein Schwert. vnd var wol auff mit den armen, vnd heng Im den ort oben ein. vnd Stich Im zu dem gesicht. Versetzt Er den Stich mit Sterckh. So laß Im den ort amm Schwert obenn ein hangen. vnd wind auff dein rechte seiten, vnd Stich, dz seind zway winden. an ainer seiten des Schwerts.

Or likewise, if he binds above against your right side, then wind against your right side as well with the long edge on his sword, and go well up with the arms and hang the point in above and thrust toward his face. If he displaces the thrust with strength, then let the point hang in above on his sword and wind on your left side and thrust. These are four winds from the two upper binds, from the left and from the right side.

91 | Likewise, you should know well that you should also drive four winds with all threats from the lower two binds, as from the upper; in this way, the winds from above and below will be eight. And recall that you should drive a cut or slice and a thrust from each particular wind, and these are called 'the three wonders'.

92 | One can and should drive from the eight winds twenty-four times, and you should learn to drive the eight winds from both sides. In this way, you gauge his threat with each particular wonder, nothing more than Hard or Soft on the sword, and when you have sensed these two things, drive the piece that correctly belongs to the wind. Where you do not do this, you will be struck by all winds.

Or, if he binds above against your right side, then wind against your right side as well with the long edge on his sword, and go well up with the arms and hang your point in above and thrust toward his face. If he displaces the thrust with strength, then let the point hang in above on his sword and wind on your left side and thrust. These are four winds from the two upper binds, from the left and the right side.

Likewise, now you should know that you should also drive four winds with all threats from the lower two binds, as from the upper; in this way, the winds from above and below will be eight. Recall that you should drive a cut, a thrust, a slice from each particular wind, and these are called 'the three wonders'.

One can drive from the eight winds twenty-four times, and you should learn to drive the eight winds from both sides. In this way, you step with each particular wind and gauge his threat with it, nothing more than if he is Hard or Soft on the sword, and when you have sensed these two things, drive the piece that correctly belongs to the wind. If you do not do this, you will be struck by all winds.

|Item oder ~~d~~ bind er dir oben an gen diner rechten sÿtten |So wind och gegen diner rechten sÿtten die langen schnid an sin schwert |vnd far wol vff mitt den armen |vnd heng im den ort oben ein |vnd stich im zu dem gesicht |verseczt er den stich mitt störck | 48ʳ | so lauß im den ort am schwert oben ein hangen Und wind vff din lincke sÿtten vnd stich |Daß sind vier winden vß den obern zwaÿen anbinden von der lincken vnd von der rechtten sÿtten ~

|**Ittem** nun solt du wissen dz du vß den vnndern zwaÿen anbinden auch solt trÿben vier winden mitt allen gefertenn alß vß den obern also worden der winden oben vnd vnden acht |Und gedenck dz du vß ÿedem winden besonnder triben solt ainen haẅ oder schnitt vnd ain stich Und dz haissen die ~~drÿ~~ drew wunder |Darumb daß man sÿ zu viervnndzwinczig malen vß den acht winden trÿben soll | 48ᵛ | vnd mag |Und die ach winden solt du wol lernen trÿben von baiden sÿtten Also daß du mitt ÿedem Wunder brieffest sin gefört Nicht mer ~~wann~~ dann öb er damitt waich oder hert am schwert sÿ |Unnd wenn du die zwaÿ ding enpfundenn ~~hast~~ haüst |So trÿb die stuck gerecht die zu dem winden gehören Wo du dz nicht tust |So wirst du bÿ allen winden geschlagen ~~~

Oder pindt er dir oben an gegen deiner rechten seytten, So wint auch gegen deiner rechten seÿtten die lang schneid an sein schwert, vnd far wol auff, mit den armen vnd heng Im dein ort obenn ein, vnd stich im zu dem gesicht, versetzt er den stich mit sterck, so las im dein ort am schwert oben ein hangen, vnd windt auff dein lincke seÿtt vnd stich, das sind vier winden aus den obern zwaÿn an pinden von der lincken vnd der rechten seÿtten.

Item Nu solt du wissen, das du aus den vnderen zwaÿn an pinden auch solst treiben vier winden mit allen geferten als aus den oberen Also werden der winden oben vnd vndten acht, gedenck das du auß yedem winden besunder solst treiben ain haw, ein stich, ein schnitt, das haist dreÿ wunder, darumb das man sÿ zu viervndzayntzigk malen aus den acht winden treiben mag, vnd die acht winden solt du woll leren treiben von paÿden seÿtten, also das du mit yedem winden zu schreittest, vnd damit prüeff sein gefert nit mer ob er damit waich oder hert am schwert ist, vnd wen du die zwaÿ ding enpfunden hast so treib die stuck gerecht die zu den winden gehõren ob du ~~es~~ des nit thuest, so wirst du peÿ allen winden geschlagen.

If he binds above against your right side, then wind against your right side as well with the long edge on his sword, and go well up with the arms and hang the point in above and thrust in the point above toward his face. If he displaces the thrust with strength, then let the point hang in above on his sword and wind on your left side and thrust. These are four winds from the two upper binds and the left and the right side.

Now you should know that you should also drive four winds in all threats from the lower two binds, as from the upper; in this way, the winds from above and below will be eight. And recall that you should drive a cut, a thrust, a slice from each particular wind, and these are called 'the three wonders'.

One can drive from the eight winds twenty-four times, and you should learn to drive the eight winds from both sides. In this way, you step with each particular wind and one gauges his threat, nothing more than if he is Hard or Soft on the sword, and when you have sensed these two things, drive the piece that correctly belongs to the wind. If you do not do this, you will be struck by all winds, etc.

pindt Er dir oben an gegen deiner rechten seiten, so wind auch gegenn deiner rechten seiten. die lang Schneid an sein Schwert, vnd var wol auff mit den armen, vnd heng Im den ort oben ein, vnd Stich Im den ort oben ein zu dem gesicht. versetzt Er den Stich mit sterckh. So laß Im den ort amm Schwert oben ein hangen. vnd wind auff dein linckhe seitten. vnd Stich das seind vier winden, auß den obern zwayen anpinden. vnd der Linckhen vnd rechten seitten.
Nu soltu wissenn. das du auß den vndern zwayen anpinden, auch solt treiben vier winden. in allen geferten. als auß den obern. also werdent der winden oben vnd vnden acht. vnd gedenckh das du auß winden besunder solt treiben ein haw, ein Stich. einen Schnit, das haissent drey wunder, Darumb das Mans zu vier vnnd zwaintzig malen auß den acht winden treiben mag. Und die acht winden soltu wol lernen treiben von beden seitten. Also das du mit Jedem winden zu schreitest. vnd das man prueff sein gefert. nit mer ob Er damit waich oder hert amm Schwert ist. vnd wenn du die zway ding enpfunden hast. so treib die Stuckh gerecht die zu den Winden gehören. ob du das nit thuest, So wirstu bey allen winden geschlagen etc

Dresden – SR

Glasgow – GF

93 |

Remember the *Record*;
In it is written the art of the 'long' sword, which Johannes Liechtenauer has allowed to be written with obscure and cryptic words; Master Sigmund Emring[116] has clarified and explained them, as is written and pictured in this book. And this has been done so that princes and lords, knights and soldiers, to whom the art belongs, should have great love for it. AD 1508.

22r | Merck die zettl, dar in geschriben stet, die kunst des langen schwerch die Johannes liechtenawen hat lassen schreiben mit verporgen vnd verdachten worten, die ~~w~~ selbigen wort hat Maister Sigmund Emring verklert vnd aus gelegt, als In[117] diessem puech geschriben stett vnd gemalt, vnd hat das gethon darumb daz fursten vnd herrnn Ritter vnd knecht den die kunst zu gehört dester grosser lieb darzu haben sullen Anno domani 1508[118]

Long sword

[116] This was probably meant to be "Einring" like JMR, but the scribe didn't dot the 'i'.
[117] Corrected from *am*.
[118] This explanation was placed at the end of the gloss in GF rather than the beginning like the others, but covers many of the same points as the introductory paragraph.

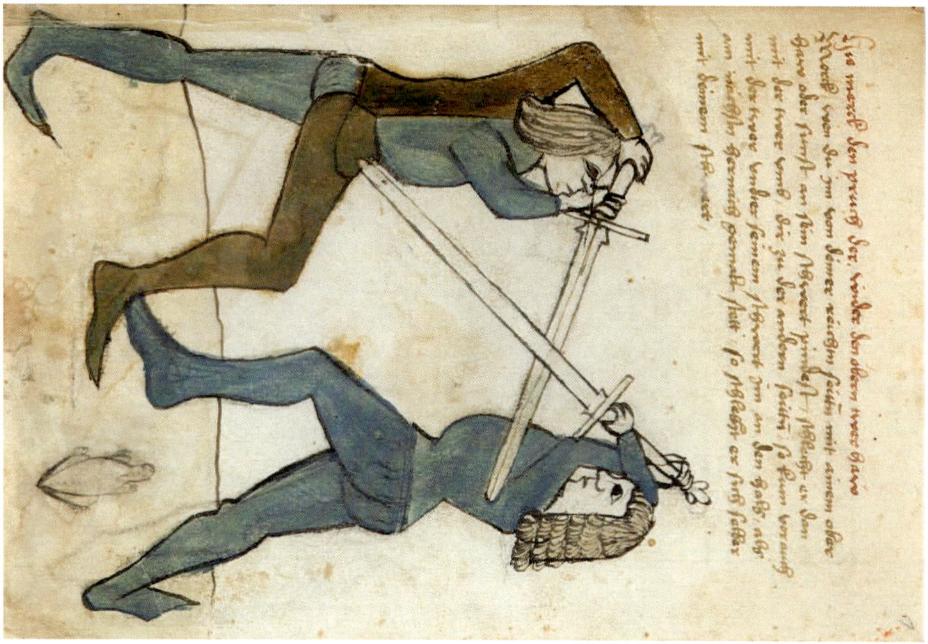

GF: 1r

GF: 1v

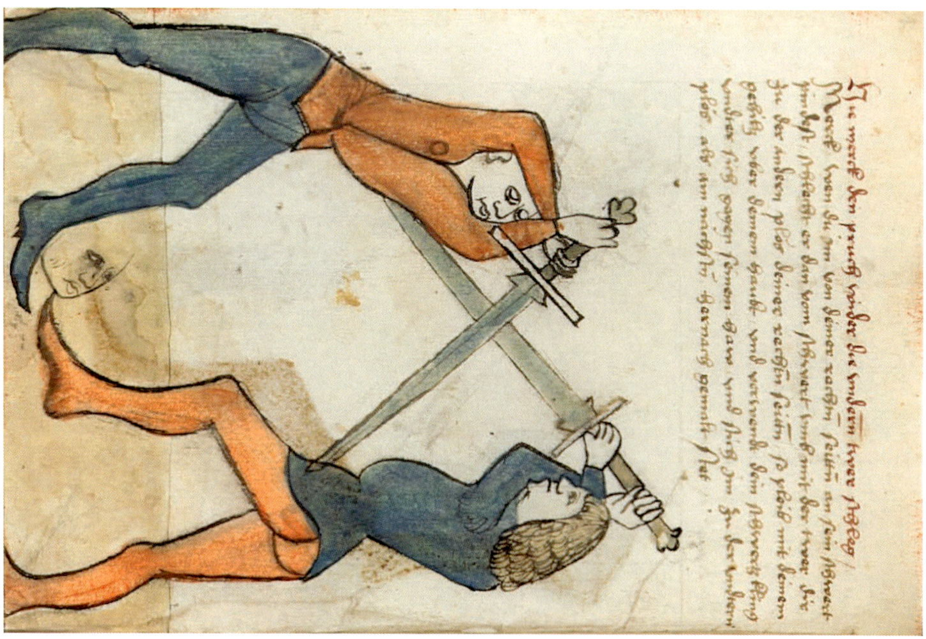

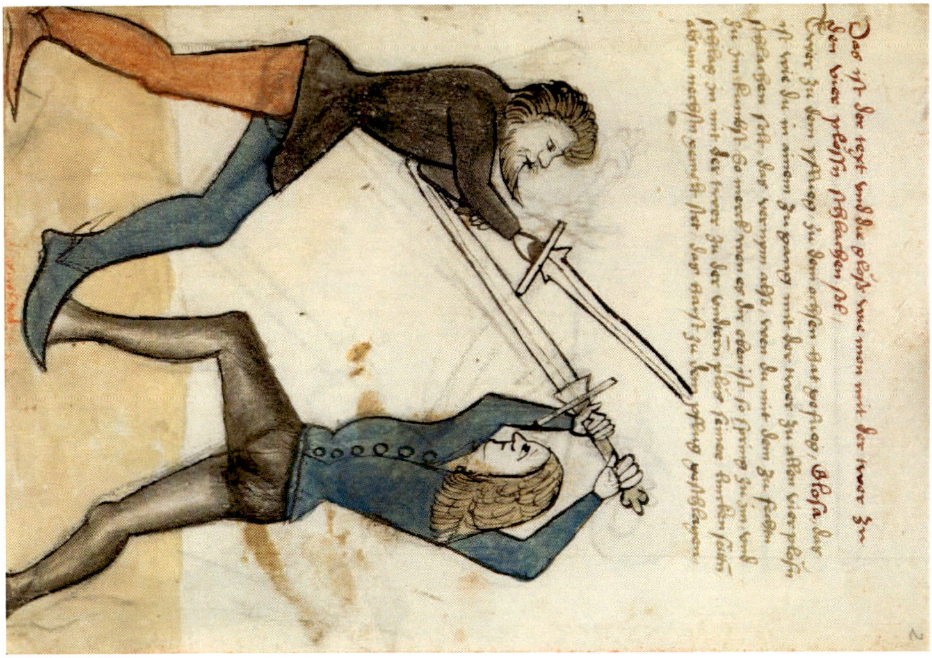

GF: 2r

GF: 2v

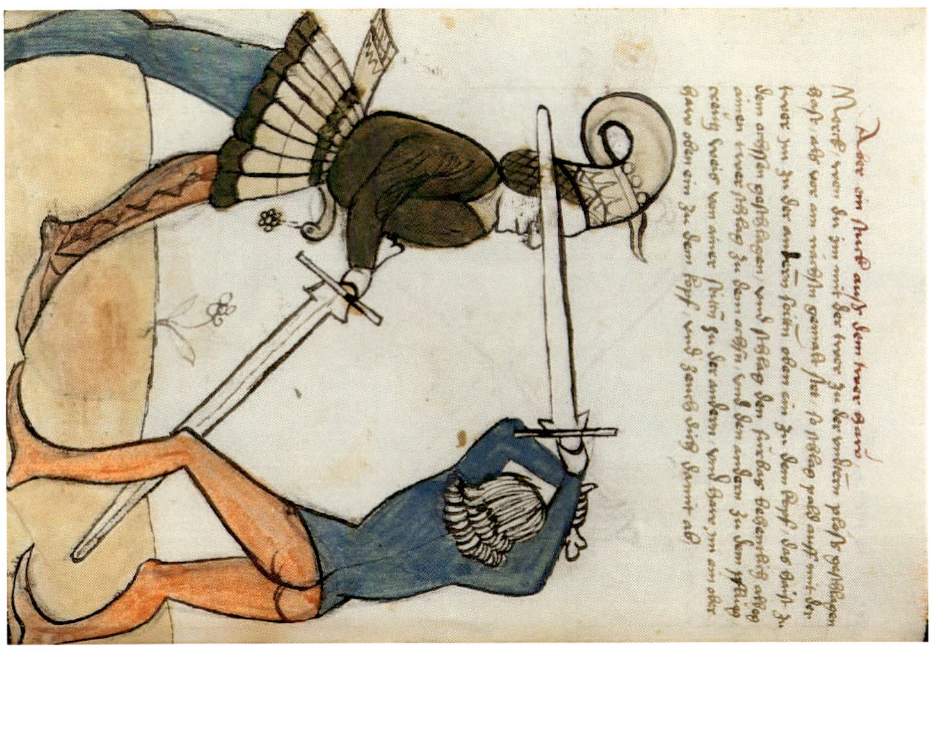

GF: 3^{v}

GF: 4^{r}

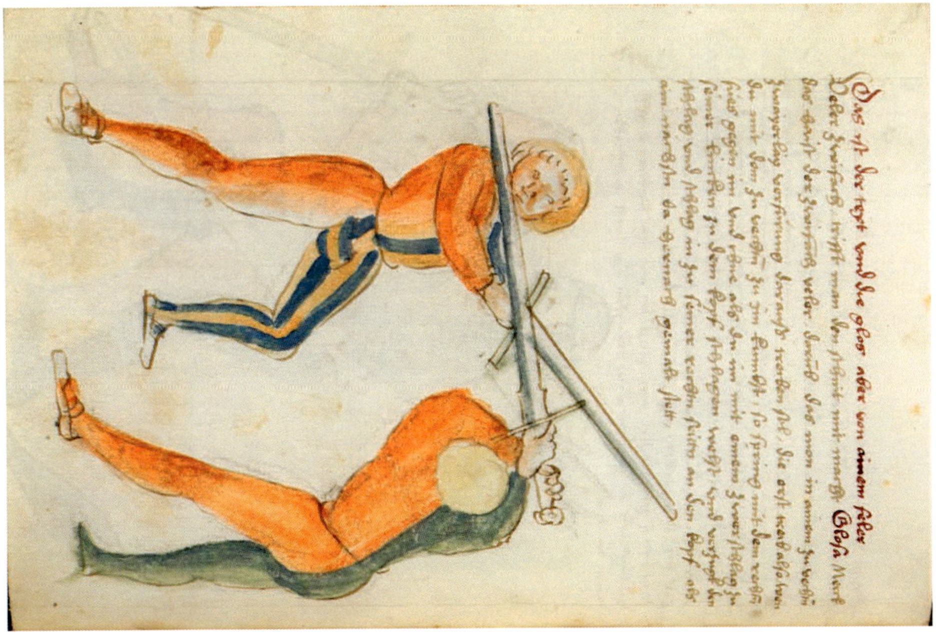

GF: 4^{v}

GF: 5^{r}

LONG SWORD

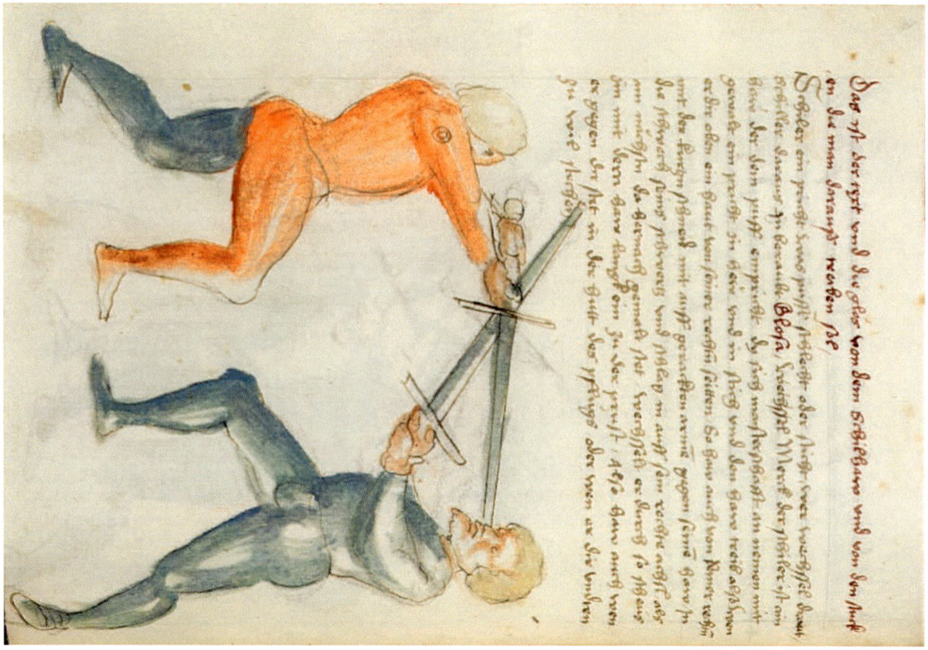

GF: 5^{v}

GF: 6^{r}

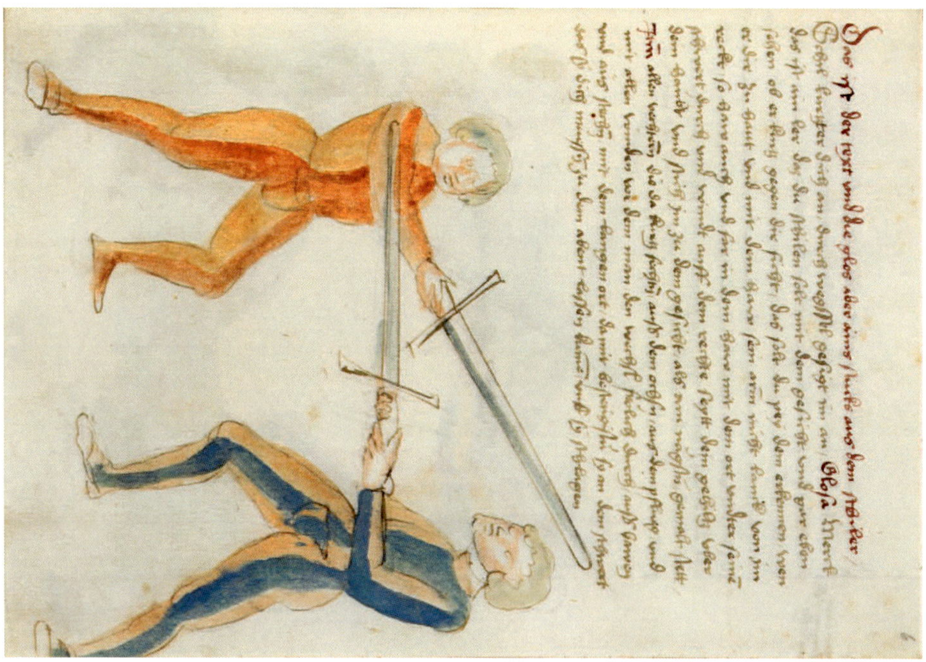

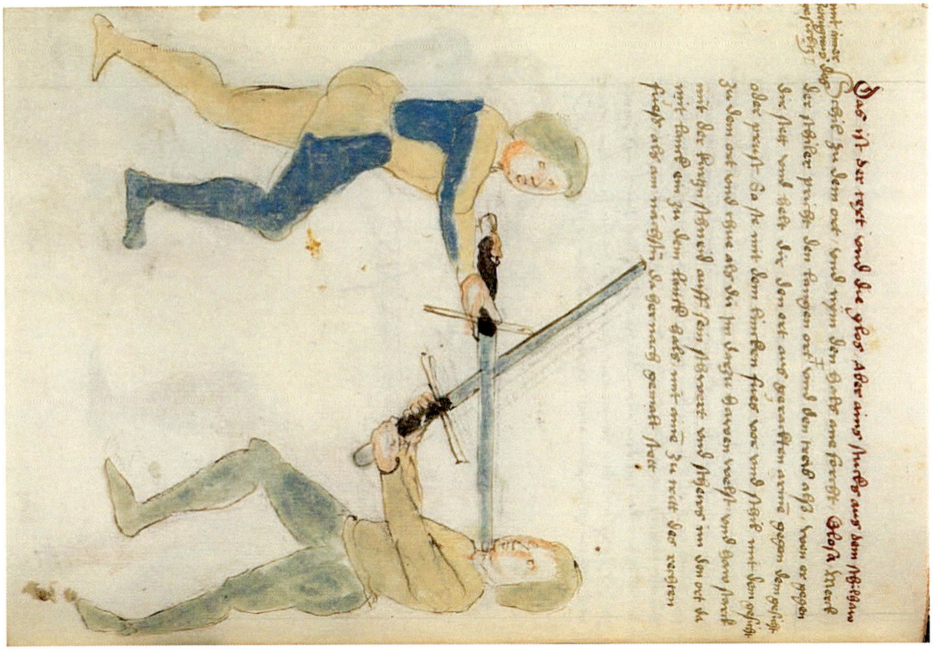

GF: 6^{v}

GF: 7^{r}

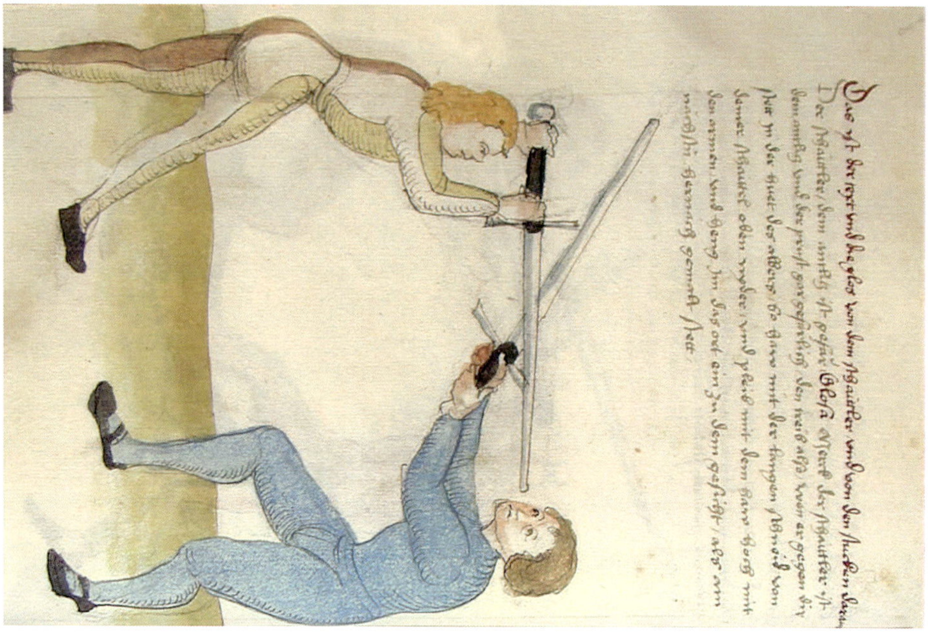

GF: 7^{v}

GF: 9^{r}

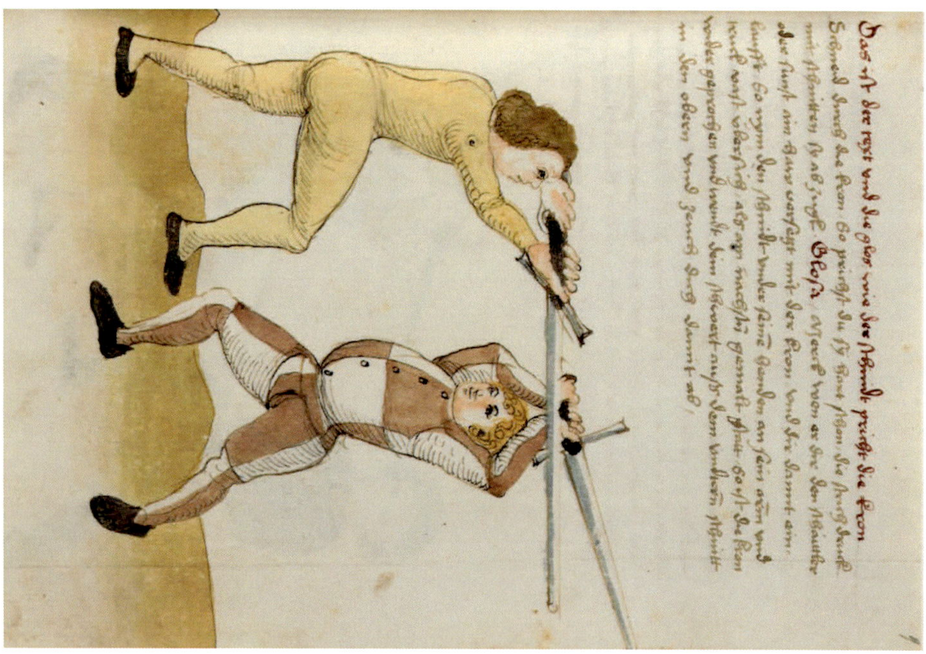

GF: 9v

GF: 10r

GF: 11r

GF: 11v

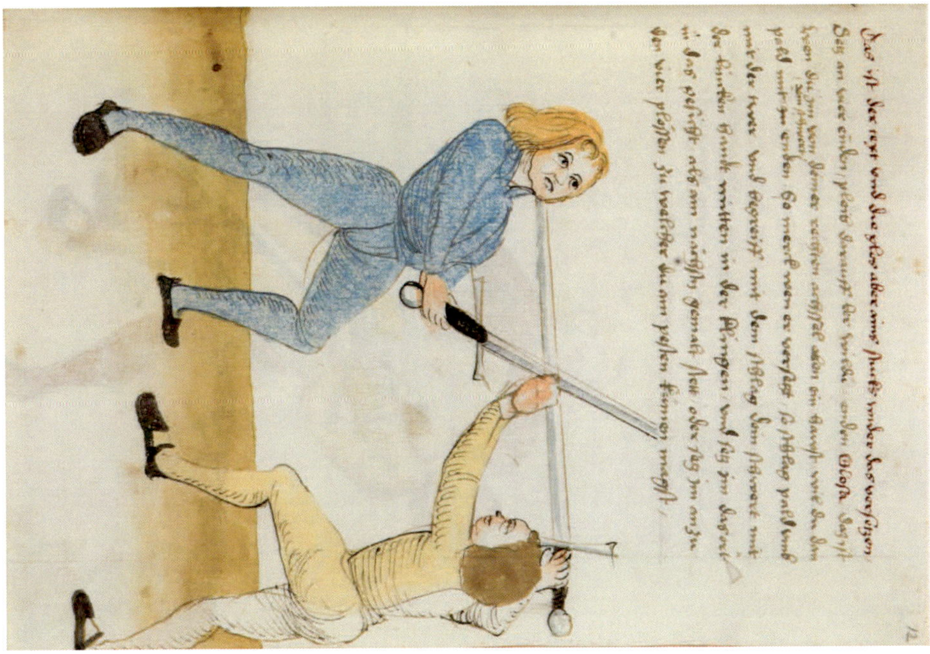

GF: 12^{r}

GF: 12^{v}

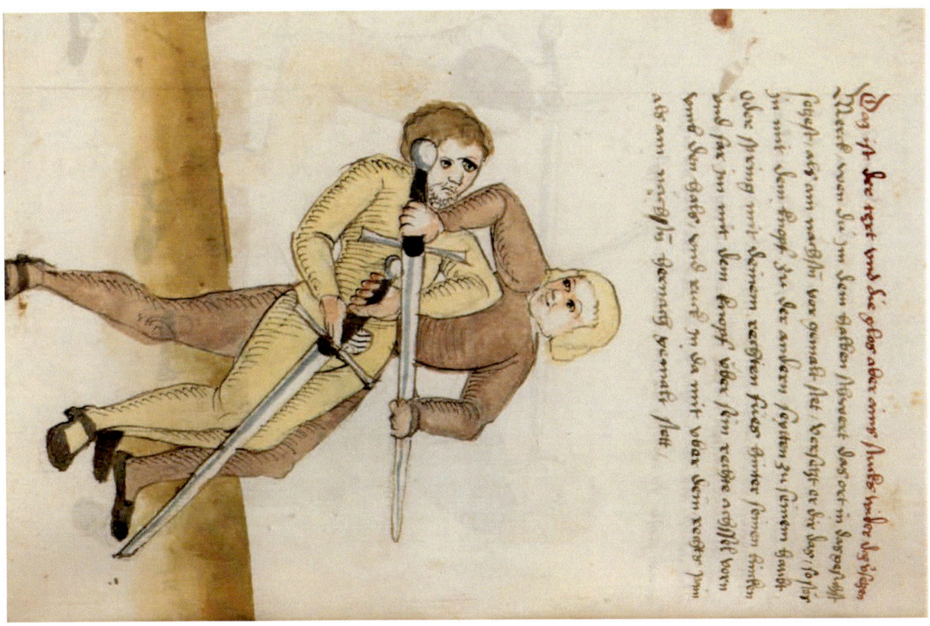

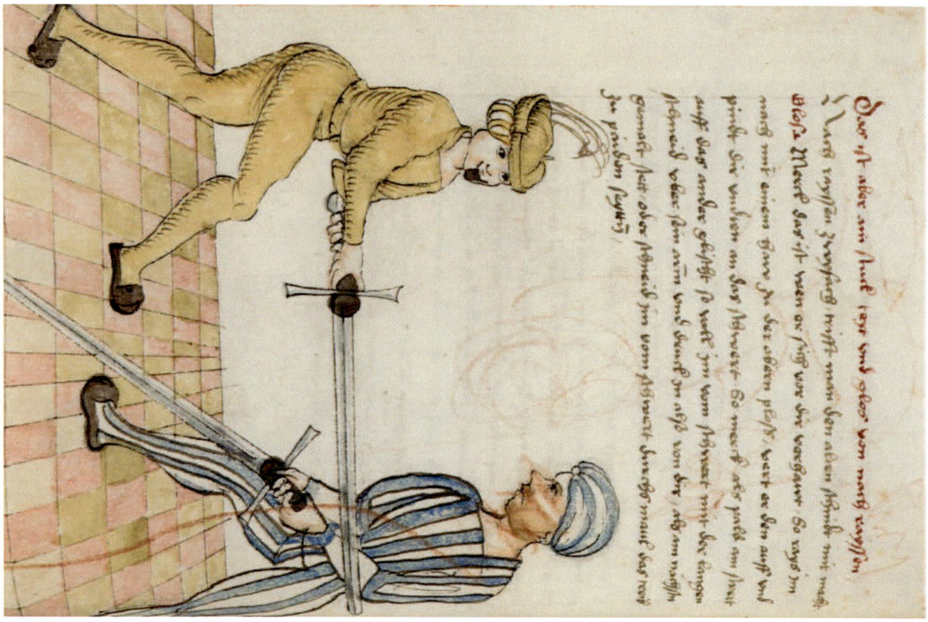

GF: 13^{v}

GF: 14^{r}

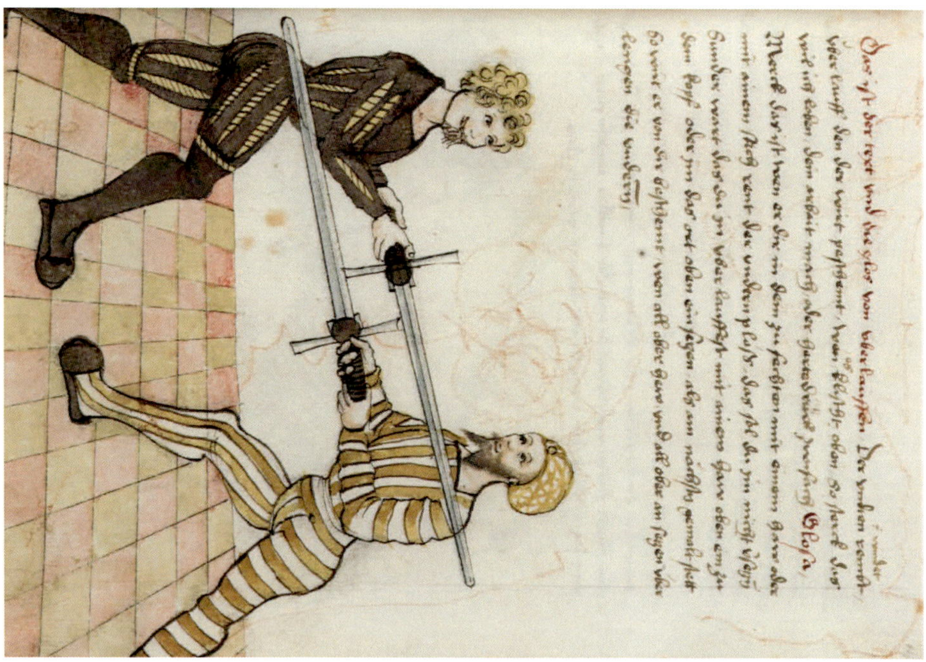

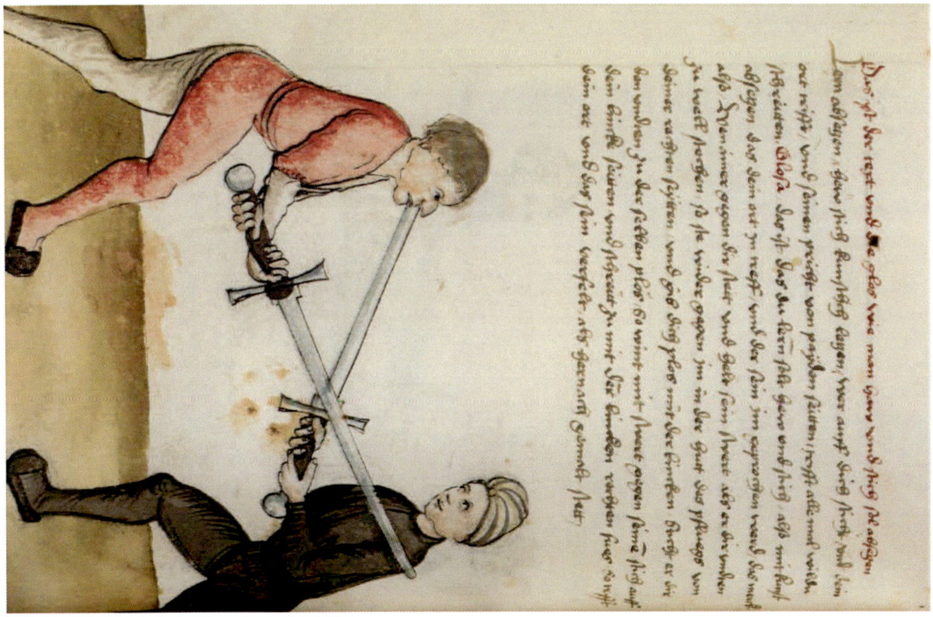

GF: 14v

GF: 15r

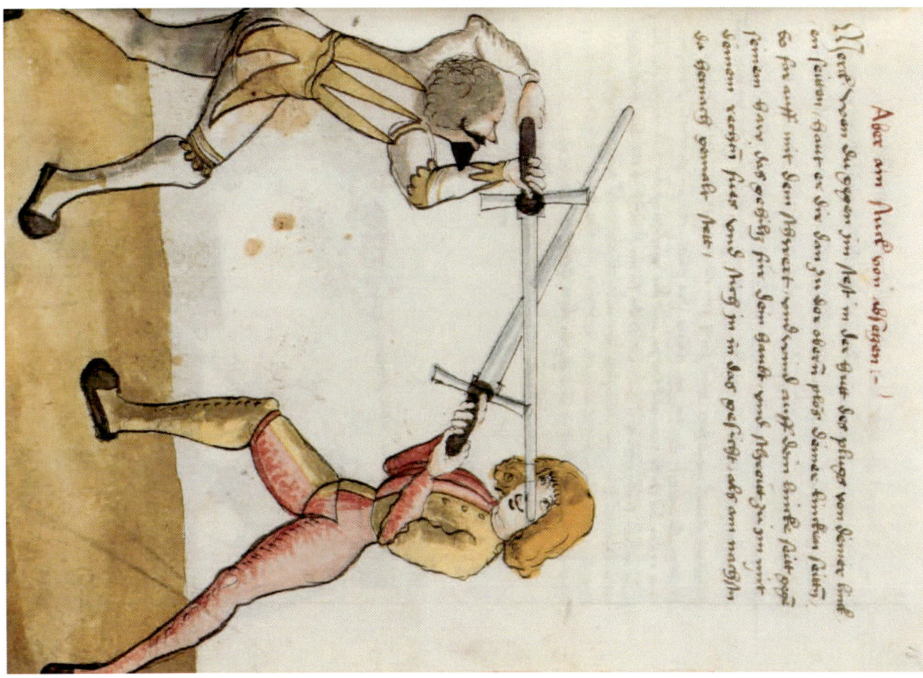

GF: 15^{v}

GF: 16^{r}

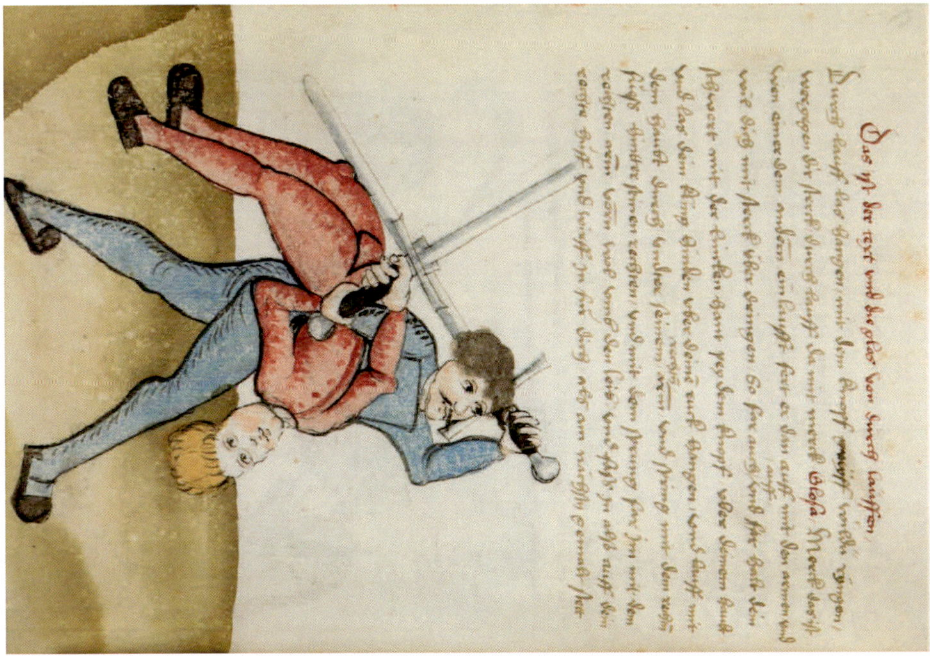

GF: 16^{v}

GF: 17^{r}

LONG SWORD

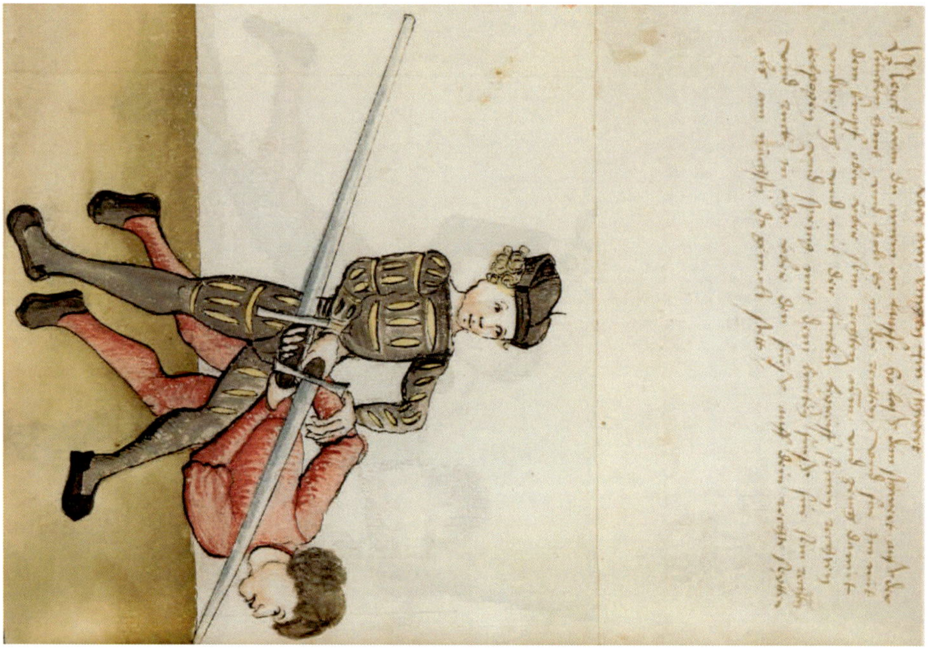

GF: 17v

GF: 19r

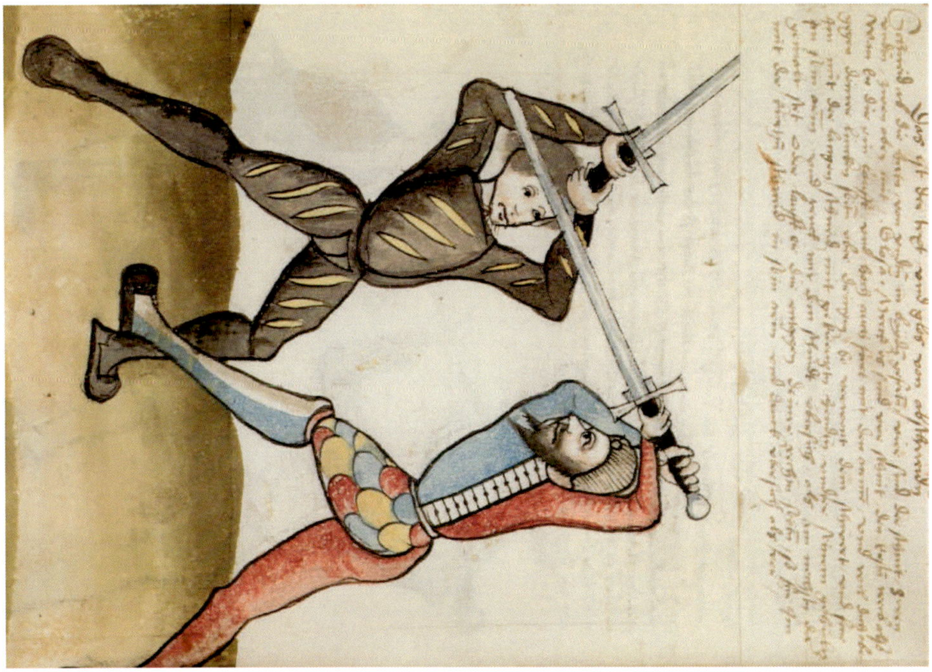

GF: 19v

In sant Jorgen
namen göpt an
die kunst

In the name of Saint George, the art begins here.

1 | The earnest duel on horse and on foot begins here:

Master Johannes Liechtenauer's fencing in dueling harness begins here, which he allowed to be written in cryptic words. They are interpreted and glossed hereafter in this booklet so that every fencer can learn the art (if he can otherwise fence).

Master Johannes Liechtenauer's fencing in dueling harness begins here, which he allowed to be written in obscure and cryptic words. They are interpreted and glossed here in this book so that any fencer can learn it (if he can otherwise fence).

2 | The foreword with the text:

41 Whoever dismounts
Begins to fence on foot;
42 He arranges his spear
In two stances to correctly defend in the beginning.

Gloss: Remember, you should understand it like this: when two want to fence with each other in harness on foot, each one should have three weapons—a spear, a sword, and a dagger—and the first occurrence should begin with the long spear. In the beginning, you should arrange yourself with it in two stances to correctly defend (as you will hear hereafter).[120]

This is the foreword with the text:

Whoever dismounts
Begins to fence on foot;
He arranges his spear
In two stances to correctly defend in the beginning.

Gloss: Remember, you should understand it like this: when two want to fence with each other in harness on foot, each one should have three weapons—a spear, a sword, and a dagger—and the first occurrence should begin with the spear. In the beginning, you should arrange yourself with it in two stances to correctly defend with harm, as you will hear hereafter.[120]

88r | In sant Jorgen namen höpt an die kunst ~

89r | Hie höpt sich an der ernstlich kampff zu roß vnd fuß

|Alhie hept sich ann |Maister |Johannsen |Liechtennawers vechten Im harnash zu kampff |Daß er hät laussen schriben mitt verborgen worten |Das stet hie nach in disem biechlin glosiert vnnd vßgelegt |Das ain ÿeder fechter vernemmen mag die kunst der ~~and~~ anderst vechten kan ~

1r | Hie hebt sich ann Mayster Johannes liechtenawers vechtenn In harnasch zu kampff das er hat lassenn schreibenn mit verporgenn vnd verdacktenn wortenn Da steet Hie Inn diesem puch glosiret vnd außlegt Das ein iglicher vechter ver nemen mag der Anders vechten kann

Die vor red mitt dem text

|Wer absinnet, vechten zu fussen beginnet, |Der schick sin sper, Zwaÿ sten an heben recht | 89v | wer ·:
Glosa |Mörck daß solt du also versten Wenn zwen zu fussen in hamasch mitt ain ander fechten wollen |So soll ÿeder man haben drÿerlaÿ wär, ain sper ain schwert vnd ain degenn |Und daß erste anheben soll geschechen mitt dem ~~lang~~ langen sper Domitt solt du dich mitt rechter wer schicken |In dem anheben In zwaÿen sten~~dt~~ alß du hernach hörn wirst ~~~~~~

Das ist die vorrede mit dem texte

Wer absynnet vechtenns zu fuß begynnet der schickt sein sper, zwey sten an hebenn recht wer
Glosa merck das soltu also vernemen wen zwen zu fuß In harnasch mit ein ander vechten wollenn So soll yderman habenn dreyerley wer Ein sper ein schwert vnnd ein degen vnnd das erst anhebenn soll geschehen mit dem Sper damit soltu dich mit rechter wer Inn dem anhebenn zw schickenn In zwen stennd Als dw hernach horen wirdest

[120] This is a close match to paragraph 2 of the Glasgow Gloss Fragment; see GF: 83r.2.

Master Johannes Liechtenauer's fencing in dueling harness begins here, which he allowed to be written in cryptic words. They are interpreted and glossed here in this book so that any fencer can learn it (if he can otherwise fence).

This is the foreword with this text:
Whoever dismounts
Begins to fence on foot;
He arranges his spear
In two stances to correctly defend in the beginning.

Gloss: Remember, you should understand it like this: when two want to fence with each other in harness on foot, each one should have three weapons—a spear, a sword, a dagger—and the first occurrence should begin with the spear. In the beginning, you should arrange yourself with it in two stances to correctly defend, as you will hear hereafter.[120]

Master Johannes Liechtenauer's fencing in dueling harness begins here, which he allowed to be written in obscure and cryptic words. They are interpreted and glossed here in this book so that any fencer can learn it (if he can otherwise fence).

This is the foreword with the text:
Whoever dismounts
Begins to fence on foot;
He arranges his spear
In two stances to correctly defend in the beginning.

Gloss: Remember, you should understand it like this: when two want to fence with each other in harness on foot, each should have three weapons—a spear, a sword, and a dagger—and the first occurrence should begin with the spear. In the beginning, you should arrange yourself with it in two stances to correctly defend with harm, as you will hear hereafter.[120]

105^{r} | **Hie hebt Sich an Meinster Johannes Liechtenawers vechtenn In harnachs[!] zw kampff das Er hatt lassen Schreybenn mitt verborgen worten Das Stett Hie In disem puch glosirett vnd außlegt Das Ein Iglicher vechter vernemen mag der anders vechten kan**

Das Ist die vorred mit disem text
Wer absinnet vechtens zw fus beginnett der Schickt Sein sper Zwey sten an hebenn recht wer
Glosa merck das Soltu also vernemen wen zwen zw fuß In harnachs mitt ein ander vechten wollen So soll Jeder man haben dreyerley wer Ein sper ein Schwertt ein degen vnnd das Erst anhebenn soll geschehenn mit dem Sper damitt soltu dich mitt rechter wer In dem anheben zw Schickenn In zwen Stend Als dw hernach horen wirdest

98^{r} | **Hie hebt sich ann meister Joannes Liechtenawers vechtenn Im harnasch tzu kampf das er hat lassenn schreiben mit verborgen vnnd verdecktenn wortten das stet hie in diesem buch glosiert vnd ausgelegt das ein ietlicher vechter vernemen mag der annders vechtenn kann.**

Das ist die vorred mit dem Texte.
Wer absinnet, fechtens zu fus beginnet, Der schick sein sper, tzwei sten anheben recht wer. Glosa.
Merck das soltu also vernemenn, wen tzwen tzu fus in harnasch mit einander fechtenn wollen, so sol Jedern habenn dreyerlei wer, Ein Sper, ein schwert, vnnd ein degenn, vnnd das erst anhebenn sol geschehenn mit dem Sper, damit soltu dich mit rechter wer |98^{v}| in dem anhebenn zu schickenn, in tzwen steendt als du hernach horenn würdest.

3| The text on two stances:
43 Spear and point
Stab the lead thrust fearlessly.
44 Leap, wind, truly set upon;
If he defends, pull; this will defeat him.

Gloss: This is the first stance with the spear: when you are both dismounted from the horses, stand with the left foot forward and hold your spear ready to throw, and step toward him like this, your left foot always remaining in front. Watch that you throw ere he does, and quickly follow after the throw toward him with the sword so he cannot throw at you safely, and grab toward the sword.

This is the text on the two stances:
Spear and point
Stab the lead thrust fearlessly.
Leap and wind, truly set upon;
If he defends, pull; this will defeat him.

Gloss: Remember, this is the first stance with the spear: when you are both dismounted from the horses, stand with the left foot forward and hold your spear toward the throw, and step toward him like this, your left foot always remaining in front. Watch that you throw ere he does, and quickly follow after the throw toward him with the sword so he cannot throw at you safely, and grab toward the sword.

4| The other stance in the spear:
Remember, if you don't want to throw your spear, then hold it by your right side at the lower guard and go toward him like this, and boldly stab from underneath toward his face ere he does toward you. If he stabs simultaneously with you or displaces, then go up with the spear into the upper guard in such a way that his point stays on your left arm, ...

This is the other stance with the spear:
Remember, if you don't want to throw your spear, then hold it by your right side in the lower guard and go toward him like this, and boldly stab from underneath toward his face ere he does toward you. If he stabs simultaneously with it or displaces, then go up with the spear into the upper guard in such a way that his point stays on your left arm, ...

Der text von zwaien stenden
|Sper vnd ort den vorstich Stich on forcht |Springe winde secz recht an, wert er zucke daß gesigt im an ·:

Glosa |Daß ist der erst stand mitt dem sper |90r| Wann ir baÿde von den rossen abgetretten sind |So stand mitt dem lincken fuß vor vnd halt din sper zu dem schuß, vnd tritt also zu im daß allweg din lincker fuß vor blÿb |Und wart dz du ee schüsst den er vnd folg bald dem schuss nach zu im mitt dem schwert |So kan er kainen gewissen schuß vff dich haben Und grÿff zu dem schwert ~

Das Ist der text von den zwenn stennden
Sper vnnd ortt denn vorstich Stich an vorcht spring vnnd wind setz recht an wert er zuck das gesigt Im ann

|1v| **Glosa** Merck Das ist der Erst Stannd mit mit dem Sper wenn Ir bayd vonn den Rossen abgedretten, seit so stee mit dem linckenn fuß vor vnnd halt dein Speer zu dem schus vnnd drytt also zw Im das albeg dein lincker fuß vor bleib vnnd wartt das du Ee scheust dann er vnnd volg pald dem schuß nach zu Im mit dem schwertt So kan er keinen gewysen schuß auff dich habenn vnnd greyff zw dem schwert

Der ander stand im sper
|Mörck ob du din sper nitt verschiessen wilt |So halt es neben diner rechten sÿtten zu der vndern hutt vnd gee also zu im |Unnd stich im kũnlich von vnden vff zu dem gesicht |Ee wann er dir |Sticht er ~~a~~ dann mitt dir glich ein oder verseetzt |90v| |So far vff mitt dem sper in die obern hut |So blÿpt dir sin ort vff dinem lincken arm

Das ist der annder Stand mit dem Sper
Merck ob du dein Sper nit verschiesen wild so halt es nebenn deiner Rechtenn seytten In der vnnderrn hut vnd gee also zw Im vnd stich Inn kunlich vonn vndenn auff zu dem gesicht Ee wenn er dir, sticht er denn...[121] |6v| ...mit der gleich ein oder versetzt So far auff mit dem Sper In die obern hut so pleibet dir sein ortt auff deinem linckenn arm

[121] Here the text in WP, AD, and JMR suddenly jumps to a different part of the treatise, but returns to this spot in the text on WP: 6v, AD: 107v, & JMR: 109r. This is probably due a scribe copying pages out of order in an earlier (lost) witness that WP and JMR were both based on.

This is the text on the two stances:
Spear and point
Stab the lead thrust fearlessly.
Leap and wind, truly set upon;
If he defends, pull; this will defeat him.

Gloss: Remember the first stance with the spear: when you are both dismounted from the horses, stand with the left foot forward and hold and hold [!] your spear with readiness to throw, and step toward him like this, your left foot always remaining in front. Watch that you throw ere he does, and quickly follow after the throw toward him with the sword so he cannot throw at you safely, and ... toward him with the sword.

Remember, if you don't want to throw your spear, then hold it by your right side in the lower guard and go toward him like this, and boldly stab from underneath toward his face ere he does toward you. If he stabs simultaneously with it or displaces, then go up with the spear into the upper guard in such a way that his point stays on your left arm, ...

This is the text on the two stances:
Spear and point
Stab the lead thrust fearlessly.
Leap, wind, truly set upon;
If he defends, pull; this will defeat him.

Gloss: Remember, this is the first stance with the spear: when you are both dismounted from the horses, stand with the left foot forward and hold your spear toward the throw, and step toward him like this, your left foot always remaining in front. Watch that you throw ere he does, and quickly follow after the throw toward him with the sword so he cannot throw at you safely, and grab toward the sword.

This is the other stance with the spear:
Remember, if you don't want to throw your spear, then hold it by your right side in the lower guard and go toward him like this, and boldly stab from underneath toward his face ere he does toward you. ~~If he stabs~~ simultaneously with it or displaces, then go up with the spear into the upper guard in such a way that his point stays on your left arm, ...

Das Ist der text von den zwen Stendenn
Sper vnnd ortt den vorstich Stich an vorcht spring vnd wend setz recht an wert er zuck das gesigt Im an

Glosa merck das[122] erst Stand mitt mutt[!] dem Sper wen Ire beid von den Rossen abgedrettenn seitt So stee mit dem linken fus vor vnd halt vnnd haltt[!] dein Sper mit dem Schus vnd dritt also zw Ime das albeg dein linker fuß vor bleib vnd |105v| wartt das dw Ee Scheust dan er vnd volg bald dem Schuß nach zw Ime mitt dem Schwertt So kan er keynen gewissen Schus auff dich habenn vnnd zw Ime mitt dem Schwertt

Merck dw ob dw dein Sper nit verschiessenn wild So haltt Es nebenn deyner rechten seyten In der vndern hutt vnnd gee also zw Ime vnd Stich in kunlich von vnnden auff zw dem gesicht Ee wen er dir, sticht er den...[121] |107v| ...mit der gleich im[?] oder versetzt So far auff mit dem Sper In die obern hutt so Bleibt dir sein ort auff deynem linken arm

Das ist der Text von den tzwen stendenn.
Sper vnnd ort, den vorstich stich one forcht. Spring, wind setz recht an, wert er tzuck das gesigt im an.

Glosa. Merck das ist der erst stand mit dem Sper, Wenn ir baid vonn den Roßen abgetrettenn seitt so stehe mit em linckenn fus vor, vnd halt dein sper tzu dem schus, vnd trit also tzu im, das alweg dein lincker fus vor bleib, vnd wart das du ee scheust dan er, vnnd volg bald dem schus nach tzu im mit dem schwert, so kan er keinenn gewisenn schuß auff dich habenn, vnnd greif tzu dem schwert.

Das ist der ander standt mit dem Sper.
Merck ob du dein Sper nit verschießenn wilt, so halt es nebenn deiner Rechtenn seitten. In der vnternn hut, vnnd gehe also tzu im, vnnd stich Im künlich von vntenn auf tzu dem gesicht, ee wen er dir. ~~sticht er denn~~[123]...[121] |109r| ...Mit dir gleich ein oder versetzt, So far auf mit dem Sper in die obernn hut, so bleibt dir sein ortt |109v| auf deinem linckenn arm,

[122] Illegible "s"; *der* would actually be expected here.
[123] This phrase is present in all copies, but for some reason was struck out in this one.

...and with that, hang the point in over his arm in toward his face. If he then goes and displaces with the left arm, then pull and set the point in underneath into the exposure of his left armpit.

5 | The text, how one should pull:

If you want with thrusts,
Learn to counter before with pulling.

Gloss: Remember, this is that you should learn to pull well. Learn it like this: when you stab toward him out of the lower guard and he displaces with the spear, so his point goes out by your side, pull through and stab toward his other side. Or, if he remains with the point in front of your face with another displacement, then do not pull and remain with the spear on his, and work with the winding toward the nearest exposure that may be yours.

in toward his face. If he goes and displaces with the left arm, then pull and set the point in underneath into the exposure of his left armpit.

This is the text, how one should pull:
If you want to pull from thrusts,
Learn to counter the defense with pulling.

Gloss: Remember in this way, this is that you should learn to pull well so that no one can set upon you while you pull through. Learn it like this: when you stab toward him out of the lower guard and he displaces with the spear, so his point goes out by your side, pull through and stab toward his other side. Or, if he remains with the point in front of your face with another displacement, then do not pull and remain with the spear on his, and work with the winding toward the nearest exposure that may be yours.

Und mitt dem so heng im den ort über sinen arm in sin gesicht fört er dann vff vnd versetzt mitt dem lincken arm |So züch vnd secz im den ort vnder sin lincke ṽchsen in die blöß ~

Der text wie man zucken soll
Wilt du mitt stechen, mitt zucken lern vor brechen ·:

Glosa |Mörck daß ist, daß du wol lernen solt daß du also zuckest Unnd vernÿm daß also wann du im vß der vndern hut zu stichst Versectz er mitt dem sper, |dz sin ort ~~besÿzt~~ besÿtz neben dir vß gat |So zuck durch vnd stich im zu der anderen sÿtten |Oder pleibt er dir mitt der andern |91r| versaczüng mitt dem ort vor dem gesicht So zucke nicht, vnd pleÿb mitt dem sper an dem sÿnen, vnd arbait mitt dem winden zu der nechsten blöß die dir werden mag ~ ~

In sein gesicht, fert er auff vnd versetzt mit dem lincken arm ~~In sein~~ so zuckt vnnd setz Im den ort vnter sein Linckes vchsenn In die plose

Das ist der Text wie mann zuckenn Sol
Wiltu zuckenn vonn stechenn mit zucken lere were prechenn

|7r| **Glosa** mercke also das ist das wol lernen solt das du also zuckest das man dir Icht ansetze die weil du durch zuckest das vernyme also wenn du Im aus der vndernn hut zu stichst versetzt er mit dem Sper das sein ort nebenn dir beseit ausgeet so zuck durch vnnd stich Im zu der andernn seittenn Oder pleibt er dir mit der andernn versatzung mit dem ortt vor dem gesicht so zuck nicht vnd pleib mit dem Sper an dem seinenn vnd arbeit mit denn winden zu der nechsten plose die dir werdenn mag

...in toward his face. If he goes and displaces with the left arm, then pull and set the point in underneath into the exposure of his left armpit.

This is the text, how one should pull:
If you want to pull from thrusts,
Learn to counter the defense with pulling.

Gloss: Remember in this way, this is that you should learn to pull well so that no one can set upon you while you pull through. Learn it like this: when you stab toward him out of the lower guard and he displaces with the spear, so his point goes out by your side, pull through and stab toward his other side. Or, if he remains with the point in front of your face with another displacement, then do not pull and remain with the spear on his, and work with the winding toward the nearest exposure that may be yours.

in seim gesicht fert er auff vnd versetzt mit dem linken arm so zuckt vnd setzt Im den ortt vnter sein links vchssen In die Blose

Das Ist der text wie man zuckenn Soll
Wiltu zuckenn vonn Stechenn mit zuckenn lere wer Brechenn
Glosa merke also das Ist das wol lernenn solt das dw also zwckes das man dir Icht an setz dieweill dw durch zwckes das vernim also wen du Im auß der vntern hutt zw Stichs versetzt er mit dem Sper das sein ortt nebenn dir beseitt auß gett so zuck durch vnd Stich Ime zw der andern seyten Oder pleibtt er dir mit der andern versatzung mit dem ortt vor dem gesicht so zuck nicht vnd Bleib mit dem Sper an dem Seynenn vnd arbeitt mitt dem windenn zw der nechsten plose die dir werdenn mag

...and with that, hang the point in over his arm in toward his face. If he then goes and displaces with the left arm, then pull and set the point in underneath into the exposure of his left armpit.

This is the text, how one should pull:
If you want from thrusts,
Learn to counter the defense with pulling.

Gloss: Remember in this way, this is that you should learn that you pull well so that no one can set upon you while you pull through. Learn it like this: when you stab toward him out of the lower guard and he displaces with the spear, so his point goes out by your side, pull through and stab toward his other side. Or, if he remains with the point in front of your face with another displacement, then do not pull and remain with the spear on his, and work with the winding toward the nearest exposure that may be yours.

vnnd mit dem so heng im denn ort vber seinenn arm in sein gesicht, fert er denn auf vnnd versetzet, mit dem linckenn arm, so tzuck vnnd setz im den ort vnter sein lincke vchsenn in die plos.

Das ist der Text wie mann tzuckenn sol.
Wiltu vonn stechen mit tzucken, lere were prechen.

Glosa. Merck also, das ist das du wol lernenn solt, das du also tzuckest das man dir nicht ansetze, dieweil du durchtzuckst. Das vernim also wenn du im aus der vnternn hut zu stichst, versetzt er mit dem sper, das sein ort neben dir beseit ausgehet, so tzuck durch, vnnd stich im tzu der andernn seittenn, Oder bleibt er dir mit der andernn versatzung mit dem ort vor dem gesicht, so tzuck nicht, vnnd pleib mit dem sper an dem seinenn, vnnd arbeit mit dem winden zu der nechstenn plos, die dir werdenn mag.

6| The text:

46 Remember, if he wants to pull away
From injury, and wants to flee,
47 You should close with him;
Yes, watch the capture surely.

Gloss: Remember, this is how you should pursue with the spear. Remember, when you come forth with the thrust and he displaces and wants to pull away on the spear, follow after him with the point. If you meet with it, then crowd him in front of you. Then, if he wants to flee backward from the point and turns one side toward you, then see that you run in on him toward that same side and surely grab him with wrestling and with arm breaks (as you will find written hereafter).

This is the text:
Remember, if he wants to pull away
From injury, and wants to flee,
You should close with him;
Yes, watch the capture reprehensibly.

Gloss: Remember, this is how you should pursue with the spear. Remember, when you come forth with the thrust and he displaces and wants to pull away from the spear, follow after him with the point. If you meet with it, then crowd him in front of you. Then, if he wants to flee backward from the point and turns one side toward you, then see that you run in on him toward that same side and surely grab him with wrestling and with arm breaks (as you will find written hereafter).

Der text
|Mörch will er ziechen, vo<u>n</u> schaid<u>en</u> vn<u>d</u> will fliechenn |So —[124] solt du im nachen Ja wÿßlich wart deß faehen
Glosa |Merck daß ist wie du solt nachraisen mitt dem sper Mörck wan<u>n</u> du vorkumst mitt de<u>m</u> stich verseczt er vn<u>d</u> will sich am sper abziechen |So volg im nach mitt de<u>m</u> ort, triftest du in do mitt |So dring in für dich Will er dan<u>n</u> vß dem ort hindersich fliechen vnd wendt dir zu ein sÿtten |So wart dz du im zu derselben sÿtten ein lauffest Vnd in wÿsslich begrÿffest mitt ring<u>en</u> |91v| vn<u>d</u> mitt armbrüchen alß du hernach geschriben fündest ~

Das ist der Text
Merck will er ziehenn vonn schaiden vnd wil er fliehenn so soltu Im nahen Ja werflich wart des vahenn

Glosa Merck das ist wie du solt nach rayssenn mit dem Sper merck wen du vor kumbst mit dem stich versetzt er vnd will siech vom Sper abziehenn so volge Im nach mit dem ortt Triffstu In da mit so tring Inn fur dich wil er dann auß dem ortt hinter sich fliehen vnd wennt dir zu ein seyttenn So wart das du Im zu der selbigenn seittenn einlauffest vnnd Inn weyslich begreuffest mit ring |7v| vnnd mit arm pruchenn Als du hernach geschribenn vindest

[124] Illegible deletion.

This is the text:
Remember, if he wants to pull away
From injury, and wants to flee,
You should close with him;
Yes, watch the capture reprehensibly.

Gloss: Remember, this is that you should pursue with the spear. Remember, when you come forth with the thrust and he displaces and wants to pull away from the spear, follow after him with the point. If you meet with it, then crowd him in front of you. Then, if he wants to flee backward from the point and indicates one side to you, then see that you run in on him toward that same side and surely grab him with wrestling and with arm breaks (as you will find written hereafter).

This is the text:
Remember, if he wants to pull away
From injury, and wants to flee,
You should close with him;
Yes, watch the capture surely.

Gloss: Remember, this is how you should pursue with the spear. Remember, when you come forth with the thrust and he displaces and wants to pull away from the spear, follow after him with the point. If you meet with it, then crowd him in front of you. Then, if he wants to flee backward from the point and turns one side toward you, then see that you run in on him toward that same side and surely grab him with wrestling and with arm breaks (as you will find written hereafter).

108r | **Das Ist der text**
Merck will er ziehenn von scheydenn vnnd will er fliehenn so soltu Im nahen Ja werflich ward des vahenn
Glosa merck das Ist das dw solt nachreysenn mit dem Sper merck wen dw vor komst mit dem stich versetzt er vnnd will sich vom Sper abziehenn So volg Ime nach mit dem ortt Triffstu In damit So tring In fur dich will Er dan auß dem ortt hinter sich fliehenn vnd meint dir zw ein seytten So wartt das dw Ime zw der selbigen seytten ein lawffest vnd In weißlich begreiffest mit ring vnnd mit arm pruchenn als dw her nach geschribenn vindest

110r | **Das ist der Text**
Merck wil er ziehenn, vonn schaidenn vnnd wil er fliehenn, so soltu im nahenn, Ja weislich wartt des fahenn.
Glosa. Merck das ist wie du solt nachreissen mit dem Sper, merck wenn du vor kumbst mit dem stich, versetzt er vnnd wil sich vonn sper abtziehenn, so volg im nach mit dem ort, trifstu in damit, so tring in fur dich, wil er denn aus dem ort hintersich fliehenn, vnnd wendt dir tzu ein seitten, So wart das du im tzu derselbigenn seitten einlaufst, vnnd in weislich begreiffest mit ring, vnnd mit armpruchenn, als du hernach geschriben findest.

7| The text on wrestling in the duel:

48 If you want to wrestle,
Learn to leap behind the leg correctly;
49 Throw a latch in place
Which skillfully encloses the front leg.

Gloss: This is when you approach to wrestle with him, you should know how you should leap in front or behind the leg, and should occur with no more than one step.

Likewise remember to it do like this: when you grab him with wrestling and he against you, whichever foot is set forward: if he has set the left forward then strike his left hand away with your right, and with this striking away, leap with your right foot behind his left and press in with the right knee in behind his left joint, and pull him over the same knee with both hands.

Likewise remember another one: when you leap with the right foot behind his left, step with the left between both his legs and hold his left knee between both your knees and hold fast with it, and push him on the front of his headgear[125] with the left hand and pull him back to the side with the right.[126]

This is the text on wrestling in the duel:
If you want to wrestle,
Learn to leap behind the leg correctly;
Throw a latch in place
Which skillfully encloses the front leg.

Gloss: Remember, this is when you approach to wrestle with him, you should know how you should leap in front or behind the leg which he has set forward, and this should occur and with no more than one step.

Likewise remember to do it like this: when you grab him with wrestling and he against you, see whichever foot he sets forward. If he has set the left foot forward then strike his left hand away with your right, and with this striking away, leap with the right foot behind his left foot and press in with the right knee in his left joint, and pull him over the same knee with both hands.

Likewise remember another one: when you leap with the right foot behind his left, step with the left between both his legs and hold his left knee between both your knees and hold fast with it, and push him on the front of his headgear[125] with the left hand and pull him back to the side with the right.[126]

Der text von ringen zu kampffe
|Ob du wilt ringen, hinder pein recht lern springen
|Rigel für schiessen daß vorbain künstlich beschliessen ·:

Glosa |Daß ist |Wenn du mitt im kumst zu ringen |So solt du wissen wie du fornen oder hinden für daß bain springen solt Und soll geschechen nicht mer dann mit ainem zu tritt

|Item mörck den thu also Wenn du in angriffest mitt ringen, vnd er dich wider, |Welchen fuß vor secz, hat er den lincken vor so schlach im sin lincke hand vß mitt diner rechten Unnd mitt dem vßschlagen so spring mitt dinem rechten fuß hinden sinen lincken |92r| vnd truck in mitt dem rechten knÿ ~~hinden~~ hinden In sin linck knÿckel vnd ruck in mitt baiden henden über daß selbig knÿ

|Item mörck ain anderß wenn du springst mitt dem rechten fuß hinder sinen lincken So schrÿtt mitt dem lincken hin nach zwischen sine baide bain vnd faß sin linckes knÿ zwischen din baÿde knÿ vnd halt es domitt föst Vnnd stoß in mitt der lincken hand vornen an die hawben vnd mitt der rechtz zeüch in hinden vff die sÿtten ~~

Das ist der Text vonn ringenn zw kampff
Ob du wilt ringenn hinter pein Recht lere springenn rigel fur schiessenn das vorpein kunstlich beschliessenn

Glosa Merck das ist wenn du mit Im kompst zu ringenn So soltu wissenn wie du vorn oder hindenn fur das pein springen solt das er hat vor gesetzt vnnd das sol geschehen vnnd nit mer den mit einem zw drit

Item Merck dem thu also wenn du In angreyffest mit ringenn vnnd er dich wider So sich welhenn fuß er vorsetzt Hat er den linckenn fuß vor so schlag Im die lincken hannt auß mit deiner Rechten vnd mit dem ausschlahenn so spring mit dem Rechten fuß hinter sein linckenn fuß vnnd druck Inn mit dem Rechtenn knye Inn sein lincke knyekell vnnd Ruck Inn mit bayden henden vber das selbig knye

Item Merck ein anders wenn du Springst mit dem rechtenn fuß hinter sein lincken so schreit mit dem linckenn hinnoch zwischenn sein baiden baynen vnnd faß sein linckes knye zwischenn dein bayde knye vnnd halt |8r| es da mit vest vnnd stoß Inn mit der lincken hannt fornn an die hauben mit der rechten zeuch in hintenn auff die seyttenn

[125] GRIMM specifies that a *Haube* is specifically a low-status metal or leather head protection worn by common soldiers, contrasted with the more elaborate and protective knightly '*Helm*=helmet'.

This is the text on wrestling in the duel:
If you want to wrestle,
Learn to leap behind the leg correctly;
Throw a latch in place
Which skillfully encloses the front leg.

Gloss: Remember, this is when you approach to wrestle with him, you should know how you should leap from or behind the leg which he has set forward, and this should occur and with no more than one step.

Likewise remember to do it like this: when you grab him with wrestling and he against you, see whichever foot he sets forward. If he has set the left foot forward then strike then strike[!] his left hand away with your right, and with this striking away, leap with your right behind his left foot and press in with your right knee in his left joint, and pull him over the same knee with both hands.

Likewise remember another one: when you leap with the right foot behind his left, step with your left between both his legs and hold his left knee with both your knees and hold fast with it, and push him on the front of his headgear[125] with the left hand and pull him back to the side with the right.[126]

This is the text on wrestling in the duel:
If you want to wrestle,
Learn to leap behind the leg correctly;
Throw a latch in place
Which skillfully encloses the front leg.

Gloss: Remember, this is when you approach to wrestle with him, you should know how you should leap in front or behind the leg which he has set forward, and this should occur with no more than one step.

Likewise remember to do it like this: when you grab him with wrestling and he against you, see whichever foot he sets forward. If he has set the left foot forward then strike his left hand away with your right, and with this striking away, leap with the right foot behind his left and press in with the right knee in in his left joint, and pull him over the same knee with both hands.

Likewise remember another one: when you leap with the right foot behind his left, step with the left between both his legs and hold his left knee and hold fast with it, and push him on the front of his headgear[125] with the left hand and pull him back to one side with the right.[126]

Das Ist der text vonn Ringenn zw kampff
Ob dw wilt ringen hinter pein Recht lere Springen rigel fur schiessenn das vorpein kunstlich beschliessen

Glosa merck das Ist wen dw mit Ime komest zw Ringenn So soltu wissen wie dw von oder hinden fur das pein springenn solt das er hatt vor gesetzt vnd das sol geschehenn vnnd nit mer dan mit eynem zw dritt

Item merck dem thw also wen dw Ine angreiffest mit Ringenn vnd er dich wider So sich welchenn fus er vorsetzt hatt er den linkenn fus vor so schlag So schlag Im den linken ~~fus auß~~ hand auß mit deyner Rechten vnd mit dem auß schlahenn so spring mit deynem rechten hinter seynen linken fus vnnd druck In mit deynem rechten knye In sein linke | 108v | kniekyl vnnd Ruck Im | mit beiden hennden vber das selbig knye

Item merck ein anders wen dw Springst mit dem Rechten fus hinter sein linkenn so schreit mit deim linken hinnoch zwischenn sein Beyde beynen vnnd fas sein linckes knie dein beyde knye vnnd halt es damit vest vnnd Stoß Inn mit der linckenn hand fornn an die hawbenn mit der Rechten zeuch In hinten auff die seyten

Das ist der Text von ringen tzu kampf.
Ob du wilt ringenn, hinter pain recht lere springen Rigel fur schießenn, das vorpain künstlich beschließen.

Glosa. Merck das ist wenn du mit im kombst zu ringenn, so soltu wissenn, wie du vorn oder hintenn fur das pain springenn, ...[127] | 103r | ...Solt das er hat vorgesetzt, vnnd das sol geschehenn nit mer denn mit einem zu trit.

Item merck den thu also, wenn du in angreiffest mit ringen, vnnd er dich wider, so sihe welchen fus er vorsetzt. Hatt er denn linckenn fus vor, so schlag im sein lincke handt aus, mit deiner Rechtenn, vnd mit dem auschlag so spring mit dem Rechtenn fus hinter sein linckenn, vnnd druck in mit dem Rechten knie, in sein lincke kniekel, vnnd Ruck in mit baiden hendenn, vber daßelbig knie.

Item merck ein anders, wenn du springst mitt dem Rechtenn fus hinter sein lincken, so schreit mit dem lincken hinach zwischen sein baiden painenn, vnd fas sein lincks knie, vnd halt es damit fest, vnd stos in mit der linckenn handt, vorn an die hauben, mit der Rechten zeuch in hinten auf ein seitten.

[126] This is a close match to paragraph 16 of the Pseudo-Danzig gloss; see GF: 84r.4.
[127] Here the Ainring section in JMR ends, but the treatise continues on JMR: 103r.

8 | The text:
50 From both hands,
If you want to end artfully.

Gloss: This is that you should know how to drive all wrestling from both sides, so that you will end artfully as soon as you approach him. Learn it like this: when you leap with the right foot behind his left, if he steps backward with his left foot during the leap, then follow after him quickly toward the other side with your left foot behind his right. Throw him over the knee or lock his knee with your legs (as was written previously).

9 | This is the text on how one should arrange oneself with the spear against the sword:
51 If it comes to pass
The sword is drawn against the spear;
52 Perceive the thrust,
Leap, catch, wrestle, rush toward him.

Gloss: Remember, this is when you have thrown your lance and he holds onto his.

This is the text:
From both hands,
If you want to end artfully.

Gloss: This is that you should know how to drive the wrestling from both sides, so that you will end artfully as soon as you are approached. Learn it like this: when you leap with the right foot behind his left, if he steps backward with the left foot during the leap, then follow after him quickly toward the other side with the left foot behind his right. Throw him over the knee or enclose his knee with your legs (as was written previously).

This is the text on how one should arrange oneself with the spear against the sword:
If it comes to pass,
The sword is drawn against the spear;
Perceive the thrust,
Leap, catch, wrestle, rush toward him.

Gloss: Remember, this is when you have thrown your lance and he holds onto his. Perceive precisely how he holds it: whether he lets the point go long or short ... his forward hand, or whether he wants to stab toward you above or below with it.[128]

Der text
|Von baÿden henden, ob du mitt kunst gerst zu enden ·:
Glosa |Daß ist daß du alle ringen solt wissen zu tribenn von baiden sÿtten |Ist daß du mitt kunst enden wilt dar nach |92ᵛ| alß du an in kumst Und dz vernÿm also |Wann du mitt dem rechten fuß springst hinder sinen lincken |Tritt er dann im sprüg mitt sinem lincken fuß zu rucke |So volg im bald nach zu der andern sÿtten mitt dinem lincken fuß hinder sinen rechten vnd wirff in über dz knÿ |Oder verschlaiß im ~~dz~~ sin knÿ mitt dinem bain alß vor geschriben stät ~

Das ist der Text
Vonn baydenn hennden ob du mit kunst gerest zu enden
Glosa Das ist das du das ringen solt wissen zu treibenn vonn baidenn seytten Ist das du mit kunst enden wild darnach als du an dich kumpst vnnd das vernyme also wenn du mit dem rechten fuß Springst Hinter seinenn lincken drit er denn Im sprung mit dem lincken fuß zu ruck so volge Im pald nach zu der andern seitten mit dem linckenn fuß Hintter sein rechten vnnd wurff Inn vber das knye ober beschleus Im sein knye mit denn peynen als vor geschribenn stet

Daß ist der text wie man sich sol schicken mitt dem sper wider daß schwert ~
|Ob er sich ver ruckt, daß schwert gegen sper würd gezückt |Der stÿch war nÿm |Spring fahe ringens eÿl zu im ·:
Glosa |Mörck daß ist wann du din gleffen ver schossen häst Vnd er behelt die sinen

Das ist der Text wie man sich soll schickenn mit dem Sper wider das schwert
Ob es sich verruckt das schwertt gegem sper wirt gezuckt Der stich war nym spring vach ringe ~~En~~ Eil zw Im
Glosa Merck das ist wenn du dein gleffen verschossenn hast vnnd er behelt die seinen |8ᵛ| So nym ebenn war wie er sy geuast hab ob er denn ortt lanck oder kurtz sein vor getzte Hannt lest fur gen oder ob er dir da mit obenn oder vnden wil zu stechenn

[128] This is a close match to paragraph 20 of the Pseudo-Danzig gloss; see GF: 84ᵛ.3.

This is the text:
From both hands,
If you want to end artfully.

Gloss: This is that you should know how to drive the wrestling from both sides, so that you will end artfully as soon as you are approached. Learn it like this: when you leap with the right foot behind his left foot (backwards), then follow after him quickly toward the other side with the left behind his right. Throw him over the knee or enclose his knee with your legs (as was written previously).

This is the text on how one should arrange oneself with the spear against the sword:
If it comes to pass,
The sword is drawn against the spear;
Perceive the thrust,
Leap, catch, wrestle, rush toward him.

Gloss: Remember, this is when you have thrown your lance and he holds onto his. Perceive precisely how he holds it: whether he lets the point go long or short ... his forward hand, or whether he wants to stab toward you above or below with it.[128]

Das Ist der text
Von beydenn hendenn ob dw mit kunst gerest zw endenn
Glosa das Ist das dw das Ringenn solt wissen zw treybenn von beydenn seyttenn Ist das dw mit kunst endenn wild darnach als dw an dich kumst vnd das vernim also wen dw mit dem Recht<u>en</u> fus Springst hinter seynen linkenn fus zw Ruck So volg Im bald nach zw der andern seytt<u>en</u> mit dem linkenn hinter seynen Recht<u>en</u> vnnd wurff Ine vber das knye oder beschleus Im sein knie mit denn beynenn als vor geschribenn Stett

Das ist der text wie man sich soll Schickenn mit dem Sper wider das Schwertt
Ob es sich verruckt das Schwert gegenn sperr wirt gezuckt der Stich warnim Spring vach ringe eyl zw Im

Glosa merck das Ist wen dw dein gleffen verschossenn hast vnnd er behelt dye seynenn So Nim eben war wie er sye gefast hab ob er den ortt lanck oder kurtz sein vorgesatzte hand lest fur gen oder ob er dir damit oben oder vnnden will zw Stechenn

This is the text:
From both hands,
If you want to end artfully.

Gloss: This is that you should know how to drive the wrestling from both sides, so that you will end artfully as soon as you approach him. Learn it like this: when you leap with the right foot behind his left, if he steps backward with the left foot during the leap, then follow after him quickly toward the other side with the left foot behind his right. Throw him over the knee or lock his knee with your legs (as was written previously).

This is the text on how one should arrange oneself with the spear against the sword:
If it comes to pass,
The sword is drawn against the spear;
Perceive the thrust,
Leap, catch, wrestle, rush toward him.

Gloss: Remember, this is when you have thrown your lance and he holds onto his. Perceive precisely how he holds it: whether he lets the point go forth long or short in front of his forward hand and whether he wants to stab toward you above or below with it.[128]

Das ist der Text.
|103v| Vonn baiden henden, ob du mit kunst gerest zu endenn.
Glosa. Das ist, das du das Ringen solt wissen zu treibenn vonn baidenn seittenn, Ist das du mit kunst endenn wilt, darnach als du ann Inn kumbst, Unnd das vernim also. wen du mitt dem Rechtenn fus springst hinter sein linckenn, trit er dan im sprung mit dem lincken fus tzu Ruck, so folge im bald nach zu der anderrn seittenn mit dem lincken fus hinter sein Rechten vnnd wirf in vber das knie, oder verschleus im sein knie mit denn painenn als vorgeschriben stehet.

Das ist der Text wie man sich sol schicken mit dem Sper wider das schwert.
Ob es sich verruckt, das schwert gegenn sper wirt getzuckt, der stich war nim, spring, fahe ringens eil tzu im.

Glosa. Merck das ist, wenn du dein glefen verschoßenn hast, vnnd er behelt die seinen, So nim ebenn war, wie er sie gefast hab, ob er denn ortt |104r| langk oder kurtz fur sein vorgesatzte handt lest fur gehenn, vnnd ob er dir damit obenn oder vntenn, wil tzu stechenn,

DRESDEN – SR

Arrange yourself against him with the sword like this: grab the middle of the blade with the left hand and lay the sword in front of you in the guard with the middle on your left knee, or hold it by your right side in the lower guard.

Likewise remember, if he stabs in toward you above with the lance, then go up and set aside the thrust in front of your left hand, with the sword on your left side, and leap toward him and look for the setting upon. If you cannot then approach, then let your sword fall and look for the wrestling.[131]

Likewise, if he stabs toward you with the lance when you are in the lower guard, then set aside the thrust in front of your left hand, with the sword on his right side, and look for the setting upon or the wrestling.

|So |93ʳ| schick dich also gegen im mitt dem schwert Griff mitte der lincken hand mitten in die clingen vnd leg das schwert für dich mitten vff ~~dz knÿ~~ din linckes knÿ In die hut oder halt es neben diner rechten sÿtten In der vndern hut ~~~ ~:~

|Item mörck |Sticht er dir dann mitt der gleffen oben eÿn zu |So far vff vnd secz im den stich ab vor diner lincken hand mitt dem schwert vff din lincke sÿtten vnd spring zu im vnd wart des anseczents |Magst du denn zu nicht komen |So lauß din schwert fallen vnnd wart der ringen

|Item sticht er dir zu mitt der glefen wann du stäst in der vndern hut |So secz im den stich ab mitt dem schwert vor |93ᵛ| diner lincken hand vff sin rechte sÿtten vnd wart deß anseczents oder der |Ringen ~~~

NÜRNBERG – WP

If he holds it short and wants [missing verb] in the upper guard, then lay the sword in front of you in the guard on your left knee.[129]

Or likewise, if he stabs toward you underneath, then set aside the thrust in front of your left hand and leap with the right behind his left, and go in with the pommel in front over his right shoulder around the neck, and throw him over the knee.[130]

If he stabs toward you above with the lance, then go up and set aside the thrust in front of your left hand, with the sword on the left side, and leap toward him and look for the setting upon. If you cannot come to it, then let your sword fall and look for the wrestling.[131]

hat er sy dann kurtz geuast vnd wil da mitt Inn die obernn hut so leg das schwertt fur dich auff dein linckes knye In die hut

Item oder sticht er dir vnden zw so setz den stich ab vor deiner linckenn handt vnd spring mit dem Rechten fuß hinter sein lynckenn vnnd far Im mit dem knopff fornn vber sein rechte achsel vmb den hals vnnd wurff In vbers knye

Sticht er dir dann mit ~~dem~~ der gleuen oben zu so far auff vnnd setz Im den stich ab vor deiner lincken handt mit dem schwert auff sein lincke seytten vnnd spring zw Im vnnd wart des ansetzen Magstu darzu nit kumen So laß dein schwertt vallenn vnnd wart der ringen

ARMORED FENCING

[129] This is a close match to the first half of paragraph 21 of the Pseudo-Danzig gloss; see GF: 84ᵛ.4–85ʳ.1.

[130] This is similar to the second half of paragraph 22 of the Pseudo-Danzig gloss; see GF: 85ʳ.2.

If he holds it short and wants [missing verb] in the upper guard, then lay the sword in front of you in the guard on your left knee.[129]

Or likewise, if he stabs toward you underneath, then set aside the thrust in front of your left hand and leap with the right behind his left, and go in with the pommel in front over his right shoulder around the neck, and throw him over the knee.[130]

If he stabs toward you above with the lance, then go up and set aside the thrust in front of your left hand, with the sword on the left side, and leap toward him and look for the setting upon. If you cannot come to it, then let your sword fall and look for the wrestling.[131]

If he holds it short and wants [missing verb] in the upper guard with it, then lay the sword in front of you in the guard on your left knee.[129]

Or likewise, if he stabs toward underneath, then set aside the thrust in front of your left hand and leap with the right foot behind his left, and go in with the pommel in front over his right shoulder around the neck, and throw him over the knee.[130]

If he stabs toward you above with the lance, then go up and set aside the thrust in front of your left hand, with the sword on his left side, and leap toward him and look for the setting upon. If you cannot come to it, then let your sword fall and look for the wrestling.[131]

hatt er sye dan kurtz geuast vnd will |109ʳ| In die obern hutt so leg das Schwertt fur dich auff dein lincks knye In die hutt

Item oder Sticht er dir vnndenn zw So setz den Stich ab vor deyner linkenn hand vnd Spring mit dem Recht<u>en</u> hinter seynen linkenn vnd far Im mit dem knopff forn vber sein Rechte achssell vmb den hals vnd wurff In vbers knye

Sticht er dir dan mit der gleuen oben zw so far auff vnd setz Im den Stich ab vor deyner linkenn hand mit dem Schwertt auff linke seytt<u>en</u> vnd Spring zw Im vnd wartt des ansetzenn magstu dar zw nit kumen So las dein Schwertt vallen vnnd wartt der Ring<u>en</u>

hat er sie dann kurtz gefast, vnd wil damit in die obernn hut, so leg das schwert fur dich auf dein lincks knie in die hut.

Item oder sticht er dich vntenn tzu, so setz denn stich ab vor deiner linckenn handt, vnnd spring mit dem Rechten fus hinter sein linckenn, vnnd far im mit dem knopf vornn vber sein Rechte achsel vmb den hals, vnnd wirf in vbers knie.

Sticht er dir dann mit der glefenn oben tzu, so far auf, vnnd setz im denn stich ab, vor deiner linckenn handt mit dem schwert auf sein lincke seittenn vnnd spring tzu im, vnnd wart des ansetzen, magstu dartzu nit kummenn, so las dein schwert fallenn vnnd wart der ringenn.

[131] This is similar to the second half of paragraph 21 of the Pseudo-Danzig gloss; see GF: 85ʳ.1.

10| The text on setting aside with empty hand:
53 Strike long from the left hand,
Leap surely and then catch.
54 If he wants to pull back
From injury, catch and press him
55 So that in the exposure,
You harass with the point of the sword.

Gloss: When you stand in the lower guard, if he stabs toward you above with the spear and has it held so that the point extends out of the hand, and stabs toward you with it above, then strike his spear away to the side with the left hand, and quickly grab your sword again with the left hand in the middle of the blade and leap toward him, and set upon him, etc.

Likewise, if he stabs underneath toward your groin with the spear, then catch his spear in the left hand and hold firmly with it, and stab underneath toward his groin with the right. If he pulls his spear toward himself firmly and wants to wrench it out of your hand, then go above him and release the spear from the hand so he gives an exposure. Quickly grab your sword with the left hand again in the middle of the thing and follow after him swiftly, and set upon him, etc.

This is the text on setting aside with empty hand:
Strike long from the **left** hand,
Leap surely and then catch.
If he wants to pull back
From injury, catch and press him
So that in the exposure,
You harass with the point of the sword.

Gloss: Remember, this is when you stand in the lower guard: if he stabs toward you above with the spear and has it held so that the point goes out of the hand, then strike him away to the side with the sword, and quickly grab your sword again with the left hand in the middle of the blade and leap toward him, and set upon him.

Or **likewise,** if he stabs underneath toward your groin with the spear, then catch his spear in the left hand and hold firmly with it, and stab underneath toward his groin with the right. If he pulls his spear toward himself firmly and wants to wrench it out of your hand, then go high and release the spear from the hand so he gives an exposure. Quickly grab his sword with the left hand again in the middle of the blade and follow after him swiftly, and set upon him.

Der text von abseczen mitt lerer hand ~
|Lincke lanck von hand schlache Spring wÿßlich vnd den fache |Ob er will zucken von schaiden, faͧch vnd truck in |Daß in die blöß, mitt schwertes ort verdröß ~:·

Glosa |Wenn du staüst in der vndern hut |Sticht er dir an oben zu mitt dem sper vnd hat dz gefasst daß im der ort lang für die hand vsß gat vnd sticht dir domitt oben zu |So schlach in mitte der lincken hand sin sper beseÿcz abe, vnd begrÿff din schwert bald wider mitt der lincken hand mitten in der clingen vnd spring zu im vnd secz im an etc

|94r| |Item sticht er dir mitt dem sper vnden zu dem gemächt |So fahe sin sper in die lincken hand vnd halt es domitt vast vnd stich im mitt der rechten vnden zu den gemächt |Und ruckt er dann sin sper fast an sich vnd will dir daß vsß der hand rÿssen |So lauß daß sper über in vß der hand far So gibt er sich blöß So begriff din schwert bald mitt der lincken hand wider mitten in der clingen vnd volg im bald nach vnd secz im an ~~~

Das ist der Text von absetzenn mit lerer hannt
Linck lanck von hannt schlag spring weyslich vnnd denn vach ob er will zuckenn von schaidenn vach vnnd druck In das In die plos mit schwertes ort ver droß

|9r| **Glosa** Merck das ist wenn du steest In der vnntern hut Sticht er dir dann obenn zw mit der Sper vnnd hat das gefast das Im der ortt fur die handt aus geet so schlag Im mit der schwer beseitt ab vnnd begreyff dein schwert bald wider mit der linckenn hannd mitten In der clingenn vnnd spring zw Im vnnd setz Im ann

Item Oder sticht er dir mit dem Sper vnden zu dem gemecht So vach sein Sper In die lincken Handt vnd halt es da mit vaste vnd stich Im mit der Rechten vndenn zw dem gemecht vnnd Ruckt er denn sein Sper vast an sich vnd wil dir das aus der handt reyssen So las das Sper vrbering aus der handt farnn So gib er sich plos So begreiff sein Schwert pald mit der linckenn hanndt wider mitten In der clingen vnd volge Im pald nach vnd setze Im ann

This is the text on setting aside with empty hand:
Left long from the hand,
Leap surely and then catch.
If he wants to pull back
From injury, catch and press him
So that in the exposure,
You harass with the point of the sword.

Gloss: Remember, this is when you stand in the lower guard: if he stabs toward you above with the spear and has it held so that the point goes out of the hand, then strike him away to the side with the sword, and quickly grab your sword again with the left hand in the middle of the blade and leap toward him, and set upon him.

Or likewise, if he stabs underneath toward your groin with the spear, then catch his spear in the left hand and hold firmly with it, and stab underneath toward his groin with the right. If he pulls his spear toward himself firmly and wants to wrench it out of your hand, then go high and release the spear from the hand so he gives an exposure. Quickly grab his sword with the left hand again in the middle of the blade and follow after him swiftly, and set upon him.

This is the text on setting aside with empty hand:
Strike long from the left hand,
Leap surely and then catch.
If he wants to pull back
From injury, catch and press him
So that in the exposure,
You harass with the point of the sword.

Gloss: Remember, this is when you stand in the lower guard: if he stabs toward you above with the spear and has it held so that the point extends out of the hand, then strike him away to the side with the sword, and quickly grab your sword again with the left hand in the middle of the blade and leap toward him, and set upon him.

Or likewise, if he stabs underneath toward your groin with the spear, then firmly his spear in the left hand and hold tightly with it, and stab underneath toward his groin with the right. If he pulls his spear toward himself firmly and wants to wrench it out of your hand, then go high and release the spear from the hand so he gives an exposure. Quickly grab your sword with the left hand again in the middle of the blade and follow after him swiftly, and set upon him.

Das ist der text vonn absetzenn mit lerer hannd
Linck lanck von hand Spring weyßlich vnnd den vach ob er will zuckenn von scheydenn vach vnd druck In das In die plos mit Schwertes ortt verdroß

Glosa Merck das Ist wen dw steest In der vntern hutt Sticht er dir dan obenn zw mit dem Sper vnd hatt das gefast das Im der ortt fur die hanndt auß geet So Schlag Inne[!][132] mit der Schwert beseyt ab vnd begreiff dein Schwertt bald wider mit der linkenn hand mitten In der clingenn vnd Spring zw Im vnd setz Im ann

Item oder Sticht er dir mit dem Sper vndenn zw dem gemecht So vach sein Sper In die linkenn hand vnd halt es damit vaste vnd Sticht Im mit der Rechtenn vnnden zw dem gemecht vnd Ruck er dan sein Sper vast an sich vnd will dir das auß der hand reyssen So las das Sper vrbering auß der hand farn So gibt er sich plos So begreiffe sein Schwertt bald mit der linkenn hand wider mitten In der klingenn |109^v| vnd volg Im Bald nach vnd Setz Ime ann

Das ist der Text vonn absetzen mit lerer handt.
Linck lang vonn handt schlag, spring weislich vnnd denn fahe, ob er wil tzuckenn von schaiden, fahe vnd druck in, das in die ploß, mit schwertes ort verdroß.

|104^r| Glosa. Merck das ist, wenn du stehest in der vnternn hut, Sticht er dir denn oben tzu mit dem Sper, vnnd hat das gefast das im der ort langk fur die hanndt ausgehet, so schlag im mit der schwer beseit ab vnnd begreif dein schwert bald wider mit der linckenn handt, mittenn in der klingenn, vnd spring zu im vnd setz im an.

Item oder sticht er dir mit dem sper, vntenn tzu dem gemecht so fahe sein sper in die lincken handt vnnd halt es damit fast, vnnd stich im mit der Rechtenn vntenn tzu dem gemecht, vnnd Ruckt er den sein sper fast ansich, vnnd wil dir das aus der hanndt reissenn, so las das sper urbering aus der hanndt farenn, so gibt er sich plos, so begreif dein schwert baldt mit der lincken hannd wider mittenn in der clingenn, vnnd volge im bald nach vnd setz im ann.

[132] Correct: *Ime.*

11 | The text on the exposures:
56 Leather and gloves
And the eyes: the correct exposures to seek.

Gloss: This is when you want to set upon someone in armor: you should look for the exposures well. The first in the face, or under the armpits, or in the palm or rear of the glove, or in the knuckle, or between the legs, or in all limbs of the harness where it has joints. These are the best for defeating an opponent who faces you, and you should seek the exposures correctly so that you do not attack a farther one when you can get a closer one. Do this with all weapons that belong to the duel, etc.

This is the text on the exposures:
Leather and gloves,
Under the eyes: the correct exposures to seek.

Gloss: This is when you want to set upon someone in armor: you should look for the exposures well. The first in the face, or under the armpits, or in the palm, or in rear of the glove, or in the knuckle, or between the legs, or in all limbs of the harness where it has joints. These are the best for defeating an opponent who faces you, and you should seek the exposures correctly so that you do not attack a farther one when you can get a closer one. Do this with all weapons that belong to the duel.

Der text von den blossen
|Leder vnnd handschüch Unnd den augen die blöß recht such ·:
Glosa |Das ist wann du ainem gewapneten man an seczen wilt |So solt du der blöß eben war nemen |Der ersten |94v| |In daß gesicht, oder vnder den uchsen, oder in den tennern, oder hinden in die handtschuch |Oder in die knÿkeln oder zwischen den bainen oder in allen glidern da der harnosch sin gelenck innen hat Wann an den stetten ist der man am besten zegewinnen vnd die blossen solt du recht wissen zu suchen, dz du nach ainer nicht wÿt griffen solt wann dir ain nächere werden mag, Daß tu mitt aller were die zu dem kanpff gehörent etc

Das ist der Text vonn denn plossen
Leder vnnd hanntschuch vntter denn augen die plos recht such
Glosa Das ist wenn du einem gewappenden mann ansetzenn wild So soltu der plös |9v| ebenn war nemenn des erstenn In das gesicht oder vnnter den vchsenn oder In den tener oder In denn hantschuch oder In die knyekel oder zwischen den peinen oder In allenn glidernn da der harnasch sein gelennck Innen hat wen an den seittenn Ist der man am pestenn zu gewynnen vnnd die plosenn soltu recht wissenn zu suchenn das du nach einer nit weyt greyffenn solt wen dir ein nehere werden mag das thu mit aller weer die zu dem kampff gehorenn

This is the text on the exposures:
Leather gloves,
Under the eyes: the correct exposures to seek.

Gloss: This is when you want to set upon someone in armor: you should look for the exposures well. The first in the face, or under the armpits, or in the palm, or in the glove, or in the knuckle, or between the legs, or in all limbs of the harness where it has joints. These are the best for defeating an opponent who faces you, and you should seek the exposures correctly so that you do not attack a farther one when you can get a closer one. Do this with all weapons that belong to the duel.

Das ist der text vonn den Blossen
Leder hantschuch vnter den augenn die ploeß recht such
Glosa Das ist wan dw eynen gewappetten man ansetzenn wild So soltu der Bloeß eben war nemen des Erstenn In das gesicht oder vnter den vchssenn oder In den tener oder In den hantschuch oder In dye knyekelen oder zwischenn den peinen oder In allen glidern do der harnasch sein gelenck Innen hatt wen an den Stellen Ist der mann am besten zw gewinnen vnd der Bloessen Soltu Recht wissenn zw Suchenn das dw nach einer nit weitt greiffen Solt wen dir ein nehere werden mag das thu mit aller wer die zw dem kampff gehorenn

This is the text on the exposures:
Leather and gloves,
Under the eyes: the correct exposures to seek.

Gloss: This is when you want to set upon someone in armor: you should look for the exposures well. The first in the face, or under the armpits, or in the palm, or in rear of the glove, or in the knuckle, or between the legs, or in all limbs of the harness where it has joints. These are the best for defeating an opponent who faces you, and you should seek the exposures correctly so that you do not attack a farther one when you can get a closer one. Do this with all weapons that belong to the duel.

Das ist der Text vonn denn plossenn.
|105r| Leder vnnd hantschuch, vnter denn augen die ploß recht such. Glosa.
Das ist wenn du einem gewappenden man ansetzen wilt, so soltu der plöß ebenn war nemen, des ersten in das gesicht, oder vnter die vchsenn, oder in denn tener, oder hintenn in denn handschuch, oder in die kniekel, oder tzwischenn denn painenn, oder in allenn glidernn, da der harnasch sein gelenck Innen hatt, wenn an denn stettenn ist der man am bestenn zugewinnenn, vnnd die blössenn soltu recht wissen zu suchenn, das du nach einer nit weit greiffenn solt, wenn dir ein nehere werdenn mag, das thu mit aller wer die tzu dem kampf gehorenn.

12| The text of the obscure wrestling:

57 The forbidden wrestling
Learn to surely bring
58 To a close; find
The strength to overcome with it.

Gloss: This is when someone runs in on another: let your sword fall and look for the sure wrestling that belongs to the duel. And they are forbidden by all wise masters of the sword, that no one should be allowed to teach them nor show them in the *Fechtschule*,[133] because they belong to duel fencing. And they are the arm breaks, leg breaks, groin strikes, death strikes, knee strikes, finger dislocations, eye grabs, and more like that.

This is the text with the forbidden wrestling:
The forbidden wrestling
Learn to surely bring
To a close; find
The strength to overcome with it.

Gloss: Remember, this is when someone runs in on another: immediately let your sword fall and look for the sure wrestling that belongs to the duel. They are forbidden by wise old masters of the sword, that no one should be allowed to teach them nor show them in the *Fechtschule*,[133] because they belong to duel fencing. They are the arm breaks and leg breaks, groin strikes, death strikes, knee strikes, finger dislocations and eye grabs, and more like that.

13| You should remember the wrestling here:
Likewise, if someone grabs you above with wrestling and wants to pull you toward him or push you from him with strength, then strike the right arm outside over his left (in front by his hand) and press on your chest with both hands, and leap with the right foot behind his left and throw him off his feet over the knee.[134]

Remember the wrestling here:
Likewise, if he grabs you above with wrestling and wants to pull you toward him or push you from him with strength, then strike the right arm outside over his left (in front by his hand) and press on your chest with both hands, and leap with the right foot behind his left and throw him off his feet over the knee.[134]

Der text von dem verborgnen ringen
|Verbotten Ringen, wÿßlich zu lern bringen, |Zu schlissen finde die starcken domitt über winde ~
Glosa |Das ist wann ainer dem ~~andern~~ andern ein laufft |So lauß din schwert fallen vnd wardt |95r| domitt wÿßlich der ringen die zu dem kampff gehören, vnd verbotten sin von allen wÿsen maistern des schwerts Daß man die vff offenbaren schulen nÿemancz lernen noch sechen lasen sol darumb daß sÿ zu dem kampff fechten gehörn vnd daß sind die armbrüch, |Bainbruch, hoden stoß, mortstoß, knÿstoß, vinger la^{u}sunge, äugen griff, vnd dar zu mer ~~~:·

Das ist der Text mit denn verpottenn ringen
Verpottenn Ringen weyslich zu lere bringen zu schliessenn vinde die starcken da mit vber winde
Glosa Merck das ist wen einer dem andernn ein laufft so las dein schwert zu hannt fallenn vnnd wart da mit weysslich der ringen die zu dem kampff gehoren die verpottenn sein von alten weysen maistern des schwertz das man auff offenwaren schulenn nyemant lernenn moch sehen sol lassenn Dar vmb das sy zw dem kampff fechtenn gehorenn das sein die Arm pruch vnnd pein pruch vnd hodenn |10r| stosß vnnd mort stoß vnnd knye stoß vnd vingerlosung vnnd augen greyffen vnd dar zw mer

Hie solt du morcken die ringen
|Item grifft dich an ainer oben mitt ringen vnd will dich mitt störck zu im rucken oder von im stossen |So schlach den rechten arm vssen über sin lincken ~~fornen~~ vornen bÿ siner hand vnnd truck den mitt baiden henden an din brüst vnd spring mitt dim |95v| rechten fuß hinder sinen lincken Und wirff in über dz knÿ vß dem füß

Hie merck die ringen
Item greyfft er dich obenn an mit Ringen vnnd wil dich mit sterck zu Im Rucken oder von Im stossenn So schlag den Rechten arm aussen vber sein glinckenn forn bey seiner hant vnnd truck den mit peiden henndenn an dein prust vnnd Spring mit dem Rechtenn fuß hinter sein linckenn vnnd wurff In vber das knye auß dem fuß

[133] This probably doesn't refer specifically to tournaments, but rather to the public training days overseen by a free fencer or fencing master, which shared the same label. See VANSLAMBROUCK 2023: xxi–xxviii.

This is the text with the forbidden wrestling:
The forbidden wrestling
Learn to surely bring
To a close, and
Overcome the strong with it.

Gloss: Remember, this is when someone runs in on another: immediately let your sword fall and look for the sure wrestling that belongs to the duel. They are forbidden by wise old masters of the sword, that no one should be allowed to teach them nor show them in the *Fechtschule*,[133] because they belong to duel fencing. They are the arm breaks and leg breaks, strikes at the groin, and death strikes, and knee strikes, and finger dislocations, and eye grabs, and {more} like that.

Remember the wrestling here:
Likewise, if he grabs you above with wrestling and wants to pull you toward him or push you from him with strength, then strike the right arm outside over his left (in front by his hand) and press on your chest with both hands, and leap with the right foot behind his left and throw him off his feet over the knee.[134]

This is the text of the forbidden wrestling:
The forbidden wrestling
Learn to surely bring
To a close; find
The strength to overcome with it.

Gloss: Remember, this is when someone runs in on another: immediately let your sword fall and look for the sure wrestling that belongs to the duel. They are forbidden by wise old masters of the sword, that no one should be allowed to teach them nor open them in the *Fechtschule*,[133] because they belong to duel fencing. They are the arm breaks, leg breaks, groin strikes, death strikes, knee strikes, finger dislocations and eye grabs, and more like that.

Remember the wrestling here:
Likewise, if he grabs you above with wrestling and wants to pull you toward him or push you from him with strength, then strike the right arm outside over his left (in front by his hand) and press on your chest with both hands, and leap with the right foot behind his left and throw him off his feet over the knee.[134]

Das ist der text mit den verpotten zw Ringen
Verpottenn Ringen weyßlich zw lere bringenn zw schliessenn vnd die Starcken damit vber wind
Glosa merck das Ist wen einer dem andern ein lawfft So las dein Schwertt zw hand fallen vnd wart damit weyßlich der ringenn die zw dem kampff gehorenn die verbott<u>en</u> sein von alten weissen meinster des schwertz das man auff offenbaren Schulen nyemantz lernen noch sehenn soll lassenn Darumb das sye zw dem kampff fecht<u>en</u> gehorenn das sein die arm bruch vnd pein bruch zw hodenn stosß vnd mortt Stosß vnnd knye |110r| Stoeß vnd vinger losung vnd augenn greyffung vnd dar zw *mer*[135]

Hie merck die ringenn
Item gerifft er dich obenn an mit ring<u>en</u> vnnd will dich mit Sterck zw Ime Ruckenn oder von Ime Stossenn So Schlag den Rechtenn arm aussenn vber sein glinckenn forn bey seyner hand vnnd truckt den mit beyden hendenn an dein Brust vnd Spring mit dem Rechten fus hinter sein linken vnd wurff In vber das knye auß dem fus

Das ist der Text vonn denn verpotnenn Ringenn.
Verpottenn Ringen, weislich tzu lere bringenn zu schliessenn finde, die starcken damit vberwinde.
Glosa. Merck das ist wenn einer dem andern einlauft, so las dein schwert zu handt fallenn, vn[d] |105v| wart damit weislich der ringenn, die tzu dem kampff gehörnn, die verpotten sein, vonn altenn weisenn meisternn des schwerts, das man die auff offenn warenn schulnn niemandts lernenn noch sehenn sol lassenn, darumb das sie tzu dem kampff fechtenn gehornn. Das sein die armpruch, vnnd pain pruch, vnnd hodenstos, vnd mortstös, vnd kniestos, vnnd finger losung, vnnd augenn greiffenn vnnd dartzu mer.

Hie merck die Ringenn.
Item greift er dich obenn an mit ringen, vnd wil dich mit sterck tzu im Ruckenn, oder von im stossenn, so schlag denn Rechtenn arm, aussenn vber sein linckenn vornn bey seiner handt, vnd druck denn mit baidenn hendenn an dein prust, vnnd spring mit dem Rechtenn fus hinter seine<u>n</u> lincken vnnd wirf in vber das knie aus dem fus.

[134] This is a close match to paragraph 29 of the Pseudo-Danzig gloss; see GF: 86v.1.
[135] The last word appears to have been added next to the text block, possibly by another hand.

Likewise, if he grabs you with wrestling and does not hold you firmly, then grab his right hand with your right and pull him toward you. Grab his elbow with the left and step with the left foot in front of his right, and pull him over it like this. Or fall on his arm with the chest and break it like this.

Likewise, grab his left hand with the left hand (in front by the hand) and pull him toward you and strike your right arm with strength over his left (in the joint of the crook of the arm) and break his left over his right with the left hand, and leap with the right foot behind his right and throw him over it like this, etc.

Likewise, when he goes through under your right with the left arm (around your body), strike him strongly down from above with the right arm, outside, in the joint of his left elbow, and turn yourself from him with it.

Likewise, when he holds you by the arms (and you him in return) and he stands with a foot outstretched, bash him on that same knee, and break his foot like this.

Likewise, if he grabs you with wrestling but does not hold you firmly, then grab his fair hand with your right and pull him toward you. Grab his elbow with the left and step with the left foot in front of his right, and pull him over it like this. Or fall on his arm with the chest and break it like this.

Likewise, grab his left hand with the left hand (in front by the hand) and pull him toward you and strike the right arm with strength over his left (in the joint of the crook of the arm) and break his left over your right with the left hand, and leap with the right foot behind his right and throw him over it like this.

Likewise, when he goes through under your right with the left arm (around the body), strike him strongly down from above with the right arm, outside, in the joint of his left elbow, and turn yourself from him with it.

Likewise, when he holds you by the arms (and he you in return) and he stands strongly with the foot that he set forward outstretched, bash him on that same knee with the foot, and break his foot like this.

|Item grifft er dich an mitt ringen vnd halt er dich dann nitt ~~fast~~ vast |So begrÿff sin rechte hand mitt diner rechten vnd ruck in zu dir mitt der lincken begrÿff im den elnbogen vnd schrÿtt mitt dem lincken fuß für sinen rechten Und ruck in also darüber |Oder fall im mitt der brust vff den arm vnd brich im den also

|Item grÿff mitt der lincken hand sin lincke vornen bÿ der hand vnd ruck in zu dir vnd schlach din rechten arm mitt störck über sin lincken In das glenck der armbüge vnd brich mitt der lincke hannd sin lincke ṽber sin rechten vnd spring |96^{r}| mitt dem rechten fuß hinder sinen rechten vnd wirff in also darüber etc

|Item fört er dir mitt dem lincken arm vnder dinen rechten durch vmb din lÿbe |So schlach in mitt dem rechten arm starck von oben nÿder vsswendig in das glenck sins lincken ~~elbog~~ elnbogens vnd wend dich do mitt von Im ~ ~

|Item wann er dich fasst bÿ den armen vnd du in wider, stat er dann gestrackts mitt dem fuß So stoß in vff daß selbig knÿ So brichst im den fuß ~ ~ ~ ~ ~ ~ ~

Item greifft er dich an mit ringenn helt er dich denn nit fast So begreiff sein gerechte hannt mit deiner Rechtenn vnd Ruck Inn zw dir mit der lincken greiff Im den Elenpogenn vnnd schreyt mit dem linckenn fuß fur seinen rechten vnnd Ruck Inn also daruber oder vall Im mit der prust auff denn Arm vnd prich Im den also

Item begreyff mit der lincken hant sein linckenn forn pey der hannt vnd Ruck In zu dir vnd schlag denn rechten Arm mit sterck vber seinen lincken In das glenck |10^{v}| der armpug vnnd prich mit der lincken hannt sein lincken vber dein rechten vnd spring mit dem Rechtenn fuß hinter sein Rechtenn vnd wurff In als dar vber

Item Fert er dir mit dem lincken Arm vnter deinenn rechten durch vmb den leib So schlag Inn mit dem Rechten arm starck vonn obenn nyder auswendig In das glenck seins linckenn Elnnpogenns vnnd went dich da mit vonn Im,

Item wenn er dich fast bey den armen vnd er dich wider Stet er denn starck gestrackt mit dem fuß den er hat fur gesetzt So stoß Inn mit dem fuß auff das selbig knye So pristu Im den fuß

Likewise, if he grabs you with wrestling but does not hold you firmly, then grab his right hand with your right and pull him toward you. Grab his elbow with the left and step with the left foot in front of his right, and pull him over it like this, and fall on his arm with the chest and break it like this.

Likewise, grab his left hand with the left hand (in front by the hand) and pull him toward you and strike the right arm with strength over his left (in the joint of the crook of the arm) and break it over the right with the left, and leap with the right foot behind his right and throw him over it like this.

Likewise, when he goes through under the right with your left arm (around the body), strike him strongly down from above with the right arm, outside, on the joint of his left elbow, and turn yourself from him with it.

Likewise, when he holds you by the arms (and he you in return) and he stands strongly with the foot that he set forward outstretched, bash him on that same knee with the foot, and break his foot like this.

Item greiff er dich an mit ringenn helt er dich den nit fast So begreiff sein Rechte hand mit deyner Rechtenn vnd Ruck In zw dir mit der linkenn greiff Im den Elebogenn vnd Schreytt mit dem linken fus fur seynen Rechten vnd Ruck In also daruber vnnd vall Im mit der prust auff den arm vnd brich Im denn also

Item begreifft mit der linkenn hand sein linken forn pey der hand vnd Ruck In zw dir vnd Schlag den Rechten arm mit sterck vber seynen linken In das glenck der armpug vnnd [p]rich mit der linkenn vber den Rechten vnd Spring mitt dem Rechten fus hinter sein Rechten vnd wurff In als das vber

Item fertt er dir mit deynem linkenn arm vnter dem Rechtenn durch vmb den leib So schlag In mit dem rechtenn arm starck von oben nyder außwendig an das glenck seins linken elpogens vnd wend dich da mit von Im

Item wen er dich fast bey den armen vnd er dich wider Stett er den Starck gestrackt mit dem fus den er hatt fur |110ᵛ| gesetzt So stoß In mit dem fus auff das selbig knye So prichstu Ime den fus

Likewise, if he grabs you with wrestling but does not hold you firmly, then grab his right hand with your right and pull him toward you. Grab his elbow with the left and step with the left foot in front of his right, and pull him over it like this. Or fall on his arm with the chest and break it like this.

Likewise, grab his left hand with the left hand (in front by the hand) and pull him toward you and strike the right arm with strength over his left (in the joint of the crook of the arm) and break his left over your right with the left hand, and leap with the right foot behind his right and throw him over it like this.

Likewise, when he goes through under your right with the left arm (around the body), strike him strongly down from above with the right arm, outside, in the joint of his left elbow, and turn yourself from him with it.

Likewise, when he holds you by the arms (and he you in return) and he stands with the foot that he set forward outstretched, bash him on that same knee with the foot, and break his foot like this.

Item greift er dich an mit ringen, helt er dich denn nit fast, so begreif sein Rechte handt mitt |106ʳ| deiner Rechtenn, vnnd ruck in tzu dir mit der lincken begreif im denn Elnpogenn, vnnd schreit mit dem linckenn fus für, seinenn Rechtenn, vnnd ruck in also daruber, oder fal im mit der prust auff denn arm, vnd brich im denn also.

Item begreif mit der linckenn handt sein lincken, vorn bey der hanndt, vnnd ruck in tzu dir, vnnd schlag den Rechtenn arm mit sterck vber seinenn lincken, in das gelenck der armpug, vnnd prich mit der linc[k]enn handt sein linckenn vber dein Rechtenn, vnnd spring mit dem Rechten fus hinter sein Rechtenn vnd wirf in also daruber.

Item fert er dir mit dem linckenn arm, vnnter deinenn Rechtenn durch vmb denn leib, so schlag in mit dem rechtenn arm starck vonn obenn nider, auswendig in das gelenck seins lincken elnpogens vnnd wend dich damit vonn im.

Item wenn er dich fast bey denn armen, vnnd er dich wider, stet er er denn gestrackt mit dem fus denn er hatt furgesetzt, So stos in mitt dem |106ᵛ| fus, auff dasselbige knie, so brichstu im den fus.

Likewise, when it's up to you, you can also bash him with the knee or at his groin with the foot. But you should watch out so he doesn't grab you by the foot, etc.

Likewise, when he grabs at you with open hands or with outstretched fingers, see whether you can grab one of his fingers and break it upward, and move him toward the ring with it.

Also, defeat him at the edge and have another great gain.[136]

Likewise, when it's up to you, you can also bash him with the knee or at his groin with the foot. Or watch out so he doesn't grab you by the foot.

Likewise, when he grabs at you with open hands or with outstretched fingers, see whether you can grab one of his fingers, and break it upward and move him in the ring with it, and defeat him at the edge and have great gain.[136]

14 | The text of a lesson:

59 Likewise, all lessons:
Turn the point against the exposure.

Gloss: This is that with all three weapons that belong to the duel, as was mentioned before, you should always thrust with the point and not do anything else (it will injure you, etc.).

This is the text of a lesson:

Likewise, all lessons:
Turn the point against the exposure.

Gloss: Remember, this is that with all faithful weapons that turn to the duel, as was mentioned before, you should always thrust with the point and not do otherwise (or it will injure you).

|Item du magst im och mitt dem knÿ oder mitt dem fuß zu den gemächten stossen, wenn es dir eben ist Aber du solt dich für sechen daß er dich bÿ dem fusse nitt begrÿffe ~

|96v| |Item wann er nach dir grÿfft mitt offen henden oder mitt gerackten fingern So wart ob du im ainen finger begrüffen mügst Und brüch im den übersich Und für in domitt zu dem kraÿß ~ ~

|Auch gewinst du Im do mitt die sÿtten an vnd sunst vill ander grosser vortail ~

Item Du magst In auch mit dem knye oder mit dem fuß zu dem gemecht stossenn wenn es dir Ebenn Ist Oder sich dich fur das er dich bey dem fuß nicht begreiffe

Item wen er nach dir greifft mit offenn, Hennden oder mit gerackten vingern So wart ob du Im ein vinger begreiffen magst vnnd prich Im denn vbersich vnd fur In da mit In dem kraiß auch gewinnstu In da mit die seitten an vnd sunst vil groß vortail

Der text von ainer lere

|Item aller lere, den ort gegen der blöß köre ·:

Glosa |Daß ist daß du mitt allen drÿ wörn die zu dem kampff gehörn allweg mitt dem ort zu den blossen stechen solt die dir vorgenant sind, vnd sunst nicht, anders es bringt dir schaden etc ~ ~ ~

|11r| **Das ist der Text vor einer lere**

Item Aller lere denn ortt gegenn der plos kere

Glosa Merck das ist das du mit allen treuen werenn Die zu dem kampff keren albeg mit dem ortt zu denn plossenn stechenn solt die dir vorgenant sein vnnd sust nit oder es pringt dir schadenn

[136] This is a close match to paragraph 34 of the Pseudo-Danzig gloss; see GF: 87r.3; it's also similar to paragraph 6 of Huntsfeld's armored fencing; see GF: 101r.1.

Likewise, when it's up to you, you can also bash him with the knee or at his groin with the foot. Or watch out so he grabs you by the foot.

Likewise, when he grabs at you with open hands or with outstretched fingers, see whether you can grab one of his fingers, and break it upward and move him in the ring with it, and defeat him at the edge and have great gain.[136]

This is the text of a lesson:
Likewise, all lessons:
Turn the point against the exposure.

Gloss: Remember, this is that with all faithful weapons that turn to the duel, as was mentioned before, you should always thrust with the point and not do otherwise (or it will injure you).

Likewise, when it's up to you, you can also bash him with the knee or at his groin with the foot. Or watch out so he doesn't grab you by the foot.

Likewise, when he grabs at you with open hands or with outstretched fingers, see whether you can grab one of his fingers, and break it upward and move him in the ring with it, and defeat him at the edge and have great gain.[136]

This is the text of a lesson:
Likewise, all lessons:
Turn the point against the exposure.

Gloss: Remember, this is that with all faithful weapons that belong to the duel, as was mentioned before, you should always thrust with the point and not do otherwise (or it will injure you).

Item dw magst In auch mit dem knye oder mit dem fus zw dem gemecht stossenn wen es dir Ebenn Ist Oder sich dich fur das er dich bey dem fus begreifft

Item wen er [nach] dir greifft mit offen henden oder mit gerackten vingern So wart ob dw Im ein vinger begreiffenn ma[g]st vnd prich Im den vbersich vnd fur In da mit In den K[r]eiß auch gewinstu In damit die seytten an vnd sunst vil groß vorteill

Das ist der text vor[!] einer lere
Item aller lere denn ortt gegen der Bloß kere
Glosa merck das Ist das du mit allen treuen weren die zw dem kamppff keren alweg mit dem ortt zw denn plosen Stechenn soltt die dir vorgenant sein vnd sust nit oder es pringt dir schaden

Item du magst im auch mit dem knie, oder mit dem fus zu dem gemecht stossenn wenn es dir ebenn ist, Oder sihe dich fur, das er dich bey dem fus nicht begreiffe.
Item wenn er nach dir greift mit offen hendenn oder mit geräcktenn fingernn, So wart ob du im ein finger begreiffenn magst, vnnd prich im denn vbersich vnnd fur in damit tzu dem krais auch gewinstu in damit die seittenn ann, vnnd sonnst vil groß vortail.

Das ist der Text von einer lere.
Item alle lere, den ort gegen der plos kere.
Glosa. Merck das ist das du mit allenn trewenn werenn, die zu dem kampf gehornn, alweg mit dem ort zu denn ploßenn stechenn, solt, die dir vorgenant sein, vnnd sonnst nit oder es bringt dir schadenn.

15 | **The text on how one should fence sword against sword in the duel:**

60 Where one, from scabbards,
Sees that both swords are drawn,
61 He should strengthen;
Remember precisely to cover correctly.

Gloss: This is if both have thrown their spears and should fence with swords: you should know and remember before all things the four guards with the 'short' sword,[137] and always thrust toward the upper exposure from them. If he thrusts in simultaneously with you or binds you on the sword, then you should immediately notice whether he is Hard or Soft on the sword, and when you have sensed it, drive the Strength against him (which you will see written hereafter).

This is the text on how one should fence with the sword against the sword:

Where one, from scabbards,
Sees that both swords are drawn,
He should stretch;
Remember precisely to cover correctly.

Gloss: This is if we both have thrown the spears and should fence with swords: you should know before all things the four guards with the 'short' sword,[137] and always thrust toward the upper exposure from them. If he thrusts in simultaneously with you or binds you on the sword, then you should immediately notice whether he is Hard or Soft on the sword, and when you have sensed it, drive the pieces against him which belong against the Soft and against the Hard.

97r | Der text wie man soll fechten Im schwert gegen schwert zu kampff ~~

|Wo man von schaiden, schwert zucken sieht von In baiden, |So soll man stercken Die schutten ~~eben~~ recht eben ~~merken~~ mörcken ·:

Glosa |Daß ist öb sÿ bai~~d~~de die sper verschossen hetenn vnd solten ~~fechten~~ vechten mitt den schwerte~~r~~n |So salt du vor allen dingen mörcken vnd wissen daß die vier huten mitt dem kurczen schwert, vnd daruß stich im allweg zu der obern blöß |Sticht er dann mitt dir glich ein oder bindt dir an dz schwert |So salt du zu hand mercken ob er hert oder waich am schwert ist |Und wenn du enpfunnden haũst |So trÿb die |97v| störck gegen im die du hernach geschriben wirst seehen ~~~ :· ~

Das ist der Text wie man soll vechtenn mit dem schwertt gegenn dem schwert

Wo man vonn schaidenn zv schwert zucken sicht vonn Inn baydenn so sol man strecken die schutten recht ebenn mercken

Glosa Das ist ob wir bayde die Sper verschossen Hettenn vnnd solten vechtenn mit denn schwerttenn So soltu vor allen dingen wissen die vier hutenn mit dem kurtzen schwert vnnd dar auß stich Im albegen zu der obern plos Sticht er denn mit dir gleich ein oder pint dir an das schwert so soltu zu hant merckenn ob er hertt oder waych am schwert ist vnnd wen du das empfunden hast So treib die stuck gegen Im die gegen der waich oder gegen der hert gehorenn

137 As with the 'long' sword, this is the literal translation, but "retracted sword" captures the intended meaning better—the sword is 'shortened' because the hand on the blade prevents it from extending as far. The synonym "half sword" probably refers to the way the hand divides the blade in half.

This is the text on how one should fence with the sword against the sword:
Where one, from scabbards,
Sees that both your swords are drawn,
He should stretch;
Remember precisely to cover correctly.

Gloss: This is if we both have thrown the spears and want to fence with swords: you should know before all things the four guards with the 'short' sword,[137] and always thrust toward the upper exposure from them. If he thrusts in simultaneously with you or binds you on the sword, then you should immediately notice whether he is Hard or Soft on the sword, and when you have sensed this, drive the pieces against him which belong against the Soft and against the Hard.

This is the text on how one should fence with the sword against the sword:
Where one, from scabbards,
Sees that both swords are drawn,
He should strengthen;
Remember precisely to cover correctly.

Gloss: This is if we both have thrown the spears and should fence with swords: you should know before all things the four guards with the 'short' sword,[137] and always thrust toward the upper exposure from them. If he thrusts in simultaneously with you or binds you on the sword, then you should immediately notice whether he is Hard or Soft on the sword, and when you have sensed this, drive the pieces against him which belong against the Soft and against the Hard.

Das Ist der text wie man soll vechten mit dem Schwertt gegen dem Schwertt
Wo man von scheyden zw schwertt zuckenn sicht vonn Ir beyden so sol man streckenn die schutten rechte eben mercken
Glosa Das ist ob wir beyde die Sper verschossenn habend vnd wollen vechten mit den Schwerten So soltu vor allen dingen wissenn die vier huten mit dem kurtzen Schwert vnd darauß stich Im alwegenn zw den[138] obern bloeß Sticht er den mit dir gleich ein oder bind dir an das Schwert so soltu zw hand merckenn ob er hert oder weich am |111r| Schwertt Ist vnd wen dw das Empffunden hast So treib die Stuck gegen Ime die gegen der weich oder gegen der hert gehorenn

Das ist der Text wie man sol fechten mit dem schwert gegen dem schwert.
|107r| Wo mann vonn schaidenn, schwert zuckenn sicht von in baidenn, so sol mann sterckenn, die schutten recht ebenn merckenn.
Glosa, das ist ob wir beide die Sper verschossenn hettenn, vnnd soltenn fechtenn mit denn schwertten so soltu vor allenn dingenn wissenn die vier huttenn mit dem kurtzenn schwert, vnd daraus stich im alwegen zu der obernn plos, Sticht er denn mit dir gleich ein oder pint dir an das schwert, so soltu zu hanndt merckenn, ob er hert oder waich am schwert ist, vnnd wen du das empfundenn hast, so treib die stuck gegenn im, die gegenn der waich oder gegen der hert gehornn.

138 Corrected, possibly from *der*.

16| The first guard in the half sword:
Likewise, hold your sword by the handle with the right hand and grab the middle of the blade with the left, and hold it over your head by your right side and let the point hang downward against the man's face, etc.[139]

Likewise, if he stands against you in the lower guard and wants to thrust toward you underneath, then thrust through down from above, between the sword and his forward hand, and press the pommel downward and wind the point in against his right side, below on the sword, and set upon him.

Likewise, thrust out of the first guard toward his face. If he defends, then pull it or go through with the thrust toward the other side as before, and when you have set upon him, then slam your sword under your right armpit, with the hilt on the chest, and crowd him away from you like this.[141]

Remember the first guard with the half sword here:
Likewise, hold your sword by the handle with the right hand and grab the middle of the blade with the left, and hold it overhead by your right side and let the point hang downward against the man's face.[139]

Likewise remember, if he stands against you in the lower guard and wants to thrust toward you underneath, then ...[140] down from above, between the sword and his forward hand, and press the pommel downward and wind the point in against his right side, below on the sword, and set upon him.

Likewise, thrust out of the first guard toward his face. If he defends widely, then pull it or go through with a thrust toward the other side as before, and when you have set upon him, then slam your sword under your right armpit, with the hilt on the chest, and crowd him away from you like this.[141]

Die erst hut In dem ha[l]ben schwert ~
|Item halt din schwert mitt der rechten hand bÿ der händhäbe vnd mitt der lincken grÿff mitten in die clingen vnd halt es neben diner rechten sÿtten über din haüpt vnd laß den ort vndersich hangen dem man gegen dem gesicht etc
|Item stätt er dann gegen dir In der vndern hut Vnd will dir vnden zu stechen |So stich durch von oben nider zwischen dem schwert vnd siner vorgesäczner hand Und truck den knopff vndersich vnd wind im den ort am schwert vnden durch gegen siner rechten sÿtten vnd secz im an ~~~~~~~~ :· ~
|98r| |Item stich Im vß der ersten hut zu dem gesicht wert ers |So zuck oder ge dürch mitt dem stich alß vor zu der anderen sÿtten |Und wenn du Im haüst angeseczt |So schlach din schwert vnder din rechte uchsen mitt demgehültz an die brüst vnd dring in also von dir hin ~

11v| **Hie merck die erst hut mit dem halben schwert**
Item Halt dein schwert mit der rechten hannt pey der hannthab vnnd mit der lincken greyff mitten In die clingen vnnd halt es nebenn deiner Rechten seytten vber dem haubt vnnd loß den ort vnntersich hangen dem man gegen dem gesicht
Item Merck Stett er dan gegen dir In der vndernn hut vnnd wil dir vnndenn zu stechenn von obenn nyder zwischen dem schwert vnnd seiner vorgesatzten Hannt vnnd truck den knopff vntersich vnd windt Im den ortt am schwert vnden[142] ~~sich~~ durch gegen seiner Rechten seyttenn vnd setz Im ann
Item Stich Im aus der Erstenn hut zu dem gesicht wert ers weit So zuck oder gee durch mit stich als vor zu der andern seyttenn vnd wen du Im hast angesetzt so slag dein schwertt vnnter dein Rechtz vchsenn mit dem gehultz an die prust vnd dring In also von dir her

[139] This is a close match to paragraph 41 of the Pseudo-Danzig gloss; see GF: 88r.2.
[140] The verb in this phrase is missing, though we can assume it was "thrust through" based on SR.

WIEN – AD

Remember the first guard with the half sword here: Likewise, hold your sword by the handle with the right hand and grab the middle of the blade with the left, and hold it over your head by your right side and let the point hang downward against the man's face.[139]

Likewise remember, if he stands against you in the lower guard and wants to thrust toward you underneath, then ...[136] down from above, between the sword and his forward hand, and press the pommel downward and wind the point in against his right side, below on the sword, and set upon him.

Likewise, thrust out of the first guard toward his face. If he defends widely, then pull it or go through with a thrust toward the other side as before, and when you have set upon him, then slam your sword under your right armpit, with the hilt on the chest, and crowd him away from you like this.[141]

ROSTOCK – JMR

Remember the first guard with the half sword here: Likewise, hold your sword by the handle with the right hand and grab the middle of the blade with the left, and hold it over your head by your right side and let the point hang downward against the man's face.[139]

Likewise remember, if he stands against you in the lower guard and wants to thrust toward you underneath, then ...[136] down from above, between the sword and his forward hand, and press the pommel downward and wind the point in against his right side, below on the sword, and set upon him.

Likewise, thrust out of the first guard toward his face. If he defends widely, then pull it or go through with the thrust toward the other side as before, and when you have set upon him, then slam your sword under your right armpit, with the hilt on the chest, and crowd him away from you like this.[141]

Hie merck die Erst hutt mit dem halben Schwertt
Item halt dein Schwertt mit der Rechten hand pey der hand hab vnd mit der linkenn greiff mittenn In die klingenn vnd halt es nebenn deyner Rechten seittenn vber dein haubt vnd los den ortt vntersich hangenn dem man gegenn dem gesicht

Item merck Stett er dan gegen dir In der vndern hutt vnd will dir vnnden zw Stechen von oben nyder zwyschen dem Schwertt vnd seyner vorgesatzten handt vnnd truck den knopff vntersich vnd wind Im den ortt vnden am Schwertt durch gegen seyner Rechten seÿtten vnd setz Im ann

Item Stich Im aus der Ersten hutt zw dem gesicht wert ers weitt So zuck oder gee durch mit Stich als vor zw der andern seytten vnd wen dw Im hast angesetzt So Schlag dein Schwertt vnter dein recht vchssenn mit dem gehultz an die prust vnd dring In also von dir hin

Hie merck die erst hut mit dem halbenn schwert.
Item halt dein schwert mit der rechtenn handt bey der handhab, vnnd mit der linckenn greif mittenn in die clingenn, vnnd halt es neben deiner Rechtenn seittenn, vber dein haupt, vnd las denn ort |107v| vnntersich hangenn, dem man gegen dem gesicht.

Item merck stet er dann gegenn dir in der vntern hut, vnnd wil dir vntenn tzu stechenn vonn obenn nider tzwischenn dem schwert, vnd seiner vorgesatztenn handt, vnnd truck denn knopf vntersich, vnnd wind im denn ort am schwert vntenn durch gegenn seiner Rechtenn seittenn, vnd setz im an.

Item stich im aus der erstenn hut zu dem gesichtt, wert ers weit so tzuck, so tzuck oder gehe durch mit dem stich als vor, zu der anderen seittenn vnnd wenn du im hast angesetzt, so schlag dein schwert vnter dein Rechte vchsenn, mit dem gehiltz an die brust, vnd dring in also von dir hin.

ARMORED FENCING

[141] This is similar to paragraph 2 of Huntsfeld's armored fencing; see GF: 100r.2.
[142] Corrected from *vnder.*

DRESDEN – SR

Likewise, remember another one: thrust toward him as before. If he displaces in front of his left hand with the sword and remains with the point in front of your face and wants to set upon you, then grab his sword by the point with the left hand and hold it fast, and thrust forcefully toward his groin with the right hand. Then, if he pulls his sword to himself firmly and wants to wrench it out of your hand, then suddenly let his sword go, and thus he gives himself an exposure. Quickly grab his sword again with the left hand in the middle of the blade and follow after him, etc.[143]

Likewise, another one: if you grab his sword and he yours, then throw his sword from the left hand and grab yours again in the middle of blade with it, and wind the point in outside over his left hand and set upon him. Or likewise, throw your sword at the feet and grab his left hand with your left and drive the arm break, or otherwise another wrestle, etc.[144]

Likewise, when you thrust toward him out of the upper guard, if he falls on your sword with the left hand in between both your hands, then go in over his left hand with the pommel, inside or outside, and wrench on your right side and set upon him...

NÜRNBERG – WP

Likewise, remember another one: thrust toward him as before. If he displaces from his left hand with the sword and remains with the point in front of your face and wants to set upon you, then grab his sword by the point with the left hand and hold it fast, and thrust forcefully toward his groin with the right. Then, if he pulls his sword to himself firmly and wants to wrench it out of your hand, then suddenly let his sword go, and thus he gives himself an exposure. Quickly grab your sword again with the left hand in the middle of the blade and follow after him.[143]

Likewise, again another one: if you grab his sword...

|Item mörck ain ~~aid~~ anders |Stich im zu alß vor, verseczt er vor siner lincken hannd mitt dem schwert vnd blipt dir mitt dem ort vor dem |98v| gesicht vnd will ~~diner~~ dir anseczen so begrÿff mitt der lincken hand sÿn schwert bÿ dem ort vnd halt daß föst vnd mitt der kerechten hand stich im kröffticlichen zu den gemächten |Zuckt er dann sÿn schwerst vast an sich vnd will dir daß vß der hand rissen |So laß im dz schert, Urbringe faren |So gibt er sich bloß |So begriff sin schwert bald mitt der lincken hand wider mitten in der clingen vnd folg im nach etc

Item ain anders Begrÿffstu sin schwert vnd er das din |So |99r| wirff sin schwert vß der lincken hand Und do mitt begrÿff daß din wider mitten inder clingen vnd wind im den ort ausen über sin lincke hand vnd secz im an ~ ~

|Item oder wirff im din schwertt für die füsß vnd begriff sin lincke hand mit diner lincken vnd trÿb den arm bruch oder sunst ander |Ringen etc

|Item wann du Im vß der obern hut zu stichst fölt er dir dann mitt der lincken hannd in din schwert zwischen dinen baiden henden |So far im mitt dem |97v| ~~kr~~ knopff vsswendig oder Inwendig über sin lincke hannd vnd reÿß vff din rerchte sÿtten vnd secz im an

Item Merck ein anders Stich Im zu als vor versetzt er vonn seiner linckenn hannt mit dem schwert vnnd pleibt dir mit dem ort vor dem gesicht vnnd will |12r| dir ansetzenn So greiff mit der lincken hannt sein schwert pey dem ortt vnnd halt das vest vnnd mit dem Rechten Stich Im krefftiglich zu dem gemecht zuckt er dan sein schwert vast an sich vnnd wil dir das aus der hannt Reyssenn So las Im das schwert vrbering farnn so gibt er sich ploß So begreiff dein schwert pald mit der lincken hannt wider mitten Inn der clingen vnnd volge Im nach

Item Aber ein anders Begreyffstu sein schwert[145]

ARMORED FENCING

[143] This is similar to paragraphs 44–45 of the Pseudo-Danzig gloss; see GF: 88r.4–88v.2.

[144] These two paragraphs are similar to paragraph 3 of Huntsfeld's armored fencing; see GF: 100r.3

WIEN – AD

Likewise, remember another one: thrust toward him as before. If he displaces from his left hand with the sword and remains with the point in front of your face and wants to set upon you, then grab his sword by the point with the left hand and hold it fast, and thrust forcefully toward his groin with the right. Then, if he pulls his sword to himself firmly and wants to wrench it out of your hand, then suddenly let his sword go, and thus he gives himself an exposure. Quickly grab your sword again with the left hand in the middle of the blade and follow after him.[143]

Likewise, again another one: if you grab his sword...

ROSTOCK – JMR

Likewise, remember another one: thrust toward him as before. If he displaces in front of his left hand with the sword and remains with the point in front of your face and wants to set upon you, then grab his sword by the point with the left hand and hold it fast, thrust forcefully toward his groin with the right. Then, if he pulls his sword to himself firmly and wants to wrench it out of your hand, then suddenly let his sword go, and thus he gives himself an exposure. Quickly grab your sword again with the left hand in the middle of the blade and follow after him.[143]

Likewise, again another one: if you grab his sword and he yours, then throw his sword from the left hand and grab yours again in the middle of blade with it, and wind the point in outside over his left hand and set upon him. Or likewise, throw your sword at his feet and grab his left hand with your left and drive the arm break, or otherwise another wrestle.[144]

Likewise, when you thrust toward him out of the upper guard, if he falls on your sword with the left hand in between both your hands, then go in over his left hand with the pommel, inside or outside, and wrench on your right side and set upon him...

Item merck ein anders Stich Im zw als vor versetzt er vonn seiner linkenn hand mit dem Schwertt vnnd pleibt dir mit dem ortt vor dem gesicht vnd will dir ansetzenn So greiff mit der linken hand sein Schwertt bey dem ortt vnd halt es vest vnd mit dem Rechten Stich Im krefftiglich zw dem gemecht zuckt er dan sein Schwertt vast an sich vnnd will dir das auß der hand Reyssenn So las Im das Schwertt |111v| vrbering farn so gibt er sich plos So begreiff dein Schwertt bald mit der linken hand wider mitten In der clingen vnd volge Im nach

Item Aber ein anders begreiffstu sein Schwertt[145]

Item merck ein annders, Stich im tzu als vor, versetzt er vor seiner linckenn handt mit dem schwert vnnd pleibt dir mit dem ort vor dem gesicht, vnd wil dir ansetzenn, so begreif mit der linckenn hanndt sein schwert bey dem ortt, vnnd halt das |108r| fest, vnnd mit der Rechtenn stich im kreftiglichenn tzu dem gemecht tzuckt er dann sein schwert fast ann sich, vnnd wil dir das aus der handt reissenn so las im das schwert vrbering farn, so gibt er sich plos, so begreif dein schwert bald mit der lincken hanndt, wider mittenn in der clingenn, vnnd volge im nach.

Item aber ein anders, begreifstu sein schwert, vnnd er das dein, so wirf sein schwert aus der linckenn hanndt, vnd begreif das dein damit wider mittenn in der clingenn, vnnd wind im denn ort außenn vber sein lincke handt, vnd setz im ann.

Item oder wirf im dein schwert fur sein fus, vnnd begreif sein lincke handt mit deiner lincken, vnnd treib den armpruch, oder sonnst ander ringenn.

Item wenn du im aus der obernn hut zu stichst felt er dir dann mit der linckenn handt in dein schwert, zwischenn deinenn beidenn hennden, so far im mit dem knopf auswendig oder inwendig, vber sein |108v| lincke hanndt, vnnd reiß auf dein Rechte seittenn, vnnd setz im ann,

ARMORED FENCING

[145] Here the text in WP and AD skips several paragraphs before resuming in the middle of a sentence. This is probably due to a missing page in the manuscript the scribe copied.

...You can also strike well out of the upper guard with the pommel, if it's up to you.

17 | **The second guard with the short sword in the duel:**

Remember, hold your sword with both hands and hold it underneath toward your right side with the handle by your right knee, and your left foot is in front and your point against your[!] face of the man.

Likewise, when you stand in the guard like this, if he stands against you in the high guard and wants to set upon you above, then thrust before him and set the point at the exposure of the palm of his forward hand, or thrust through above his forward hand and and[!] your sword, and press your pommel against the ground and set upon him toward the other side, etc.[146]

Likewise, when he thrusts above toward you, grab his sword with the left hand in front of his left, and set your sword with the hilt on your chest with the right, and set upon him, etc.[147]

aüch magst du vsß der obern hut mitt dem knopff wol schlachen wann es dir eben ist ~~

Die ander hutt mitt dem kurczen schwert zu kampff

|Merck halt din schwert mitt baiden henden vnd halt daß vndersich zu diner rechten sÿtten mitt der handhäben neben dinem rechten knÿ vnd dz din lincker fuß vor stee vnd din ort dem man gegen din ~~ge~~ gesicht ~ ~ ~ : ~

|100r| |Item wenn du also steest In der hut |Stet er dann gegen dir in der obern hut vnd will dir oben anseczen |So stich du vor vnd secz im den ort für sin fürgeseczte hand zu der blöß des teners ~~v~~ oder stich im über sin vorgeseczten hand durch vnd vnd din schwert vnd truck dinen knopff gegen der erden vnd secz im an zu der andern sÿtten etc

|Item wann er dir oben zu sticht |So grÿffe mitt der lincken hand sin schwert vor siner lincken vnd mitt der rechten secz ~~I~~ din schwert mitt dem gehülcz |100v| an din brüst vnnd secz im an etc

[146] The final part of this paragraph is a close match to paragraph 29 of Huntsfeld's armored fencing; see GF: 102v.5.

...You can also strike well out of the upper guard with the pommel, if it's up to you.

This is the second guard with the short sword in the duel:
Remember, hold your sword with both hands and hold it underneath toward your right side with the handle by your right knee, and your left foot is in front and the point against the face of the man.

Likewise, when you stand in the guard like this, if he stands against you in the high guard and wants to set upon you above, then thrust before him and set the point in the exposure under his left armpit, or thrust toward the exposure of the palm of his forward hand, or thrust through above his forward hand and the sword, and press the pommel against the ground and set upon him toward the other side.[146]

Likewise, when he thrusts above toward you, grab his sword with the left hand in front of his left, and set your sword with the hilt on your chest with the right, and set upon him.[147]

auch magstu aus der obernn hut mit dem knopf wol schlagen wen̲ es dir eben̲ ist.

Das ist die ander hut mit dem kurtzen schwert zu kampff.
Merck halt dein schwert mit baiden hendenn, vnnd halt es vntersich, tzu deiner Rechten seittenn, mit der handhab nebenn deinem Rechtenn knie vnnd das dein lincker fus vor stehe, vnnd der ort dem man gegen dem gesicht.
Item wenn du also stehest, in der hut, stehet er den gegenn dir in der obernn hut, vnnd wil dir obenn ansetzenn, So stich du vor, vnd setz im den ort, vnter sein lincke vchsenn in die plos, oder stich im tzu seiner vorgesatzenn handt, zu der plos des teners. Oder stich im vber seiner vorgesatzten handt durch, vnnd dem schwert, vnd truck denn knopf, gegenn der erdenn, vnd setz im an zu der andern seitten.
|109ʳ| Item wenn er dir obenn tzu sticht, so begreif mitt der linckenn handt sein schwert vor seiner lincken vnnd mit der Rechtenn setz dein schwert mit dem gehiltz an dein prust, vnd setz im ann.

147 This is a close match to paragraph 56 of the Pseudo-Danzig gloss; see GF: 89ᵛ.2.

18 | A counter against the setting though:

Likewise, when you thrust toward him out of the lower guard, if he thrusts through at you out of the upper guard, between your forward hand and the sword, then remember: while he presses the pommel down, go up into the upper guard and set upon him.[148]

Likewise, thrust toward him out of the lower guard. If he goes through underneath your sword with the pommel and wants to set aside with it, then remain with the point strongly in front of his face and press his right hand down like this while he winds through, and set upon him.[149]

Likewise, thrust strongly toward his face out of the lower guard. If he thrusts in simultaneously with you, then grab his sword in the middle (against yours) with your inverted left hand and hold both fast together, and go through with the pommel under your sword, and pull upward with the right arm on your right side. In this way, you take his sword.[150]

...underneath the sword with the pommel and wants to set aside or wrench with it, then remain with the point strongly in front of the face and press his right hand downward like this while he winds through, and set upon him.[149]

Likewise, thrust strongly toward his face out of the lower guard. If he thrusts in simultaneously with you, then grab his sword in the middle (at yours) with your inverted left hand and hold both fast together, and go through with the pommel under your sword, and pull assuredly with the right arm on your right side. In this way, take his sword.[150]

Ain bruch wider daß durchseczen

|Item wann du im vß der vnder hut zu stichst |Stich^{t}~~s~~ er dir vß der obern hut durch zwischen diner vorgesäcztenn hand vnd dem schwert |So mörck die wil er den knopff nider truckt, so far vff zu der obern hut vnd secz im an ~~

|Item stich im zu vß der vndern hut fert er dann durch mitt dem knopff vnder din schwert vnd will domitt abseczen |So blÿb im mitt dem ort starck vor dem gesicht vnd truck Im sin gerechte hand also vnder |122r|[151] die ~~wil~~ wÿl er durch windt vnd secz im an ~~

|Item stich im starck vß der vndern hut zu dem gesicht |Sticht er dann mitt dir glÿch ÿn |So begrÿff sin schwert in der mitte zu dem dinen mitt lincker ver korter hand vnd halt sÿ baÿde föst zu samen vnd far mitt dem knopff vnden durch sin schwert vnd mitt dem rechten arm rück ubersich vff din rechten sÿtten so nÿmpst du im sÿn schwert ~ ~

vnnter dem schwert vnd will damit absetzen oder Reissenn So pleib mit dem ortt starck vor dem gesicht vnd truck Im sein Rechte hannt also vntersich die weil er durchwindet vnnd setze Im ann

Item Stich Im starck auß der vntern hut zum gesicht Sticht er dann mit dir gleich ein, so begreiff sein schwertt Inn der mitt zu dem deinem mit lincker verkerter hant vnd halt sie bede vest zu samen vnd far mit dem knopff vnder durch sein schwert vnd mit dem rechten Arm ruck vbersich auff dein Rechte seyttenn So nymbst Im das schwertt

[148] This is a close match to paragraph 61 of the Pseudo-Danzig gloss; see GF: 90r.2.
[149] This is a close match to paragraph 68 of the Pseudo-Danzig gloss; see GF: 90v.6.
[150] This is a close match to paragraph 70 of the Pseudo-Danzig gloss; see GF: 91r.2.
[151] The sheet containing ff. 122 & 125 got mixed up when SR was bound, some time before the modern →

Likewise, this is a counter against the thrusting though:
When you thrust toward him out of the lower guard, if he thrusts through at you out of the upper guard, between your forward hand and the sword, then remember: while he presses the pommel down, go up into the upper guard and set upon him.[148]

Likewise, thrust toward him out of the lower guard. If he goes through underneath the sword with the pommel and wants to set aside or wrench with it, then remain with the point strongly in front of his face and press his right hand downward like this while he winds through, and set upon him.[149]

Likewise, thrust strongly toward his face out of the lower guard. If he thrusts in simultaneously with you, then grab his thrusts in the middle (at yours) with your inverted left hand and hold both fast together, and go through with the pommel under your sword, and pull upward with the right arm on your right side. In this way, take his sword.[150]

...underneath your sword with the pommel and wants to set aside or wrench with it, then remain with the point strongly in front of his face and press his right hand downward like this while he winds through, and set upon him.[149]

Likewise, thrust strongly toward his face out of the lower guard. If he thrusts in simultaneously with you, then grab his sword in the middle (at yours) with your inverted left hand and hold both fast together, and go through with the pommel under your sword, and pull assuredly with the right arm on your right side. In this way, take his sword.[150]

Item das ist ein pruch, wider das durchstechen,
wen du im aus der vnternn hut zu stichst, Sticht er dir aus der obernn hut durch, zwischenn deiner vorgesatztenn handt vnnd dem schwert, so merck dieweil er denn knopf nider truckt, so far auf, Inn die obernn hut, vnnd setz im ann.

Item stich im tzu, aus der vnternn hut, fertt er dan durch mit dem knopf, vnter dem schwert, vnnd wil damit absetzenn oder reissenn, so bleib im mit dem ort starck vor dem gesicht, vnd truck im sein Rechte hanndt also vntersich, dieweil er durchwindet, vnd setz im an[152]

|99v| Item stich im starck aus der vntern hut zum gesicht, sticht er dann mit dir geleich ein, so begreif sein schwert In der mit zu dem deinen mit lincke[r] vekerter hanndt, vnnd halt sie baide fest tzusamen vnnd far mit dem knopf vntenn durch sein schwert, vnnd mit dem Rechtenn arm Ruck vber sich auf dein Rechte seiten, so nimbst im das schwert.

vnnter dein Schwertt vnd will damit absetzen oder Reyssenn So pleib Im mit dem ortt Starck vor dem gesicht vnd truck Ime sein Rechte handt also vntersich dieweyl er durch windett vnd setze Ime ann

Item Stich Im Starck auß der vntern hutt zwm gesicht Sticht er dan mit dir gleich ein, so begreiff sein Schwertt In der mitt zw dem deynen mit linker verkerter hand vnd halt sie bede vest zw samen vnd far mit dem knopff vnder durch sein Schwertt vnd mit dem Rechten arm Ruck versich[!] auff dein Rechte seyttenn So nimst Im das Schwertt

foliation was applied. This is their original location (see Chapter 4 for the full collation of SR).
[152] Here the text in JMR suddenly jumps to a different part of the treatise, whereas the text of the treatise continues on JMR: 99r. This is probably due to the pages the scribe copied being out of order.

Likewise, counter it like this: when someone grabs your sword in the middle (against his) with his left hand and wants to wrench it away from you, then remember, while he holds your sword firmly in the left hand against his, go up into the upper guard and set upon him.[153]

Likewise, thrust inside toward his face out of the lower guard. If he displaces, then pull and thrust out toward his face. If he displaces further, then go forward with the pommel over his right shoulder, around the neck, and leap with the right foot behind his left, and pull him over the leg with the pommel so he falls, etc.

Likewise, counter it like this: whoever goes around the neck in front with the pommel and leaps behind your left foot with the right, grab his left hand and press it firmly to the chest, and turn yourself away from him on the right side, and hold him on your left hip and throw him in front of you, etc.

Likewise, you can also strike him well out of the lower guard, if it's up to you.

Likewise, counter it like this: when someone grabs your sword in the middle (against his) with his left hand and wants to wrench it away from you, then remember, while he holds your sword firmly in the left hand against his, go up into the upper guard and set upon him.[153]

Likewise, thrust inside toward his face out of the lower guard. If he displaces, then pull and thrust outside toward his face. If he displaces further, then go forward with the pommel over his right shoulder, around the neck, and leap with the right foot behind his left, and pull him over the leg with the pommel so he falls.

Likewise, counter it like this: whoever goes around the neck in front with the pommel and leaps behind your left foot with the right, grab his right arm with your left hand and press it firmly to your chest, and turn yourself away from him on your right side, and hold him on your left hip and throw him in front of you.

Likewise, you can also strike him also out of the lower guard with the pommel, if it's up to you.

|Item also brich daß Wenn dir ainer mitt siner lincken hand begrÿfft din schwert In der |122ᵛ| mitten zu dem sinen Vnd will dir daß vsßrissen |So mörck die wil er dir daß schwert fasst in die lincken hand zu dem sinen So far vff in die obern hut vnd secz im an ~ ~

|Item stich im vsß der vndern hut in wendig zu dem gesicht, verseczt er |So ~~zuch~~ zuck vnd ~~schu~~ stich im vß zu dem gesicht, verseczt er fürbaß |So far im mitt dem knopff vornen über sin rechte achseln vmb den halß vnd spring mitt dem rechten fuß hinder sin lincken vnd ruck in mitt dem knopff über daß bain so fölt er etc

|125ʳ| |Item also brich daß wer dir mitt dem knopff vornen vmb den halß fört vnd mitt dem rechten fuß springt hinder din lincken |So begrÿff im sin lincke hand vnd truck die fast an die brüst, vnd wend dich von im an die rechten sÿtten vnd fass in vff din lincke hüffe vnd wirff in für dich etc.

|Item och magstu du im vß der vndern hut wol zu schlachen wenn es dir eben ist ~

Item Also prich das wen dir einer mit seiner |12ᵛ| lincken hand begreyff dein schwert In der mit zu seinem vnd wil dir das außreissen So merck die weyl er dein schwert vast In die linck hant zu dem seinen so far auff In die oberhut vnnd setz Im an

Item Stich Im aus der[154] vndern hut In wendig zum gesicht versetzt er so zuck vnnd stich Im aussen zum gesicht versetzt er furpas so far Im mit dem knopff vorn vber sein Rechte achselnn vmb den hals vnnd spring mit Dem Rechtenn fuß hinter sein lincken vnd Ruck Inn mit dem knopff vber das pein so felt er

Item Also prich das wer dir mit dem knopff forn vmb denn hals fert vnd mit dem Rechten fuß springt hinter dein lincken So greiff sein rechten Arm mit deiner lincken hant vnnd truck denn vast an dein prust vnd wendt dich von Im auff dein rechte seitten vnd faß In auff dein lincke huff vnd wurff In fur dich

Item Auch magstu In auß der vnntern hut auch zu schlagen mit dem knopff wen es dir eben ist

[153] This is a close match to paragraph 71 of the Pseudo-Danzig gloss; see GF: 91ʳ.3.
[154] "d" corrected from "v".

Likewise, counter it like this: when someone grabs your sword in the middle (against his) with his left hand and wants to wrench it away from you, then remember, when he holds your sword firmly in the left hand against his, go up into the upper guard and set upon him.[153]

Likewise, thrust inside toward his face out of the lower guard. If he displaces, then pull and thrust outside toward his face. If he displaces further, then go forward with the pommel over his right shoulder, around the neck, and leap with your right foot behind his left, and pull him over the leg with the pommel so he falls.

Likewise, counter it like this: whoever goes around your neck in front with the pommel and leaps behind the left foot with the right, grab his right arm with your left hand and press it firmly to your chest, and turn yourself away from him on your right side, and hold him on your right hip and throw him in front of you.

Likewise, you can also strike him also out of the lower guard with the pommel, if it's up to you.

Item Also prich das wen dir eyner mit seyner linken hand begreifft dein Schwertt In der mitt zw seynem vnd will dir das außreissenn So merck wen er dein Schwertt vast In die linck hand zw dem seynen so far auff In die ober hutt vnd setz Im ann

Item Stich Im auß der vndern hutt In wendig zwm gesicht versetzt er so zuck vnnd Stich Im aussenn zwm gesicht versetzt er furpas so far Im mit dem knopff vorn vber sein rechte achsseln vmb den hals vnd Spring mit deim Rechtenn fus hinter sein linkenn vnd Ruck In mit dem knopff vber das pein So felt er

Item Also prich das wer dir mit dem knopff forn vmb dein hals fert vnd mit dem Rechten fus Springt hinter den linken |112ʳ| So greiff sein rechtenn arm mit deyner linkenn handt vnd truck den vast an dein prust vnd wend dich von Im auf dein ~~prust~~ rechte seitten vnd faß In auff dein Rechte hufft vnd wurff In fur dich
Item Auch magstü In auß der vnntern hutt auch zw Schlagenn mit dem knopff wen es dir Eben Ist

Likewise, counter it like this: when someone grabs your sword in the middle (against his) with his left hand and wants to wrench it away from you, then remember, while he holds your sword firmly in the left hand against his, go up into the upper guard and set upon him.[153]

Likewise, thrust inside toward his face out of the lower guard. If he displaces, then pull and thrust outside toward his face. If he displaces further, then go forward with the pommel over his right shoulder, around the neck, and leap with the right foot behind his left, and pull him over the leg with the pommel so he falls.

Likewise, counter it like this: whoever goes around your neck in front with the pommel and leaps behind your left foot with the right, grab his right arm with your left hand and press it firmly to your chest, and turn yourself away from him on your right side, and hold him on your left hip and throw him in front of you.

Likewise, you can also strike him also out of the lower guard with the pommel, if it's up to you.

Item also prich das, wenn dir einer mit seiner linckenn hanndt begreift dein schwert in der mitt, tzu dem seinenn vnd wil dir das außrayßenn, So merck dieweil er dein schwert fast in der linck[en] handt, zu dem seinen, so far auf in die oberhut vnnd setz im ann.

Item stich im aus der vntern hut inwendig tzum gesicht, versetzt er so tzuck vnnd stich im außenn tzum gesicht, versetzt er furbas, so far im mit dem knopf, vorn vber sein Rechte achseln vmb den hals vnd spring mit dem Rechtenn fus hinter sein lincken, vnd ruck in mit dem knopf vber das |99ᵛ| pain so felt er.

Item also prich das. wer dir mit dem knopf, vornn vmb denn hals fert, vnnd mit dem Rechten fus springt hinter dein linckenn, So begreif sein Rechten arm mit deiner lincken handt, vnd truck den fast an dein brust, vnnd wend dich von im auf dein Rechte seittenn, vnd fas in auf dein lincke huf, vnd wirf in fur dich.
Item auch magstu im aus der vnternn hut auch tzuschlagenn mit dem knopf, wen es dir ebenn ist.

ARMORED FENCING

19| The third guard with the short sword:

Likewise, hold your sword with both hands (as is written previously) and lay it over your left knee, and counter all his pieces with displaces in it, etc.

Likewise, if he thrusts out of the upper guard toward your face, and you set aside the thrust against his right side with the sword in front of your left hand, then go up into the upper guard with the sword and set upon him.[155]

Or likewise, go up with the sword and displace the upper thrust between both of your hands, and go in with the pommel over his forward hand and pull downward with it, and set upon him.[156]

Likewise, go through below with the pommel, over his forward hand, and pull downward with it and set upon him. You can also change through below with the pommel and set aside his thrust.

Likewise, you also find it written hereafter how you should displace the strikes a man strikes out of the third guard with the pommel.

This is the third guard with the short sword in the duel:

Hold your sword with both hands (as is written previously) and lay it over your left knee, and counter all his pieces with displaces out of it.

Likewise, if he thrusts out of the upper guard toward your face, and you set aside the thrust against his right side with the sword in front of your left hand, and go up into the upper guard and set upon him.[155]

Or **likewise,** go up with the sword and and[!]displace the upper thrust between both of your hands, and go in with the pommel over his forward hand and pull downward with it, and set upon him.[156]

Or **likewise,** go through below with the pommel, over his forward hand, and pull downward with it and set upon. You can also change through below with the pommel and set aside his thrust.

Likewise, you also find it written hereafter how you should displace the strikes a man strikes out of the third guard with the pommel.

Die dritt hutt mitt dem kurczen schwert

|Item halt din schwert mitt baÿden henden alß vor geschriben stät vnd leg es über |125v|[157] din linckes knÿ Vnd daruff brich im alle sine stuck mitt verseczen etc

|Item sticht er dir vß der obern hut zu dem gesicht vnd seczt den stich abe mitt dem schwert vor diner lincken hand gegen siner rechten sÿtten |So ~~far~~ far vff mitt dem schwert in die obern hut vnd secz im an ~~

|Item oder far vff mitt dem schwert vnd versecz den obern stich zwischen dinen baiden henden vnd far im mitt dem knopff über sin vor geseczte hand vnd ruck domitt vndersich vnd secz im an ~~

|Item far vnden durch mitt dem knopff über sin vorgeseczte hand vnd ruck domitt |101r| vndersich vnd secz im an |Aüch magst du vnden durch wechlen mitt dem knopff vnd im den stich abseczen ~~~~~ : ~

|Item du vindest och hernach geschriben wie du vß der drÿtten hut die schläg die man schlächt mitt dem knopff verseczen soll

Das ist die dritt hut mit dem kurtzenn schwert zu kempffen

Halt dein schwert mit baiden henden als vorgeschribenn stet vnd leg es vber dein linckes knye vnnd darauss prich Im all |13r| sein stuck mit versetzenn

Item Sticht er dir auß der obern hut zum gesicht vnnd setzt denn stich ab mit dem schwertt vor deiner lincken hant gegen seiner rechten seyttenn vnd far auff In die ober hut vnd setz Im ann

Item Oder far auff mit dem schwert vnd vnnd[!] versetz denn obernn stich zwischen deinen beiden henden vnd far Im mit dem knopff vber sein versatzten hannt vnd Ruck da mit vntersich vnd setz Im ann

Item Oder far vnden durch mit dem knopff vber sein vorgesatzte hant vnd ruck damit vntersich vnnd setze an Auch magstu vnden durch wechselnn mit dem knopff vnd Im denn stich absetzenn

Item Auch vinstu hernach geschriben wie du aus der trittenn hut die schleg die man schlecht mit dem knopff versetzen solt

[155] This is a close match to paragraph 77 of the Pseudo-Danzig gloss; see GF: 91v.5.
[156] This is a close match to paragraph 78 of the Pseudo-Danzig gloss; see GF: 91v.6.
[157] There's a library stamp on the left edge.

WIEN – AD

This is the third guard with the short sword in the duel:
Hold your sword with both hands (as is written previously) and lay it over your left knee, and counter all his pieces with displaces out of it.

Likewise, if he thrusts out of the upper guard toward your face, and you set aside the thrust against his right side with the sword in front of your left hand, and go up into the upper guard and set upon him.[155]

Or likewise, go up with the sword and displace the upper thrust between both of your hands, and go in with the pommel over his forward hand and pull downward with it, and set upon him.[156]

Likewise, you also find it written hereafter how you should displace the strikes a man strikes out of the third guard with the pommel.

ROSTOCK – JMR

This is the third guard with the short sword in the duel:
Hold your sword with both hands (as is written previously) and lay it over your left knee, and counter all his pieces with displaces out of it.

Likewise, if he thrusts out of the upper guard toward your face, and you set aside the thrust against his right side with the sword in front of your left hand, and go up into the upper guard and set upon him.[155]

Or likewise, go up with the sword and displace the upper thrust between both of your hands, and go in with the pommel...[156]

WIEN – AD

Das ist die dritt hutt mit dem kurtzenn Schwertt zw kempffenn
Halt dein Schwertt mitt beiden henndenn als vorgestriben[!] steet vnd leg es vber dein linkes knye vnd daraüß prich Im alle sein Stuck mit versetzenn

Item Sticht er dir aus der obern hutt zwm gesicht vnd setzt den Stich ab mit dem Schwertt vor deyner linken handt gegenn seyner rechten seytten vnd far auff In die obern hutt vnd setz Im an
Item oder far auff mit dem Schwertt vnd versetz den obern stich zwischen deinen beyden hendenn vnd far Im mitt dem knopff vber sein versatzte hand vnd Ruck da mit vntersich vnd setz Im ann

Item auch vinstu hernach geschribenn wie dw auß der dritten hutt die Schleg die man Schlecht mit dem knopff versetzen solt

ROSTOCK – JMR

Das ist die drit hut mit dem kurtzen schwert zu kempffenn.
Halt dein schwert mit beiden henden, als vorgeschribenn stet, vnnd leg es vber dein linckes knie vnd daraus prich im al sein stuck mit versetzen.

Item sticht er dir aus der obernn hut zum gesicht vnnd setzt denn stich ab mit dem schwert vor deiner lincken handt, gegen seiner Rechten seitenn vnd far auf in die oberhut, vnd setz im an.
|100r| Item oder far auf mit dem schwert, vnd setz denn obernstich zwischenn deinen baiden henden, vnd far im mit dem knopf...[158]

ARMORED FENCING

[158] Here the text in JMR suddenly jumps to a different part of the treatise, whereas the text of the treatise continues on JMR: 102r; in the process, it skips several paragraphs. This is probably due to the pages the scribe copied being out of order (and missing a page).

20| The fourth guard with the short sword in the duel:

Likewise, hold the sword with both hands (as is written previously) and hold it with the handle under your right armpit, and set the hilt in front on the right chest, so that the point is against the man. And in all three aforementioned guards, you should come into the guard. Thus, when you thrust toward his exposure and the point is stuck in his harness, always wind your hilt onto your chest and crowd him away from you like this, and do not let him come away from the point. In this way, he can neither thrust nor cut nor strike against you.[159]

Likewise, when you have set upon him, if he is longer when[!] you are, then crowd him away from you so that your point goes up above and is set into the rings well. [160]

But if he is shorter than you are, then let your sword sink down to your right hip and with the pommel at your right side, so that your point is upward and set into his rings as before, and then crowd him in front of you and do not let him come away from the sword.[161]

This is the fourth guard with the short sword in the duel:

Hold the sword with both hands (as is written previously) and hold it with the handle under your right armpit, and set the hilt in front on the chest, so that the point is against the man. And you should come into the guard out of all three guards. Thus, when you thrust toward his exposure and the point is stuck in his harness, always wind your hilt onto your chest and crowd him away from you like this, and do not let him come away from the point. In this way, he can neither thrust nor strike against you.[159]

Remember, all that you want to set upon, set upon the face, or upon the neck, or upon his left shoulder, or under his left armpit; and when you have set upon him, if he is longer than you are, then crowd him away from you there so that your point goes up above and is set into the rings well.[160]

Or **likewise,** if he is shorter when[!] you are, then let your sword sink down to your right hip and with the pommel at your right side, so that your point is upward and set into his rings as before, and then crowd him in front of you and do not let him come away from the sword.[161]

Die vierd hutt mitt dem kurczen schwert zu kampff

|Item halt dz schwert mitt baiden henden alß vor geschriben stät vnd halt es mitt der handhäbe vnder din rechte uchsen vnd secz daß gehülcz vornen an die rechte brüst daß der ort gegen dem man stee |Und in die hut solt du vff allen |drÿen vorgenäpten | 101^{v} | hutten kumen |Also wenn du im stichst zu der blöß vnd dz der ort hafft in dem harnasch |So winde allwegen din gehülcz an din brust vnd dring in also von dir hinweg ~~vnl~~ vnd lauß in nitt von dem ort abkomen |So mag er weder stechen noch hawen noch schlachen ~

|Item wann du im hast angeseczt Ist er dann lenger wann/du bist So dring in also von dir daß din ort übersich vff gee vnd im wol in die ringe geseczt sÿ

|Ist er aber kürczer dann du bist |So lauß din schwert mitt dem knopff zu diner rechten | 102^{r} | sÿtten vnder sich ab sincken bisß vff die rechten hüffe vnd daß din ort übersich stande Und im in die ring wol geseczt sÿ alß vor vnd dring in also für dich vnd lauß in von dem schwert nicht abkomen ~~~~:~

Das ist die vierd hut mit dem kurtzenn schwert zu kempffen

Halt das schwert mit beiden henden als vorgeschribenn stet vnd halt es mit der hanthab vnter dein Rechte vchsen vnd setz das gehultz forn an die prust das der ortt gegenn dem man stee vnnd In die hut soltu auß allenn dreyen hutten kumen | 13^{v} | Also wen du Im stichst zu der ploß vnd das der ortt hafft In dem harnasch So wind albeg dein gehultz an dein prust vnd dring In also vonn dir hinweg vnd las In nicht abkummen vonn dem ortt So mag er weder stechenn noch geschlagenn

merck alles das du wilt ansetzenn das setz an das gesicht oder an den hals oder an sein lincke achsell oder vnter sein lincke vchsen vnd wen du Im hast angesetzt Ist er denn lenger den du pist so dring Inn also von dir hin das dein ortt vbersich auff gee vnd Im wol In die ring, gesatzt sey

Item Oder Ist er kurtzer wen du pist so las dein schwert mit dem knopff zu deiner rechten seittenn vntersich absincken bis auff die rechte huff vnd das dein ortt vbersich stee vnd Im In die Ring woll gesetzt sey als vor vnnd dring In also fur dich vnd las In vonn dem schwert nit abkumen

[159] The second half of this paragraph is a close match to paragraph 84 of the Pseudo-Danzig gloss; see GF: 92^{r}.6.

[160] This is a close match to paragraph 85 of the Pseudo-Danzig gloss; see GF: 92^{v}.1; the first half is also similar to paragraph 24 of Huntsfeld's armored fencing and the second half is similar to paragraph 26; →

This is the fourth guard with the short sword in the duel:
Hold the sword with both hands (as is written previously) and hold it with the handle under your right armpit, and set the hilt in front on the chest, so that the point is against the man. And you should come into the guard out of all three guards. Thus, when you thrust toward his exposure and the point is stuck in his harness, always wind your hilt onto your chest and crowd him away from you like this, and do not let him come away from the point. In this way, he can neither thrust nor strike against you.[159]

Remember, all that you want to set upon, set upon the face, or upon the neck, or upon his left shoulder, or under his left armpit; and when you have set upon him, if he is longer than you are, then crowd him away from you there so that your point goes up above and is set into the rings well.[160]

Or likewise, if he is shorter when[!] you are, then let your sword sink down to your right hip and with the pommel at your right side, so that your point is upward and set into his rings as before, and then crowd him in front of you and do not let him come away from the sword.[161]

Das ist die vierd hutt mit dem kurtzen schwertt zw kempffenn
Halt das Schwertt mit beyden henndenn als vorgestriben[!] steet vnd halt es mit der hand hab vnter dein Rechte vchssenn vnd setz das gehultz forn an die Brust das der |112v| ortt gegenn dem man Stee vnnd In die hutt soltu auß allenn dreien hutten kumen Also wen dw Im Stichst zw der ploes vnd das der ortt hafft In dem harnaschs So wind alweg dein gehultz an dein prust vnd dring In also von dir hinweg vnd las In nicht abkomenn von dem ortt So mag er weder Stechen noch Schlagen
merck alles das dw wild ansetzen das setz an das gesicht oder an den hals oder an sein linke achsseln oder vnter sein linke vchsseln vnd wen dw Im hast angesetzt Ist er den lenger den dw pist so dring In also von dir hin das dein ortt vbersich auff gee vnd Im wol In die ring gesetzt sey
Item Oder Ist er kurtzer wen dw pist So las dein Schwertt mit dem knopff zw deiner Rechten seytten vntersich ab sincken bis auff die Rechten huff vnd das dein ortt vber sich Stee vnd Im In die ring wol gesetzt sey als vor vnnd dring In also fur dich vnd las In von dem Schwertt nit abkomen

see GF: 102r.4 & 102v.2.

161 This is a close match to paragraph 86 of the Pseudo-Danzig gloss; see GF: 92v.2; it is also similar to paragraph 27 of Huntsfeld's armored fencing; see GF: 102v.3.

21 | The text of the Before and After:
62 Before and After, these two things[162]
Correspond[!] surely; learn to leap away.

Gloss: This is that first of all, you should know the Before and the After, when from these two things goes all the art of the duel. Remember to always come forth ere he does in such a way that he must displace, be it with a strike or with a thrust, and as quickly as he binds the sword with the displacement, nimbly drive a piece so he cannot come forth with {his} piece in front of your work, and this is called 'the Before'.

Likewise, remember the After here: it is the counter to all pieces that he drives against you. Learn it like this: if he comes forth such that you must displace, then with the displacement, immediately seek the nearest exposure with the point. In this way, you win the work with your displacement, and this is called 'the After'.

Likewise, you should remember here that in the duel fencing, there should be not more when[!] one step out and one step in. Learn it like this: if he rushes at you such that you cannot come to displace, you should only make one step back with left foot, and see that you set upon him surely with a step forward of the left foot, or grab him with wrestling.

This is the text of the Before and of After:
Before and After, these two things[162]
Gauge surely; learn to leap away.

Gloss: This is that first of all, you should know the Before and the After, when from these two things goes all the art of the duel. Remember to always come forth ere he does in such a way that he must displace, be it with a strike or with a thrust, and as quickly as he binds the sword with the displacement, nimbly drive a piece so he cannot approach with his piece in front of your work, and this is called 'the Before'.

Likewise, remember here what is called 'the After': it is the counter to all pieces that he drives against you. Learn it like this: if he comes forth such that you must displace, then with the displacement, immediately seek the nearest exposure with the point. In this way, you win the work with your displacement, and this is called 'the After'.

Likewise, you should remember here that in the duel fencing, there should be not more than one work and one step in. Believe it like this: if he rushes at you such that you cannot come to displace, you should only make one step back with the left foot, and see that you set upon him surely with a step forward of the left foot, or grab him with wrestling.

Der text von den vor vnd nach
|Vor vnd nach die zwaÿ ding, brieffe[!] wÿßlichen lere mitt ab spring ·:
Glosa |Das ist dz du vor allen sachen solt wissen daß vor vnd daß nach wann vß den zwaÿen dignen ~~gätt~~ gett alle kunst zu kampffe |Doch |So gedenck daß du allweg ~~y for~~ vor |102v| komest ee dann er |Es sÿ mitt ainem schlag oder mitt ainem stich So muß er dir verseczen vnd alß balde alß er mitt der ~~vez~~ versaczung an daß schwert bindt |So trÿbe din stuck behentlich |So mag er vor ~~diner~~ # sinen[163] stucken vor diner arbaÿt nicht komen Und daß haisst daß vor ~~~~~
|Item hie mörck daß nach |Daß sind die/brüch wider alle stuck die er vff dich ~~trÿbt~~ trÿpt |Und daß ver ~~nem~~ nÿm also |kumpt ~~v~~ er vor dz du im verseczen must So such zu hand mitt der versäczüng mitt dem ort die nechsten blöß |So gewinst du |103r| mitt diner versaczüng die arbait Vnnd dz haisst daß nach ~~
|Item hie solt du mörcken daß in dem kampff fechten nicht mer soll sin wann ain abtritt vnd ain zutritt Und daß vernÿm also ~~vberilt er dich~~ überÿlt er dich daß du zu kainer versaczung komen magst |So solt du nur ain tritt zu ruck thon mitt lincken fuß vnd wart wÿßlich daß du im mitt ainem zu tritt deß lincken fuß wider an seczest oder mitt ringen begrÿffest ~~~~~~

Das ist der Text vonn dem vor vnd von nach
Vor vnnd nach die zwey ding pruff weisßlich lere mit abspringk
Glosa das ist das du vor allen sachen solt wissen das vor vnd das nach wen aus den zweien dingenn geet alle kunst...[164] |1v| ...zv kampff Doch so gedenck das dw alweg vor kombst denn er Es sey mit einem schlag oder mit einem ~~seh~~ stich So muß er dir versetzen vnnd als pald als er mit der versatzung ann das schwertt pindt So treib deine stuck behenndiglich So mag er vor deiner arbeit zw seinen stuckenn nit kummen vnd das heist das vor
Item Hie merck was da heist das nach das |2r| sind die pruch wider alle stuck die er auff dich treibt vnnd das vernyme also kompt er vor das dw Im versetzenn must so such zw hannd mit der versatzung mit dem ortt die nachst plos so gewinstu die Arbeit mit deiner versatzung vnnd das haist das nach
Item Hie soltu merckenn das Inn dem kampff vechtenn nicht mehr sol sein denn ein abtrit vnnd ein zw tryt vnnd das vernyme also vbereilt er dich das du zu keiner versatzung komenn magst So soltu newer ein trit zu ruck thun mit dem linckenn fuß vnnd wartt weislich das du Im mit einem zw drytt des linckenn fuß wider ansetzest oder begreiffest mit ringenn

[162] This line is the same as the first line of long sword couplet 17, in the common lesson.
[163] *sinen* has been added to the edge of the page with an insertion mark.

Wien – AD

This is the text of the Before and of After:
Before and After, these two things[162]
Gauge surely; learn to leap away.

Gloss: This is that first of all, you should know the Before and the After, when from these two things goes all the art of the duel. Remember to always come forth ere he does in such a way that he must displace, be it with a strike or with a thrust, and as quickly as he binds the sword with the displacement, nimbly drive a piece so he cannot approach with his piece in front of your work, and this is called 'the Before'.

Likewise, remember here what is called 'the After': it is the counter to all pieces that he drives against you. Learn it like this: if he comes forth such that you must displace, then with the displacement, immediately seek the nearest exposure with the point. In this way, you win the work with your displacement, and this is called 'the After'.

Likewise, you should remember here that in the duel fencing, there should be not more than one work and one step in. Believe it like this: if he rushes at you such that you cannot come to displace, you should only make one step back with the left foot, and see that you set upon him surely with a step forward of the left foot, or grab him with wrestling.

Das ist der text vonn dem vor vnd von Nach
Vor vnnd nach die zwey ding pruff weißlich lere mit abspring
Glosa das Ist das dw von allen Sachenn solt wissenn das vor vnd das nach wen auß denn zweyenn dingen gett alle kunst...[164] |105ᵛ| ...zw kampff Doch So gedenck das dw alweg vor kompst den er es sey mitt eynem Slag oder mit eynem Stich So muß er dir versetzen vnnd als bald er mit der versatzung an das Swertt bindt So treib dein stuck behendiglich So mag er vor deyner arbeitt zw seynen Stucken nit komen vnd das heist das vor
Item hie merck was da heist das nach das sind die bruch wider alle Stuck die er auff dich treibt vnnd das vernim also kompt er vor das dw Ime versetzen must So such zw hand mit der versatzung mit dem ortt die nachst Blos so gewinstu die arbeitt mit deyner versatzung vnd das heist das nach
Item hie soltu merckenn das In dem kampff vechten nicht mer soll sein denn ein arbeitt vnd ein zw tritt vnd das vermein also vbereilt er dich das dw zw keyner versatzung komen magst So soltu newer Ein tritt zw Ruck thün mit dem linkenn fus vnd wartt weyßlich das dw Ime mit eynem zw tritt des linkenn fus wider ansetzest oder begreiffest mitt Ringen

Rostock – JMR

...of the duel. Remember to always come forth ere he does in such a way that he must displace, be it with a strike or with a thrust, and as quickly as he binds the sword with the displacement, nimbly drive a piece so he cannot approach with his piece in front of your work, and this is called 'the Before'.

Likewise, remember here what is called 'the After': it is the counter to all pieces that he drives against you. Learn it like this: if he comes forth such that you must displace, then with the displacement, immediately seek the nearest exposure with the point. In this way, you win the work with your displacement, and this is called 'the After'.

Likewise, you should remember here that in the duel fencing, there should be not more than one step out and one step in. Learn it like this: if he rushes at you such that you cannot come to displace, you should only make one step back with the left foot, and see that you set upon him surely with a step forward of the left foot, or grab him with wrestling.

|102ʳ| ...zu kampf, doch so gedennck das du alweg vor kumbst es den er es sey mit einem schlag oder mit einem stich, so mus er dir versetzenn, vnnd als bald er mit der versatzung an das schwert bindt, so treib deine stuck behendiglich so mag er vor deiner arbeit tzu seinenn stuckenn nit kummen, vnd das heist das vor.
Item hie merck was da heist das nach das sind die prüch wider alle stuck die er auf dich treibt, vnnd das vernim also, kumbt er vor das du im versetzen must, so such tzu handt mit der versatzung mit dem ort die nechst plöß, so gewinstu die arbeit mitt deiner versatzung, vnd das heist das nach.
Item hie soltu merckenn, das in dem kampfechtenn nicht mer sol sein den ein abdrit, vnd ein tzu trit, vnnd das vernim also. Ubereilt er dich das du tzu keiner versatzung kummen magst, so soltu nur ein drit zu Ruck thun mit dem lincken fus, vnn[d] Wart weislich das du im mit einem tzu trit des |102ᵛ| linckenn fus, wider ansetzest oder begreiffest mitt ringenn.

164 Here the Ainring section ends in WP and AD, but the remainer of the treatise continues on WP: 1ᵛ & AD: 105ᵛ. This is probably due to the pages the scribe copied being out of order.

22 | The text of the pursuit with the sword in the duel:
63 Follow in all meetings
If you want to meet the strong.[165]
64 If he defends, pull back,
Stab; if he defends, pull forward.
65 If he seems extended,
You are artfully prepared.

Gloss: This is the pursuit. You should drive it against the strong that fence long and wide, and otherwise don't hold well to the correct art.

Likewise, arrange yourself against him like this: lay yourself against him in a guard; if he also lays himself in one, then remember precisely: when he pulls his sword to himself, wanting to thrust, or rises up and wants to strike, quickly follow after him toward the nearest exposure with the point ere he brings the thrust or the strike. If he sees the setting upon and he displaces widely with the sword but seeks no exposure with the point, then pull and through again, and always drive the pulling as often as he goes after the sword with a displacement. With this, you also come to the arm breaks and to other breaks, and this is the art against all those that fence long and wide, toward the sword and not toward the man.

This is the text of the pursuit with the sword in the duel:
Follow in all meetings
If you want to confound the strong.[165]
If he defends, pull back,
Stab; if he defends, pull forward.
If he fences extended,
You are artfully prepared.

Gloss: Remember the pursuit. You should drive it against the strong that fence long and wide, and otherwise don't hold well to the correct art.

Likewise, arrange yourself against him like this: lay yourself against him in a guard; if he also lays himself in one, then remember precisely: when he pulls his sword to himself and wants to thrust, or rises up and wants to strike, quickly follow after him toward the nearest exposure with the point ere he stabs or brings the strike. If he sees the setting upon and he displaces widely with the sword but seeks no exposure with the point, then pull and through again, and always drive the pulling as often as he strikes after the sword with the displacement. With this, you also come to the arm breaks and to other pieces, and this is the art against all those that fence long, toward the sword and not toward the man.

103^{v} | Der text von Nachraÿsen mitt dem schwert zu kampff

|Volge allen treffen, den starcken wilt du sÿ treffen Wört er so zucke, stich wert er zu im rucke |Öb er langk sicht, So bÿß du künstlich bericht :·

Glosa |Daß ist die nachraÿsen solt du ~~tr~~ trÿben gegen den starcken die do lang vnd wÿt ~~fachten~~ fechten vnd wöllend sunst von rechter kunst nicht halten ~

|Item gegen den schick dich also |Leg dich gegen im in ain hut Legt er sich dann och in aine |So mörck eben wann | 104^{r} | er sin schwert an sich zücht, will stechen oder vff hept vnd will schlachen |So folg im ~~bald~~ bald nach mitt dem ort zu der nechsten blösse |Ee wenn er den stich oder den schlag verbringt Wirt er dann des anseczens gewar, vnd verseczt aber wÿt mitt dem schwert vnd sucht an dir kain blösß mitt dem ort |So zuck aber durch vnd daß zuchen trÿb allwegen alß offt er dir mitt versäczung nach dem schwert fert, Do mitt so ~~kumst~~ kompst du och zu den arm brüchen | 104^{v} | vnd zu andren brüchen Unnd daß ist die kunst wider alle die, die do lang vnd wÿt fechten zu dem schwert vnd nicht zu dem man ~~~~~ :· ~

Das ist der text von nach reyssen mit dem schwertt zu kempffen

Volg allenn treffenn, denn starcken wiltu sy effenn, wertt, er so zuck stich wertt er io zu Im ruck ob er langk ficht So bistu kunstlich bericht

Glosa Merck die nachreissenn soltu treibenn gegenn den starckenn die da langk vnd weit vechtenn vnnd wollenn sunst von rechter kunst nit haltenn

| 2^{v} | **Item** Gegenn dem schick dich also leg dich gegenn Im In ein hut: legt er sich dan auch In ein So merck ebenn wen er sein schwert an sich zeucht vnnd wil stechen oder auff hebt vnnd wil schlahen so volg Im pald nach mit dem ortt zu der nachsten plose Ee wenn er denn stich oder den schlach verpringt wirt er den des[166] ansetzens gewar vnnd versetzt aber weyt mit dem schwert vnnd sucht an dir kein ploß mit dem ort So zuck aber durch vnnd das zuckenn treib alweg Als offt er dir mit versatzung nach dem schwert seit Damit kumpstu auch zw denn Arm pruchenn vnd zw andernn stuckenn vnnd das ist die kunst wider alle die da langk vechten zu dem schwert vnnd nicht zu dem mann

[165] This couplet is very close to long sword couplet 35, in the section on the wrathful cut, and 90, in the section on the overrunning. The piece described is nearly identical to the First Pursuit in the long sword, even though the verb used is '*nachfolgen*=follow' rather than '*nachreisen*=pursue'.
[166] Corrected from *den*.

This is the text of the pursuit with the sword in the duel:
Follow in all meetings
If you want to confound the strong.[165]
If he defends, pull back,
Stab; if he defends, pull forward.
If he fences extended,
You are artfully prepared.

Gloss: Remember the pursuit. You should drive it against the strong that fence long and wide, and otherwise don't hold well to the correct art.

Likewise, arrange yourself against him like this: lay yourself against him in a guard; if he also lays himself in one, then remember precisely: when he pulls his sword to himself and wants to thrust, or rises up and wants to strike, quickly follow after him toward the nearest exposure with the point ere he stabs or brings the strike. If he sees the setting upon and he displaces widely with the sword but seeks no exposure with the point, then pull and through again, and always drive the pulling as often as he strikes at the sword with a displacement. With this, you also come to the arm breaks and to other pieces, and this is the art against the long fencing toward the sword and not toward the man.

Das Ist der text von nach Reyssen mit dem Swert zw kempffen
|106ʳ| Volg allem treffen, denn Starcken wiltw sie effen wertt er So zuck stich wertt er io zw Im Ruck Ob er lanck ficht So bistu kunstlich bericht
Glosa Merck die nachreÿsen soltẅ treibenn gegen den Starcken die da lanck vnd weitt vechten vnnd wollen sunst von rechter kunst nit halten
Item Gegen dem Schick dich also, leg dich gegen Ime In ein Hut legt er sich dan auch In ein, So merck eben wen Er sein swertt an sich zeucht vnd will Stechenn oder auff hebt vnnd will Schlagen, so volg Im bald nach mit dem ortt zw der nechsten plose Ee wen er Sticht oder den Slag verbringt Wirt Er dan des ansetzens gewar vnd versetzt aber weytt mit dem Swertt, vnd sucht an dir kein bloß mit dem ort So zuck aber durch vnd das zücken treib alweg, als offt er dir mit versatzung nach dem Swert sleitt damit kümstü auch zw denn arm bruchenn vnd zw andern Stückenn vnd das Ist die kunst ~~die~~ wider die lanck vechten zw dem Swertt vnd nicht zw dem mann

This is the text of the pursuit with the sword in the duel:
Follow in all meetings
If you want to confound the strong.[165]
If he defends, pull back,
Stab; if he defends, pull forward.
If he fences extended,
You are artfully prepared.

Gloss: Remember the pursuit. You should drive it against the strong that fence long and wide, and otherwise don't hold well to the correct art.

Likewise, arrange yourself against him like this: lay yourself against him in a guard; if he also lays himself in one, then remember precisely: when he pulls his sword to himself and wants to thrust, or rises up and wants to strike, quickly follow after him toward the nearest exposure with the point ere he stabs or brings the strike. If he sees the setting upon and he displaces widely with the sword but seeks no exposure with the point, then pull and through again, and always drive the pulling as often as he goes after the sword with the displacement. With this, you also come to the arm breaks and to other pieces, and this is the art against all those that fence long and wide, toward the sword and not toward the man.

Das ist der Text vonn Nachreissen mit dem schwert tzu kampf.
Folg allen treffenn, den starcken wiltu sie effenn Wert er so tzuck, stich wert er Jo zu im ruck, ob er lanngk ficht, so bistu kinstlich bericht.
Glosa. Merck die Nachraysenn soltu treiben gegen denn starckenn, die da lang vnnd weit fechtenn, vnnd wollenn sonst vonn Rechter kunst nit halten,
Item gegenn denn schick dich also, leg dich gegen Im in ein hut, legt er sich den auch in ein, so merck ebenn, wenn er sein schwert an sich tzeucht, vnnd wil stechenn oder aufhebt, vnnd wil schlagen so folg im bald nach mit dem ort, tzu der nechsten ploß ee wenn er denn stich oder den schlag verpringt, Wirt er denn des ansetzens gewar, vnnd versetzt aber weit mit dem schwert, vnnd sucht ann dir kein ploß mit dem ort, so tzuck aber durch, vnnd das tzuckenn treib alweg, als oft er dir mit der |103ʳ| versatzung nach dem schwert fert, damit kumbstu au[ch] tzu denn armpruchenn vnd tzu andern stucken, vnnd das ist die kunst, wider alle, die da lang vnd weit fechtenn tzu dem schwert, vnd nicht tzu dem man...[167]

[167] Here the text in JMR suddenly jumps to a different part of the treatise, whereas the text of the treatise continues on JMR: 101ʳ. This is probably due to the pages the scribe copied being out of order.

23| The text of the setting upon:
66 If he also grabs on strongly,
The shooting in wins him.

Gloss: This is when someone has set upon you and crowds you: stab him in the palm of the hand by which he holds the sword in the middle. Or, if he has turned the hand around, stab down from above in toward the same guard again, or stab at his arm behind at the back of his glove, and run forward when the thrust sticks, so you win on his side and another great advantage. Or stab through down from above, over his forward hand and your sword, and set your hilt on your chest and set upon him, etc.

Likewise, if he has set upon your left shoulder, then step backward with the left foot so his point goes off and yours sticks. You can also extend your sword when you pull your pommel to your chest with the wrestling.

This is the text of the setting upon:
If he also grabs on strongly,
The shooting in defeats him.

Gloss: Remember, this is when someone has set upon you and crowds you: stab him in the palm of the hand by which he holds the sword in the middle. Or, if he has turned the hand around, stab down from above in toward the same hand again, or stab at his arm behind in toward the back of his glove, and run forward when the thrust sticks, so you win on his side and another great advantage. Or stab through down from above, over his forward hand and the sword, and set your hilt on your chest and set upon him, etc.

Likewise, if he has set upon your left shoulder, then step backward with the left foot so his point goes off and yours sticks. You can also extend your sword when you pull your pommel to your chest with the crowding.

Der text von anseczen
|Grÿfft er och starcke an, daß schiessen sigt im an :·
Glosa |Das ist ~~wenn~~ wenn dir ainer hat angeseczt vnd dich dringet |So stich In in den tener der hand do er daß schwert in der mitti bÿ helt oder hat er die hand vmb gewändt |So stich von oben nÿder aber in die selbigen hut |Oder |105r| stich ~~v~~ im vff sinen arm hinden in den händschuch Und wann der stich hafft So lauff für dich So gewinst du im die sÿtten an vnd sunst ain groß vortel |Oder stich im durch über sin vorgesaczten hand vnd din schwert von oben nider vnd secz din gehülcz an din brűst vnd secz im an etc

|Item hat er dir dann angeseczet an din lincke achsel So tritt zu rucke mitt dem hncken fuß, So gatt sin ort abe |Vnd der din haffte |Auch magst du din schwert erlengen wann du mitt dem ringen dinem |105v| knopff für dich zuckest an din brűst ~ ~ ~ ~ :· ~ ~ : ~

Das ist der text vonn ansetzenn
Greyfft er auch sterck an das schiessen gesigt Im ann
Glosa Merck wenn dir ainer hatt angesetzt vnnd dich dringkt So stich Im In den tener der hant do er das schwert In der mit bey helt Oder hat er die hanndt vmb gewannt, so stich vonn obenn nider aber |3r| Inn die selbig hanndt oder stich Inn auff seinenn Arm hindenn In den hanttschuch vnnd wan der stich hafft so lauff fur dich so gewinstu Im die Seytten ann vnd sust ain groß fortail Oder stich In durch vber seiner vngesatztenn hannt vnnd dem schwert vonn obenn nyder vnnd setz dein gehultz ann dein prust vnd setz Im ann

Item Hat er dir denn angesetzt an dein Lincke achselnn so tritt zw Ruck mit dem linckenn fuß So get sein ort an vnd der dein der hafft Auch magstu dein schwert erlengenn wenn dw mit dem dringenn deinenn knopff fur dich Ruckst an dein prust

WIEN – AD

This is the text of the setting upon:
If he also grabs on strongly,
The shooting in defeats him.

Gloss: Remember, this is when someone has set upon you and crowds you: stab in toward the palm of the hand by which he holds the sword in the middle. Or, if he has turned the hand around, stab down from above in toward the same hand again, or stab at his arm behind at the back of his glove, and run forward when the thrust sticks, so you win on his side and ~~an~~-other {great} advantage. Or stab through down from above, over his inset hand and the sword, and set your hilt on your chest and set upon him.

Likewise, if he has set upon your left shoulder, then step backward with the left foot so his point goes off and yours sticks. You can also extend your sword when you pull your pommel to your chest with the crowding.

ROSTOCK – JMR

This is the text of the setting upon:
If he also grabs on strongly,
The shooting in defeats him.

Gloss: Remember, this is when someone has set upon you and crowds you: stab him in the palm of the hand by which he holds the sword in the middle. Or, if he has turned the hand around, stab down from above in toward the same hand again, or stab at his arm behind in toward the back of his glove, and run forward when the thrust sticks, so you win on his side and another great advantage. Or stab through down from above, over his forward hand and the sword, and set your hilt on your chest and set upon him, etc.

Likewise, if he has set upon your left shoulder, then step backward with the left foot so his point goes off and yours sticks. You can also extend your sword when you pull your pommel to your chest with the crowding.

WIEN – AD

Das Ist der text vom Ansetzenn
Greifft Er auch sterck an, das schiessen gesigt Im an
Glosa Merck wenn dir eyner hatt angesetzt vnd dich druckt, So stich In den tener der hand do Er das Swertt In der mit behelt oder hatt Er die hand vmbgewant so Stich von oben nider aber In die selbig hand, oder stich In auff seynen arm |106ᵛ| Hinden an der hantschüch vnd wan der stich hafft, So lauff fur dich so gewinstu Ime die Seyten an vnnd süst **grossen**[168] ~~ain~~[169] forteill Oder stich In durch vber seyner vngesatztenn hanndt vnd dem Swert von oben nider vnd setz dein gehultz an dein prust vnnd setz Ime ann

Item Hatt er dir denn angesetzt an dein lincke achßelnn so trit zw Rück mit dem lincken fus So get sein ort ab vnd der dein der hafft aüch magstü dein Swert Erlengenn wen dw mit dem dringen deinenn knopff fur dich Ruckst an dein brust

ROSTOCK – JMR

101ʳ| **Das ist der Text vonn ansetzenn.**
Greift er auch starck an, das schießen gesigt im an.
Glosa. Merck wenn dir einer hat angesetzt, vnnd dich dringt, so stich in in den tener der handt, da er das schwert In der mitt bey helt, Oder hat er die handt vmbgewant, so stich vonn obenn nider aber in die selbig hanndt. Oder stich in auf seinenn Arm, hintenn in denn handtschuch, vnnd wen der stich hafft, so lauf fur dich, so gewinstu im die seitten, an, vnd sonst ein groß vortail. Oder stich in durch vber seiner vorgesatzenn hand vnnd dem schwertt vonn oben nider, vnd setz dein gehiltz an dein brust vnnd setz im ann.[170]

Item hat er dir denn angesetzt, an dein lincke achselnn so trit tzu Ruck mit dem lincken fus, so gehet sein ort ab, vnd der dein hafft. Auch magstu dein schwer[t] erlengenn wenn du mit dem dringenn, deinenn knopf fur dich ruckst an dein prust.

ARMORED FENCING

[168] Added in darker ink in the margin.
[169] This word is present in all copies, but for some reason was struck out in this one.
[170] Illegible deletion.

24 | The text, how one should displace strikes:
67 With the bashing point,
He covers himself and meets without fear.
68 With both hands,
Learn to turn the point toward the eyes.

Gloss: Remember, the pommel is the bashing point. If he wants to overrun you with strong blows with it, then hold your sword over your left knee and in the guard. If he strikes toward your head and is a strong man, then sweep the strike away against his right side with the sword in front of your left hand, and go up on the sword into the upper guard. Or if he is weaker than you are, then step in enthusiastically and catch the strike on your sword between both hands, and set the point into his face.[171]

Likewise, when you catch the strike with the pommel in the middle of your sword, go over his sword with the pommel, in front by the hilt, and pull upward with it on your right side. In this way, you take his sword.

This is the text, how one should displace:
With the bashing point,
He covers himself and meets without fear.
With both hands,
Learn to turn the point toward the eyes.

Gloss: Remember the pommel—the bashing point. If he wants to overrun you with strong blows with it, then hold your sword over your left knee and stand in the guard. If he strikes toward your head and is a strong man, then sweep the strike away against his right side with the sword from your left hand, and go up on the sword into the guard again. Or if he is weaker than you, then step in boldly and catch the strike on the sword between both hands, and set the point into his face.[171]

Likewise, when you catch the strike with the pommel in the middle of your sword, go over his sword with the pommel, in front by the hilt, and pull downward with it at your right side. In this way, you take his sword.

Der text wie man die schlege verseczen soll ~
|Mitt sinem schlachenden ort, schüczt er sich trifft one forcht Mitt baiden henden, den ort zu den augen lere wenden ·:~
Glosa |Mörck der knopff ist der schlaehent ort, Will er dich domitt überlauffen mitt starcken schlegen |So halt din schwert über din lincke knÿ vnd in der hut |Schlecht er dir dann zu dem haupt Vnd ist ain starcker manne |So streÿch im den schlag ab mitt dem schwert vor diner lincken hand | 106r | gegen siner rechten sÿtten, Und far vff am schwert in die öbern hut |Oder ist er schwecher dann du bist |So tritt im frischlich ein vnd fach den schlach zwischen baÿden henden in din schwert vnd secz im den ort Inn daß gesicht ~ ~
|Item wenn du den schlag mitt dem knopff fächst mitten in din schwert so far mitt dem knopff über sin schwert vornen bÿ dem gehülcz vnd ruck domit übersich vff din rechte sÿtten |So nÿmst du Im ~~im~~ im sin schwert ~ ~

Das ist der text wie man die versetzen sol
Mit Seinem Slagennden ortt schutzt er sich trifft anforcht mit beyden hendenn, denn ortt zw denn Augenn lere wendenn
Glosa Merck der knopff der schlagend ort wil er dich da mit vberlauffenn mit starckenn schlegenn So halt dein schwert | 3v | vber dein lingkes knye vnnd stee Inn der hut Schlecht er dir denn zw dem haubt vnd Ist ein starck mann So streich Im denn schlag ab mit dem schwert vonn deiner linckenn hannt gegen seiner Rechtenn seyttenn vnnd far auff am schwertt Inn die aber hut oder Ist er schwecher denn dw so trytt im kunlich ein vnnd vach denn Schlag zwischenn bedenn henndenn Inn dem schwertt vnnd setz Im denn ortt Inn das gesicht
Item wenn dw denn schlag mit dem knopff vechst mittenn Inn dein schwert so var mit dem knopff vber sein schwert vorn bey dem gehultz vnnd Ruck da mit vntersich auff dein Rechte seytten So nymbstu Im sein schwertt

[171] This is similar to paragraphs 100 and 102 of the Pseudo-Danzig gloss; see GF: 94v.1 & 3.

This is the text, how one should displace:
With the bashing point,
He covers himself and meets without fear.
With both hands,
Learn to turn the point toward the eyes.

Gloss: Remember the pommel—the bashing point. If he wants to overrun you with strong blows with it, then hold your sword over your left knee and stand in the guard. If he strikes toward your head and is a strong man, then sweep the strike away against his right side with the sword from your left hand, and go up on the sword into the guard again. Or if he is weaker than you, then step in boldly and catch the strike on the sword between both hands, and set the point into his face.[171]

Likewise, when you catch the strike with the pommel in the middle of the sword, go over his sword with the pommel, in front by the hilt, and pull downward with it at your right side. In this way, you take his sword.

This is the text, how one should displace strikes:
With the bashing point,
He covers himself and meets without fear.
With both hands,
Learn to turn the point toward the eyes.

Gloss: Remember the pommel—the bashing point. If he wants to overrun you with strong blows with it, then hold your sword over your left knee and stand in the guard. If he strikes toward your head and is a strong man, then sweep the strike away against his right side with the sword from your left hand, and go up on the sword into the upper guard. Or if he is weaker than you, then step in boldly and catch the strike on your sword between both hands, and set the point into his face.[171]

Likewise, when you catch the strike with the pommel in the middle of the sword, go over his sword with the pommel, in front by the hilt, and pull downward with it at your right side. In this way, you take his sword.

Das Ist der text wie man die versetzen sol
Mitt Seinem Slaendenn ortt, schützt Er sich trifft on forcht mit beyden hendenn denn ortt, zw den augen ler wenden

Glosa Merck der knopf der Slagend ortt will er dich damit vberlawffenn mit Starckenn slegenn So halt dein Schwert vber dein linkes knie vnd stee In der hutt Slecht er dir dann zw dem hawbt vnd Ist ein Starck man So streich Im den Slag ab mit dem schwert von deyner linken hand gegen seyner Rechten Seyten vnd far auff am Schwert In die aber hutt oder Ist er Schwecher denn dw so tritt Im kunlich Ein vnd vach den Schlag zwischen beyden henden In dem Schwertt vnd setz Im den ort In das gesicht

Item wenn dw denn schlag mit dem knoppff vechst mitten In dem Swertt so var mit dem knopff vber sein Swertt |107r| vorn bey dem gehultze vnd Rucke damit vnter sich auff dein Rechte Seytenn So Nimstu Im sein Swertt

Das ist der Text wie mann die schleg Versetzenn sol.
|101v| Mit seinem schlagendenn ort, schutzt er sich trif one forcht, mit baiden henden, denn ort tzu den augen lere wendenn.

Glosa. Merck der knopf der schlagendt ort, will er dich damit vberlauffenn mit starcken schlegen, so halt dein schwert vber dein linckes knie, vnd stee in der hut, schlecht er dir denn tzu dem haupt, vnd ist ein starck mann, so streich im den schlag ab, mit dem schwert vonn deiner linckenn handt, gegenn seiner Rechtenn seittenn, vnnd far auff am schwert in die oberhut. Oder ist er schwecher dann du, so trit im künlich ein, vnd fahe denn schlag tzwischenn baiden henden in dein schwertt vnd setz im den ort in das gesicht.

Item wenn du denn schlag mit dem knopf fechst mittenn in dem schwert, so far mit dem knopf vber sein schwert vornn bey dem gehiltz, vnnd ruck damit vntersich auf dein Rechte seittenn, So nimbstu im sein schwert.

Likewise, if he strikes toward the pommel with your left knee, then catch the strike on the sword between your hands (so the pommel hangs toward the ground), and go through with the pommel below his sword by the hilt and pull upward on your right side. In this way, you pull his sword from his hands.

Or likewise, if he strikes with the pommel and toward your foot, then throw your sword against his strike, with the pommel at your left side toward the ground, and leap toward him with it, and look for the wrestling.

Likewise, when you hold your sword by your right side in the lower guard, if he strikes toward your point with the pommel and lets it run widely around, then meanwhile, leap boldly toward him in such a way that he strikes over you with the pommel so it does not harm you, and set upon him.

Likewise, you can also pursue and set upon him wherever you want while he lets his pommel run around widely.

Likewise, if he strikes toward your left knee with the pommel, then catch the strike on the sword between your hands (so the pommel hangs toward the ground), and go through with the pommel below his sword by the hilt and pull upward on your right side. In this way, you pull his sword from his hands.

Or **likewise,** if he strikes below toward your foot with the pommel, then throw your sword against his strike, with the pommel at your left side toward the ground, and leap toward him with it, and look for the wrestling.

Likewise, when you hold your sword by your right side in the lower guard, if he strikes toward your point with the pommel and lets it run widely around, then meanwhile, leap boldly toward him in such a way that he strikes over you with the pommel so it does not harm you, and set upon him.

Likewise, you can also pursue and set upon him wherever you want while he lets his pommel run around widely.

|Item schlecht er dir zu dem |106ᵛ| kopff mit dem lincken knÿ So fach den schlag zwischen dinen henden In daß schwert daß der knopff zu der erden hang |Und far mitt dem knopff vnden durch sin schwert vornen bÿ dem gehülcz vnd ruck übersich vff din rechten sÿtten So ~~ruck~~ ruckst im sin schwert vsß den henden ~ ~

|Item oder schlecht ~~er~~ er dir mitt dem knopff, vnd nach dem fuß wirff din schwert mitt dem knopff zu diner lincken sÿtten in die erden gegen sinen schlag vnd spring do mitt zu im Unnd wart der |107ʳ| ringen ~ ~

|Item wann du häst din schwert neben diner rechten sÿtten In der vndern hut |Schlecht er dir dann mitt dem knopff nach dem ort vnd lausst in wÿt vmb dich lauffen |So spring die wil künlich zu im |So schlecht er mitt dem knopff überdich daß ist dir nicht schädlich Vnd secz Im an ~~

|Item och magst du Im nach raÿsen vnd an seczen wo du wilt die wil er den knopff also wÿt lasst vmb sich lauffen ~~~ ~

Item Schlecht er dir mit dem knopff zu dem lynckenn knye So vach denn schlag zwischenn deinen henden In das schwertt das der knopff zu der Erden Hange vnnd far mit dem knopff vnden durch sein schwertt forn pey dem geHultz vnnd Ruck vbersich auff dein rechte seytenn So ruckestu Inn seynn |6ʳ| [172] schwertt aus denn Henndenn

Item Oder schlecht er dir mit dem knopff vnden noch dem fuß So wurff dein schwert mit dem knopff zu deiner linckenn seyten Inn die erdenn gegenn seinen schlag vnnd Spring da mit zu Im vnnd wart der ringen

Item wenn du hast dein schwertt neben deiner rechtenn seyttenn Inn der vndern hutt schlecht er dir dann mit dem knopff nach dem ortt vnnd lest In weyt vmb sich lauffen So spring die weyl kunlich zw Inn so schlecht er mit dem knopff vber dich das ist dir nit schedlich vnnd setz Im ann

Item Auch magstu Im nachraissenn vnd an setzenn wo du wild dieweil er den knopff also weyt lest vmb sich lauffenn

[172] The text on WP: 4–5 contains unrelated writings on dagger and mounted fencing; the pages were mixed up before binding.

Likewise, if he strikes toward your left knee with the pommel, then catch the strike on the sword between your hands (so the pommel hangs toward the ground), and go through with the pommel below his sword by the hilt and pull upward on your right side. In this way, you pull his sword from his hands.

Or likewise, if he strikes below toward your foot with the pommel, then throw your sword against his strike, with the pommel at your left side toward the ground, and leap toward him with it; look for the wrestling.

Likewise, when you hold your sword by your right side in the lower guard, if he strikes toward your point with the pommel and lets it run widely around, then meanwhile, leap boldly toward him in such a way that he strikes over you with the pommel so it does not harm you, and set upon him.

Likewise, you can also pursue and set upon him wherever you want while he lets his pommel run around widely.

Likewise, if he strikes toward your left knee with the pommel, then catch the strike on the sword between your hands (so the pommel hangs toward the ground), and go through with the pommel below his sword by the hilt and pull upward on your right side. In this way, you pull his sword from his hands.

Or likewise, if he strikes below toward your foot with the pommel, then throw your sword against his strike, with the pommel at your left side toward the ground, and leap toward him with it, and look for the wrestling.

Likewise, when you hold your sword by your right side in the lower guard, if he strikes toward your point with the pommel and lets it run widely around, then meanwhile, leap boldly toward him in such a way that he strikes over you with the pommel so it does not harm you, and set upon him.

Likewise, you can also pursue and set upon him wherever you want while he lets his pommel run around widely.

Item Slecht er dir mit dem knopff zw dem linken knye So vach den Schlag zwyschenn deynen hendenn In das Schwert das der knoppff zw der Erden hange vnd far mit dem knopff vnden durch sein Swertt vorn bey dem gehultze vnd Ruck vbersich auff dein Rechte seytten So Ruckstu Im sein Schwertt aus den henden

Item oder Schlecht er dir mit dem knopff vnden nach dem fus So wirff dein Schwertt mit dem knopff zw deyner linkenn seytten Inn die Erden gegen seynem Schlag vnd Spring damit zw Ime wartt der ringen

Item wenn dw hast dein Schwertt neben deyner Rechten seitten In der vntern hutt schlecht er dir dann mit dem knopff nach dem ortt vnd lest Inn weitt vmb sich lawffen So springe dieweille kunlich zw Ime So Schlecht er mit dem knopff vber dich das ist dir nit schedlich vnd setz Im an

Item auch magstu Im nachreisen vnd an Setzen wo dw wilt dieweil er den knopff also weit lest vm sich lawffen

102ʳ| Item schlecht er dir mit dem knopf zu dem lincken knie so fahe denn schlag zwischenn...[173] |100ʳ| ...deinenn hendenn in das schwert, da[s] der knopf tzu der erdenn hange, vnd far mit dem knopf vntenn durch sein schwert, vorn bey dem gehiltz vnnd ruck vbersich auf dein Rechte seittenn, So ruckstu im sein schwerdt aus denn hendenn.

Item oder schlecht er dir mit dem knopf, vnten nach dem fus, so wirf dein schwert mit dem knopff, tzu deiner linckenn seittenn, in die erdenn gegenn seinem schlag, vnd spring damit zu im, vnd wart der ringenn.

Item wen du hast dein schwert neben deiner Rechtenn seitten in der vntern hut, schlecht er dir dan mit dem knopf nach dem ort, vnd lest inn weit vmb sic[h] laufen, so spring dieweil künlich tzu im, so schlecht er mit dem knopf vber dich, das ist dir nit schedlich vnd setz im an.

Item auch magstu im nachraißen vnd ansetzen, wo d[u] wilt, dieweil er den knopf also weit lest vmbsich lauffenn.

[173] Here the text in JMR suddenly jumps to a different part of the treatise, whereas the text of the treatise continues on JMR: 100ʳ. This is probably due to the pages the scribe copied being out of order.

25 | The text of the strikes with the pommel:
69 The forward foot
You must defend with striking.

Gloss: When you want to strike with the pommel, you should very precisely aim at his forward members. Learn it like this: when you want to strike, hold your sword in the guard over your head and do as if you want to set upon him in the face, but let your sword go from the right hand and come to the middle of the blade with it to help the left hand, and strike toward his forward foot with the pommel, or toward his forward hand in which he holds the sword in the middle. In this way, you can also strike out of the lower guard from the right side.[174]

Likewise, you should also very precisely foresee when he strikes toward your forward knee or toward your forward hand with the pommel, so that in the strike, you displace with the counters written previously in such a way that he cannot harm you, etc.

Hereafter is the text of the strikes with the pommel:
The forward foot
You must defend with striking.

Gloss: Remember, this is when you want to strike with the pommel, you should very precisely aim at his forward members. Learn it like this: when you want to strike, hold your sword in the guard over the head and do as if you want to set upon him in the face, but let your sword go from the right hand and come to the middle of the blade with it to help the left hand, and strike toward his forward foot with the pommel, or toward his forward hand in which he holds the sword in the middle. In this way, you can strike in out of the lower guard from the same right side.[174]

Likewise, you should also very precisely foresee when he strikes toward your forward knee with the pommel, or toward your forward hand, so that in the strike, you displace with the counters written previously in such a way that he cannot harm you.

107ᵛ | Der text von den schlegen mitt dem knopff ~

|Deß[ß] fordern fuß, mitt schlegen du hietten must
Glosa |Wenn du mitt dem knopff schlachen wilt |So solt du do mitt gar eben remen siner vorgesäczten glider Und daß vernÿm also |Wann du schlachen wilt |So halt din schwert in der hut über din haüpt vnd thu alß du im in daß gesicht an wöllest seczen Vnnd lauß din schwert farn vß der rechten hannd, vnd kom do mitt der lincken zu hilff mitten in die clingen vnd schlach in mitt dem knopff zu dem |108ʳ| fürgeseczten fuß oder zu siner fürgeseczter hand do er daß schwert in der mitte da mitt helt Also magst du vß der vndere hut von der rechten sÿtten auch schlaehen ~~~ ~ ~
|Item du solt och gar eben fürsechen wann er dir mitt dem knopff zu dinem vorgesaczten knÿ oder zu diner fürgesäczter hand schlecht Daß du im den schlag also versecztest mitt den vor geschribnen brüchen daß er dir nitt schaden müge etc

Das hiernach ist der text von denn slegen mit dem knopff
Des fordernn fuß mit schlegenn du huten must
Glosa merck das ist wenn du mit dem knopff schlagenn wild So soltu da mit gar ebenn remenn seiner vorgesatzten glider vnnd das vernyme[175] also wen du schlahen wild So halt dein schwert Inn der hut vber dem |6ᵛ| haubt vnnd thu als du Im In das gesicht wollest an setzenn vnnd las dein schwert farn auß der rechtenn hannt vnd kome domit der linckenn zu hilff mitten In dye[176] clingenn vnnd Schlag In mit dem knopff zu dem vorgesatztem fuß oder zu seiner vorgesatzten hannt da er das Schwert In der mitte mit helt Also magstu aus der vndernn hut vonn der selbenn rechten seittenn auff schlagenn
Item Auch soltu dich gar eben fur sehenn wenn er dir mit dem knopff schlecht zu deinem vorgesatztem knye oder zu deiner vorgesatztenn hanndt das du In denn schlag also versetzt mit den vorgeschriben pruchenn das er dir Icht schaden mug

[174] This is a close match to paragraph 109 of the Pseudo-Danzig gloss; see GF: 95ʳ.4.
[175] The first up and down strokes of an original "w" were crossed out to correct it to "v".
[176] *dye* corrects a previously-written word that cannot be clearly identified.

Hereafter is the text of the strikes with the pommel:
The forward foot
You must defend with striking.

Gloss: Remember, this is when you want to strike with the pommel, you should very precisely aim at his forward members. Learn it like this: when you want to strike, hold your sword in the guard over the head and do as if you want to set upon him in the face, but let your sword go from the right hand and come to the middle of the blade with it to help the left hand, and strike toward his forward foot with the pommel, or toward his forward hand in which he holds the sword in the middle. In this way, you can strike in out of the lower guard from the same right side.[174]

Likewise, you should also very precisely foresee when he strikes toward your forward knee with the pommel, or toward your forward hand, so that in the strike, you displace with the counters written previously in such a way that he cannot harm you.

Hereafter is the text of the strikes with the pommel:
The forward foot
You must defend with striking.

Gloss: Remember, this is when you want to strike with the pommel, you should very precisely aim at his forward members. Learn it like this: when you want to strike, hold your sword in the guard over your head and do as if you want to set upon him in the face, but let your sword go from the right hand and come to the middle of the blade with it to help the left hand, and strike toward his forward foot with the pommel, or toward his forward hand in which he holds the sword in the middle. In this way, you can also strike out of the lower guard from the right side.[174]

Likewise, you should also very precisely foresee when he strikes toward your forward knee with the pommel, or toward your forward hand, so that in the strike, you displace with the counters written previously in such a way that he cannot harm you.

Das hiernach Ist der text von den Schlegen mit dem knopff
Des fordern fus mit schlegen dw huetten must
Glosa Merck das Ist wenn dw mit dem knopff Schlagen wild So Soltu da mit gar ebenn remenn seyner vorge satzten glieder vnd das vernim also wen dw Slagen wild |107v| do halt dein Schwert In der hutt vber dem haubt vnd thü als dw Im In das gesicht wollest ansetzen vnnd las dein Schwert vorn auß der Rechten hand vnnd kom domit der linken zw hilff mitten In die klingenn vnnd Schlag Ime mit dem knopff zw dem vorgesatzten fus oder zw seyner vorgesatzten hanndt do er das Schwert In der mitt mit helt also magstu auß der vndern hutt von der selben Rechten seyten auff Schlagen
Item Aüch soltu dich gar ebenn fursehenn wen er dir mit dem knopff schlecht zw deynem vorgesatzten knye oder zw deyner vor gesatzten hand das dw In dem Schlag also versetzt mit den vor geschribenn Bruchenn das er dir nicht schaden mag

100v| **Das hiernach ist der Text von den schlegen mit dem knopf**
Des fordern fus, mit schlegen du huten must. Glosa Merck das ist, wenn du mit dem knopf schlagenn wilt, so solstu damit gar ebenn remen seiner vorgesatztenn glider, vnnd das vernim also, wenn du schlagenn wilt, so halt dein schwert in der hut vber dein haupt, vnnd thu als du im in das gesicht wollest ansetzenn, vnnd las dein schwert farn aus der Rechtenn hanndt, vnnd kum damit der linckenn tzu hilf mittenn in die clingen. vnnd schlag in mit dem knopf zu dem vorgesatzten fus, oder zu seiner vorgesatztenn handt, da er das schwert in der mitt mit helt, Also magstu aus der vnternn hut, von der Rechten seitten auch schlagen.

Item auch soltu dich gar eben fürsehen, wen er dir mit dem knopf schlecht tzu deinem vorgesatzten knie, oder tzu deiner vorgesatztenn handt, das du im denn schlag also versetzt mit denn vorgeschriben pruchen, das er dir nicht schadenn mug.

1. Hunt from your chest toward his right hand.

2. Turn around with the horse; pull his right hand with your left.

3. With scraping, take the saddle-strap or weapon.

4. Set upon high, swing, move through, or counter the sword.

5. The cover goes before all meetings, hewing, stabbing.

6. Grab onto the strong with both hands.

7. Here someone begins to seek the belt pouch cut.

8. Turn the right hand to him, set your point toward his face.

9. Whoever defends against the thrust, turn, catch his right hand in the left.

10. Press firmly, push from the reins and seek his knife.

11. Seek the exposure: arms, leather, gloves, under the eyes.

12. Learn to sweep against all weapons with empty hands.

74ʳ | 1 Jag von deiner prust zw seiner rechten handt	2 Vmb ker mit dem roß, zeuch sein rechte handt mit deiner lincken	3 Mit Straÿffen satel riem, oder wer nymbt
4 Setz an hoch, schwing, durchfar, oder schwert prich	5 Das schutten, vor geng allen treffen, hawen stechen	6 Greyff an mit paiden henden die sterck
7 Hie heb an den man taschen haw zw suechen	8 Went im die recht handt setz dein ortt zw seinem gesicht	9 Wer den stich went dem vach sein rechte handt in dem glincke.
10[177] Drück vast stoß vom zaüm vnd suech sein messer	11[155] Suech die plos armm leder handt schuech vndter den augen	12 Mit lerer handt lern straich gegen aller were.

[177] These two figures have their numbers reversed compared to the diagram on PD: 7ᵛ.

13. The bucket hold defends against all grips, wrestle under the eyes.

14. Turn around again at length, thus the horse steps back.

15. Up close, catch the hand, reverse his face to where the nape is.

16. Catch the weapon at length when riding against.

17. Hunt toward the right hand with your art.

18. If you hunt left, fall upon the sword's pommel, push under the eyes.

19. Set the point against the face.

20. Cover against all meetings that ever happen.

21. Arrange yourself in the strength in the beginning.

22. Now this is the spear, run to the other meetings under the eyes.

23. The unnamed hold takes the weapon or fells him.

24. If someone hunts you from both sides, turn around on the left so he comes on the right.

74v | 13 Der Schaffgriff werdt alle griff ringes vndter augen	14 An der weidt ker vmb, so die roß hyn hurttñ.	15 In der nech fach die handt verker sein antlutz do der nacke ist
16 Die weer fach in der weitt in dem wider reÿtten	17[178] Jag zw der rechten handt, mit iren kunstñ	18[156] Jagstu linck fall aüfs schwertz knopf stos vndter augen
19 Setz an den ortt gegen dem gesicht	20 Schutt gegen allen treffen die ymmer werden	21 Die stercke in dem an heben daryn dich recht schick.
22 Das ist nun der sper lauff der dem anderen begegnet vndter augen.	23 Der vngenant griff weer nymt oder felt in	24 Ob man dich jagt von paiden seÿtten ker vmb linck so er recht kumbt

[178] These two figures have their numbers reversed compared to the diagram on PD: 8r.

25. Think about taking and holding the knife.

26. Grab over whoever falls on you, or do it riding against him.

1 | Here begins the chivalric art of fencing on horse
Here begins Master Johannes Liechtenauer's fencing on horse, which he has allowed to be written with obscure and cryptic words, and it is interpreted and glossed hereafter so that any fencer can learn it well (if he otherwise can or knows how to fence).

Here begins Master Johannes Liechtenauer's horse fencing, which he has allowed to be written with words obscure and cryptic; it is thus interpreted and glossed here in this book so that any fencer can learn it well (if he otherwise can fence).

2 |

1 Direct your spear,
Make it useless to ride against [you].

This is the text:
Direct your spear,
Make it useless to ride against [you].

Gloss: This is when you have a lance and another also has one, and wants to ride together with you: you should know to arrange yourself like this with your lance so that you divert his, and you land with the thrust and he does not. And you should know how to drive the pieces with the spear out of two guards which will be named hereafter, etc.

Gloss: Remember, this is when you have a lance and another also has one, and wants to ride together with you: you should know to arrange yourself like this with your lance so that you divert his with it, and you land in with the thrust and he does not. And you should know how to drive the pieces with the lance out of two guards which will be named hereafter.

The 21st figure speaks about this: "The strength in the beginning", etc.

75r | 25	26
Der messer nemen vnd behalden gedenck	Über greÿff wer dich an felt oder thue im wider reÿttens.[179]

109r | Hie hept sich an die ritterlich kunst das fechten zuo roß ~
|Hie hept sich an maister |Johannsen |Liechtennawers fechten zu roß |Das er hat lassen schriben mitt verborgen vnd verdöcten worten |Vnnd stät her nach glosiert vnnd vßgelegt daß es ain ÿettlicher fechter wol vernemen mag der anderst fechten kan oder waist ~~ ~

Hye hebt sich an Maister Johannes Liechtenawers Roß vechten, das er hat lassen schreibenn mit verporgen worten vnd verdackten das stett alhÿe in diesem puech also auß gelegt vnd glosiert das es ein yegklicher vechter wol vernemen mag der anderst vechten kan.

|Din sper berichte, gegen rÿtten macht zu nichte ·:
Glosa |Daß ist Wann du hast ain glefen | 109v | vnd ain anderer och aine vnd mitt dir zu samen will ~~tretten~~ rÿttenn |So solt du dich mitt diner glefen also wissen zu schicken ~~faß~~ daß du im die sinen ablaÿtest vnd treffest mitt dem stich vnnd er dich ~~nitt~~ nicht |Vnd die stucke solt du wissen zu trÿben mitt dem sper |Uß zwaÿen huten die dir her nach genennt werden etc

Das ist der text
Dein sper bericht gegen reytten mach zu nicht.
Glossa Merck das ist wen dw hast ein gläffen, vnd ain ander auch aine, vnd mit dir zw samen wil reytten, so soltu dich mit deiner gleffen also wissen zw schicken, das dw im da mit die sein ableyttest vnd in trefst, mit dem stich, vnd er dich nit, vnd die stuck soltu wissen zw treiben mit der gleffen auß zwein huetten, die dir hernach genant werden, darumb spricht die xxj figur, die sterck in dem an heben etc

[179] The circles drawn around these 26 figures may indicate they were copied from a conventional tree presentation (see PD: 7v–8r) but the scribe didn't understand the significance of the arrangement.

3| The text of the pieces out of the first guard:
2 If it falls down,
The end undoes him.

This is when you ride together with, hold your lance under your arm at the thrust and when he comes to the meeting, do as if it is too heavy for you and allow it to sink down in front with the point on your left side. Then, if he rides upon you with a thrust, then raise your lance upward onto his with strength so that you meet with him and he doesn't you, since his lance goes off beside your own, etc.

This is the text of the pieces out of the first guard:
If it falls down,
Your end undoes him.

Gloss: Remember, this is the piece out of the first guard: when you ride together with him, hold your lance under the arm at the thrust and when he comes to the meeting, do as if it is too heavy for you and allow it to sink down in front with the point against your left side. Then, if he rides upon you with a thrust, then raise your lance upward onto his with strength so that you meet him and he doesn't you, since his lance goes off by the side of your own.

The first figure speaks about this: "Hunt from the chest", etc.

4| A piece out of the second guard:
Likewise, when you ride together with, hold your lance with both hands in the middle. If he rides upon you with a thrust, then strike his lance away from you on your right side with the front part of yours and wind your lance with it under your right arm, and in this way, you meet him and he doesn't you.

This is the piece out of the second guard:
Remember, when you ride together with him, hold your lance crosswise in front of you on the saddle bow with both hands in the middle. If he rides upon you with a thrust, then strike his lance away from you on your right side with the front part of yours and wind your lance with it under your right arm, and in this way, you meet him and he doesn't you.

The 17th figure speaks about this: "Hunt to the", etc.

Der text von den stucken vß der ersten hutt ~

|Ob es enpfalle, dem ende Im abschnalle
|Daß ist wenn du mitt zu samen rÿttest |So halt din glefen vnder dinem arm zu dem stich vnd wenn es kumpt zu dem treffen |So thu alß sÿ dir |110r| ze schwer sÿ vnd lauß sÿ mitt dem ort vornen nider sincken vff din lincke sÿtten |Rÿtt er dann vff dich mitt ainem stich so heb din glefen übersich mitt störck an die sinen |So trifft du in mitt vnnd er dich nitt |Wann sin glefe get neben dir besÿcz ab, etc

75v| Das ist der text von dem Stuck auß der ersten huet
Ob es empfall, dein endt im ob schnall.
Glosa Merck das ist das stuck auß der ersten huet, wen du mit Im zu samen reyttest, so halt dein gleffen vndter dem arm zw dem stich, vnd wen es kumbt zw dem treffen so thue als sÿ dir zw schwer seÿ, vnd laß sy mit dem ort voren nÿder sincken gegen deiner lincken seytten, Reytt er dan auff dich mit einem stich, so heb dein gleffen v̈ber sich mit sterck auff an die sein, so trifstu in, vnd er dich nit, wen sein gleffen get neben die seytt ab, dar vmb sprich die erst figur, Jag von der prust etc

Ain stuck vß der anderen hutt
|Item wann du mitt zu samen rÿttest |So halt din glefen mitt baÿden henden in der mitten |Rÿtt er dann vff dich mitt ainem stich |So schlach mitt dem ~~ford~~ vordem tail diner gleffen die sine vff din rechte sÿtten von dir/abe vnd wind din glefen |110v| do mitt/vnnder dinen rechten arm |So triffest du in vnd er dich nitt ~ ~~[180]

Das ist das stuck auß der andern huett
Merck, Wen du mit im zw sammen reyttest, so halt dein gleffen mit baiden henden in der mitt fur dich twergs auff dem satelpogen, reytt er dan auff dich mit einem stich, so schlag mit dem vorder tail deiner gleffen die sein, auff dein rechte seytten von dir ab vnd windt dein gleffen da mit vndter dein rechten armm, so trifstu in vnd er dich nit, darumb vmb sprich die xvij figur, Jag zw der etc

[180] At this point, SR changes over to the Pseudo-Danzig gloss without any indication that the source has changed, and then terminates abruptly in the middle of a sentence a few folia later.

5| **This is the text:**
3 Hew in, don't draw
From the scabbard, pull him to the left.
4 Grab on his right,
Thus you catch him without fencing.

Gloss: This is if both miss with the lance in the riding in: immediately let yours fall and draw neither sword nor knife, and ride in on him and turn yourself with your left side on his right and drive the wrestling written hereafter.

Likewise, when you approach with the left side on his right, if he grabs you with the right hand in front and wants to wrestle, then grab his right arm with the right (in front by the hand), and drive the unnamed hold or the obscure hold.

Or likewise, if he has drawn his sword and goes up with it and wants to strike, then grab his right elbow with the left hand and push him from you with it, and lift his right foot with the left foot; in this way he falls. Or, when you have grabbed his right elbow with your left hand, then grab his sword-pommel with your right and pull it toward you; in this way, you take his sword.[181]

Das ist der text

Haw drein nicht zuck, von schaiden linck zw im ruck, Greyff in sein rechten, so fechst du in ane fechten.

Glosa. Das ist ob in in dem an reÿtten bede verfelt mit der gleffen so laß zw hant die dein fallen, vnd zeüch weder swert noch messer |76ʳ| vnd reit an in vnd wendt dich mit deiner lincken seytten an sein rechte vnd treib die hernach geschriben ringenn

Item Wen du kumbst mit der lincken seÿtten an sein rechte greifft er dich den mit der rechten handt voren an, vnd wil ringen, So begreiff mit der rechten seinen rechten armm vorn peÿ der handt, vnd treib den vngenanten grieff, oder den verporgen Item. oder hat er sein schwert gezogen vnd fert damit auff, vnd wil schlagen, so begreyff mit der lincken handt sein rechten Elpogen vnd stos in damit von dir, vnd mit dem lincken fueß erheb seinen rechten so felt er, oder wen du Im mit der lincken handt seinen rechten elpogen begriffen hast, so begreiff mit der rechten sein schwert knopf vnd ruck da mit zu dir so nÿmbstu Im das swert

[181] This is a close match to paragraphs 46–47 of the Lew gloss; see JMR: 43ᵛ.4–5.

Dresden – SR

6 |

Glasgow – GF

Here remember another, that is, if you can't approach on his right side with the left, then hold yourself with the right on his right. If he grabs onto you with wrestling, then drive the bucket hold or the sun show.

Likewise, when you approach with your right side on his right, if he has his sword drawn and has gone up with it and wants to strike, then go down from above with your right arm, outside over his right arm, and press the arm into your right side and ride forward; in this way, you take his sword. Or, go with the right arm from inside, up over his right, and press the arm in front on your chest and ride forward; in this way, you take his sword again.[182]

Likewise, grab his right hand with your left and pull it in front of your chest, and turn your horse from him; in this way, he falls.[183]

Or **likewise,** grab his right elbow with the right hand and lift his right foot with the right foot; in this way, he must fall.[184]

The second figure speaks about this: "Turn around with the horse", etc.

Hye merck ein anders. das ist, ob du mit der linck<u>e</u>n seytt<u>e</u>n nicht magst kumen an sein rechte, so halt dich mit der recht<u>e</u>n an sein rechte greyfft er dich dan an, mit ringen, so treib den schaff grieff, oder das sunen zaichen

Item. Wenn du kümbst mit deiner recht<u>e</u>n seÿtt<u>e</u>n an sein rechte, hat er den sein schwert gezog<u>e</u>n vnd fert da mit auff, vnd wil schlag<u>e</u>n, so far mit deinem recht<u>e</u>n arm<u>m</u>, von oben nÿder auswendig v̈ber sein recht<u>e</u>n, vnd drück den|arm in dein rechte seÿtt<u>e</u>n, vnd reyt fur dich, so ny<u>m</u>bst du im schwert, Oder far mit dem recht<u>e</u>n arm von inwendig auff v̈ber seinen recht<u>e</u>n, vnd drück den arm<u>m</u> vorn an dein prust, vnd reyt fur dich, so nÿmbstu im aber sein schwert, ~~Darum<u>b</u> sprich die xij figur, mit lerrer hant~~

|76^v| **Item.** Begreiff sein rechte handt mit deiner lincken vnd ruck sÿ fur dein prust, vnd wen dein Roß von im so felt er.

Item. Oder begreÿff mit der recht<u>e</u>n handt, sein recht<u>e</u>n Elpog<u>e</u>n vnd mit dem recht<u>e</u>n fueß erheb im sein recht<u>e</u>n fueß, so mueß erfallen, Darub spricht die Ander figür, vmb ker mit dem roß etc

[182] This is a close match to paragraphs 24–25 of the Pseudo-Danzig gloss; see SR: 112^v.3–113^r.2.
[183] This is a close match to paragraph 57 of the Lew gloss; see JMR: 44^v.3.
[184] This is a close match to paragraph 60 of the Lew gloss; see JMR: 44^v.6.

7 5 This is the text:
The lance stabbing,
Fencing, learn to counter calmly without running.

Gloss: When someone rides upon you with a lance, you should ride against him quite calmly, and not running with the horse, you can undo all of his riding at you with the pieces written previously. Or, when you run or course, you cannot come to the art or to the pieces, and you are uncertain on the horse.

The 22nd figure speaks about this: "Now this is the spear, run", etc.

8 This is the text of the belt pouch cut:[185]
6 If it shifts,
The sword will be traded against sword.
7 Correctly hold the strong,
Seek and remember the belt pouch cut.

Gloss: This is if you both have come from the lance and you should fence with sword, then remember to lay your sword off[186] your left arm in the guard[187] and ride straight at him under the eyes toward his right side...

Das ist der text
Das gleffen Stechen vechten, sÿtigkleich an hurt lere prechen

Glosa. Wen ainer auff dich reitt, mit einer gleffen, so soltu gar sytigklich gegen im reytten, vnd mit hurrtten mit dem Roß so magstu mit den vorgeschriben stucken Im alle seins an reÿtten wol zu nicht machen, Oder wend du hurts oder renst, so kanstu zu der kunst oder zu den stucken nit kumen, vnd pist selber vnsicher auff dem roß, darumb spricht die xxij figur, das ist nun der sper lauff, etc

Das ist der text von dem Taschn haw
Ob es sich verwandlt das Schwert gegen schwert wirt gehandlt, Recht fast dye sterck, Taschn haw du suech vnd merck.

Glosa. Das ist ob ir pede von der gleffen kumen seyt, vnd solt fechten mit schwerten, So leg dein schwert auß den lencken armm in die huet vnd reÿt Im gerad vndter augen zu seiner rechten seÿtten...

[185] A *Taschen* is a belt pouch worn on the side toward the back; STEPHEN CHENEY likens it to a "fanny pack".
[186] Probably meant to be *auf* instead of *aus*, "on your left arm".
[187] This guard is labeled "first guard" in other glosses, but Ainring omits the list of guards and their numbers.

Dresden – SR

Glasgow – GF

...If he hews over you, then go up with the sword and displace the cut strongly with the long edge, and stab toward his face. If he displaces the thrust and goes up high, then hew in at the left hand or at the reins with the long edge. And if the horse will press you, then hew at his right leg while running away.[188]

The 7th figure speaks about this: "Here begins", etc.

9| This is the text:
8 Learn to strongly cover well
In all meetings without danger. Distress him,
9 Set upon without danger.
Hang in at him from stirrup to hair.[189]

Gloss: Remember, this is that you should bind boldly with the sword, be it with hewing or with stabbing, and do not pull yourself away from the sword, and constrain him with the point into the pieces written hereafter.

The 20th figure speaks about this: "Cover against", etc.

...hawt er dan auff dich ein oberen haw, so far auff mit dem swert, vnd versetz den haw starck mit der langen schneid, vnd stich im zw dem gesicht, versetzt er den stich vnd fert hoch auff, so haw im mit der langen schneid, nach der lincken handt, oder nach dem ~~zwi~~ zawm, vnd ob dich das roß vertrueg so haw im nach dem rechten pain in dem hin rennen, darumb spricht dye vij figur, Hie Heb an etc

77r| Das ist der text
Ler wol starck schütten allen treffen an far domit nott in Setz an angefar, wer strayfft henng Im zu dem har
Glosa Merck das ist, Das du albeg künlich mit dem schwert solt anpinden, Es sey mit hawenn oder mit stechen, vnd dich von dem schwert nit abziechen, vnd in zwingen mit dem ort zu den hernach geschriben stucken, Darumb spricht die xx figur, schut gegen etc

Mounted Fencing

188 This is similar to paragraphs 37–38 of the Pseudo-Danzig gloss and specifically the variant in SR and PKW; see SR: 116v.2–117r.2.

189 Grimm indicates that 'from foot to hair' was an expression equivalent to 'from head to toe', and I assume this is similar.

Likewise, watch if you can set upon him with the sword. If he displaces on his left side and rides upon you with it, then go through with the pommel from underneath his sword, over his left shoulder, around the neck, and come with the left hand behind the pommel to help, and pull him to you on the side.[190]

The 6th figure speaks about this: "Grab on with both hands", etc.[191]

Likewise, when he displaces your cut, then go with your hilt under his jaw and grab by the helmet with the left hand, and pull toward yourself with it, and push from you with the hilt, so he falls.[192]

Likewise, if he displaces your cut or thrust and steps toward you, then grab his right hand with the left, and with the right, set the point in his face.[193]

The 8th figure speaks about this: "Turn the right hand to him", etc.

Likewise, if he displaces your high cut with the crosswise, then go out high with the right hand and with the point over his sword toward his face or chest, and set upon him.[194]

Item wart ob du im mit dem schwert magst an setzen, versetzt er auff sein lincke seÿtten, vnd reytt an dich, so far mit dem knopf von vndten auff, durch sein schwert im vmb den hals, vnd kumb mit der lincken dem knopf zu hilff, vnd ruck in zu dir auff die seÿtten Darumb spricht die vj figur, Greyff an mit paiden hendten etc

Item Wen er dir den haw versetzt, so far im mit dein gehultz vndter kunpacken vnd mit der lincken handt begreiff in hindten pey dem Eÿsnhuet, vnd zeuch damit zu dir, vnd mit dem gehultz stoß von dir so felt er

Item Versetzt er dir ein haw oder stich, vnd hürt an dich, so begreyff sein rechte hant mit der lincken, vnd mit der rechten setz im dein ort in das gesicht, darub spricht die viij figur, Went im die recht hant etc

Item Versetzt er dir ein oberhaw mit der twer, so far hoch auff mit der rechten handt, vnd~~t~~ mit dem ort v̄ber sein schwert zu seinem ge~~sch~~sicht oder prust vnd setz im an, ~~darumb spricht die iiij figur, Setz an hoch schwing etc~~

[190] This is similar to the second half of paragraph 40 of the Pseudo-Danzig gloss; see SR: 117v.
[191] The Pseudo-Danzig and Lew glosses invoke the 5th figure at this point, but label it "6th".
[192] This is similar to paragraph 44 of the Pseudo-Danzig gloss; see SR: 118v.2.
[193] This is a close match to paragraph 42 of the Pseudo-Danzig gloss; see SR: 118r.3.
[194] This is a close match to paragraph 86 of the Lew gloss; see JMR: 47r.5. Note that in Lew, it states "hang the point in over his sword". It's possible the verb *hengen* was intended here too.

10|

This is the text:
10 If you want to harry,
Hunt long, that greatly troubles.
11 Whoever defends against you,
Wind, that hurts the eye.
12 If he defends further,
Catch the reins and do not let go.

Gloss: Hold your sword in the guard near your right leg[195] and ride in like this, and stab toward his face from long rightful[!] arms. If he displaces the thrust, then go up with the right hand and wind on the sword, and remain with the point in front of his face.[196]

The 4th figure speaks about this: "Set upon high", etc.

Likewise, if he displaces with the sword and runs in on you, then go with your hand inward, in over his right arm, and grab your reins under his arm with inverted left hand. In this way, you engulf his hand with the reins.[197]

The 10th figure says, "Press firmly", etc.

Das ist der text
Wiltu geruet[198] lanck jagen das seer mutt wer dir das wert so windt das aug versert, Wert ers furpaß, fach |77v| zawm vnd wer piß laß.

Glossa Halt dein swert neben deinem rechten pain in der huet, vnd reyt also zu im, vnd stich im zu dem gesicht mit lanck auß gerechten armm Versetzt er den stich, so far auff mit der rechten handt, vnd wind am swert, vnd pleib im mit dem ort vor dem gesicht, darumb spricht die iiij Setz an hoch etc

Item versetzt er den mit dem swert, vnd hurt an dich so far mit deiner handt inwenidig auff uber sein rechten armm, vnd begreyff deinen zawm mit lincker verkertter handt, vndter seinem armm, also verschlingst du Im dy handt mit dem zawm, die x figur spricht, druck vast etc

195 This guard is labeled "third guard" in other glosses, but Ainring omits the list of guards.

196 This is a close match to the first half of paragraph 53 of the Pseudo-Danzig gloss; see SR: 121r.3–121v.1.

197 This is similar to the second half of paragraph 53 of the Pseudo-Danzig gloss; see SR: 121v.1. SR ends here abruptly, so with reluctance, we'll have to switch to PD for subsequent references.

198 Corrected from *geradt*.

11 |

Here remember the setting aside with the sword on horse:

Remember, when you ride on someone and you have your sword in a guard, then notice on which side he hews toward you. If he hews from above toward your left side, then wind against his cut (also on your left side). Or if he hews at you on your right side, then wind on your right as well, such that your thumb always comes underneath. And with the displacement, always set the point in his face, and drive this against the lance like this.

The 19th figure speaks about this: "Set the point against", etc.

Likewise, when he hews in at you [or] displaces,[199] if he remains by you in this way, then go out over his right hand with the pommel, and push the hand with the hilt in front of you, toward your saddlebow, and grab his sword-pommel with your left hand and ride forward. In this way, you take the sword.[200]

Hye merck die absetzen mit dem swert zw roß

Merck, wen du reyttest an den man, vnd hast dein swert in einer huett, So merck zu welicher seytten er dir zw hawt, Hawt er dir von oben zw deiner lincken seytten so windt auch auff dein lincke seÿtten gegen seinen haw, Oder hawt er dir zw deiner rechten seytten, so wint auch auff dein rechte, das dein dawm albeg vndten kumb, vnd mit der versatzung setz im denn ort albeg in das ge, gesicht, vnd das treib gegen der gleffen ~~auß~~ auch also, dar vmb spricht die xix figur, Setz an den ort etc

Item Wen er dir ain hawt versetzt, pleibt er dan also pey dir, so far mit knopf aussen v̈ber sein rechte hant, vnd stoß die handt mit dem gehultz fur dich zu deinem satelpogen, vnd mit deiner lincken begreyff seinen swertzknopf vnd reyt fur dich so nÿmbstu Im das swert.

[199] There are two verbs in the first clause here, *ainhawet* and *versetzt*, and they're both third person—"he hews in" and "he displaces". It could be read as "When he displaces your hew", which would bring it in line with pseudo-Danzig, but that's not really grammatical and assumes it's misspelled. My solution is to turn it into a disjunction—"he hews in *or* he displaces".

[200] This is a close match to paragraph 50 of the Pseudo-Danzig gloss; see SR: 120^r.2.

Dresden – SR

12|
13

Glasgow – GF

This is the text:
Think of the exposure
Seek the knife, don't tend to the pommel.

Gloss: This is when you want to fence with someone in harness: first of all, you should know which places you can best win in.

Likewise, that is, under the face, or under the armpit, or outside on the hand in the glove, or inside in the palm of the hand, and in all joints of the harness at the arms and at the legs, and seek all the exposures with thrusting, and not with hewing. And if you cannot work with the sword when you come close to him, then work with the dagger, and if you cannot get to your dagger then see if you can take his, and work with it toward the exposure.

The 11th figure says, "Seek the exposure: arms, leather", etc.

78r| Das ist der text
Bedenck dye plos, suech messer nicht wart kloß
Glosa. das ist, Wen du mit ainem vechten solt in harnasch, so soltu vor allen sachen wissen an welchen stetten du in am besten gewinnnen magst
Item. das ist vndter dem gesicht, oder vndter den ṽchsen, oder ausswendig auff der handt in den hantschuechen, oder inwendig in dye hendt, des taners, vnd in allen gelencken des harnasch an armen vnd an paynen, vnd dye plõß suech alle mit Stechen, vnd nit mit schlagen, Und ob du mit dem swert, wen du zu nahent an in kunpst nicht ~~gera~~ arbayttten magst, so arbait mit dem tegen, vnd ob dw zu deinem degen nicht kumenn magst, so wart ob dw im den seinen genemen magst, vnd arbait domit zu der plõß, Spricht die xj figur, Such die ploß armm leder etc

Mounted Fencing

13| This is the text:
14 Learn two sweeps
Against the weapons with empty hand.

Gloss: This is that, first of all, you should learn[201] and know how you should work with a free hand on horse, and up front with the wrestling. Therefore, you should set the reins in such a way that you can change from one hand to the other, and seek your advantage with it in this way, and this is the greatest art on horse.

The 12th figure speaks about this: "With empty hand, learn", etc.

14| This is the text of the bucket hold:[202]
15 The bucket hold teaches:
Whoever turns and wrestles to you,
16 While under the eyes,
Grab him correctly with striking.

Gloss: This is the best wrestling for anyone on horse. Drive it like this: when you want to wrestle, ride equally toward the man's right side, under the eyes, and grab onto him with wrestling...

Das ist der text
Zwen strich lere, mit lerer handt gegen der were.
Glosa das ist, das du vor allen sachen wissen solt, vnd ler nemen, wie dw mit ledigen hendten solst arbaytten zw Roß, Und zu vor auß mit den ringen, darumb solt dw den zawm also an richten das dw in magst gewechsln, von ainer handt in dye ander, vnd also dein vortail damit suechen, vnd das ist die grõst kunst zu roß, darumb spricht die xij figur, Mit lerer handt lere, etc

Das ist der text von dem Schaff griff
Der schaff griff lert wer sich ringes zu dir kertt Als vndter augen, greyff in recht mit schlaugen.
|78[v]| Glosa. das ist der pestn ringen ains zu roß, das treirb[!] also, Wen dw ringen wilt, reyt dem man gleich vndteraügen zw seiner rechtten seÿtten, vnd greyff yn an mit ringen, ...

[201] *Ler nemen* was probably intended to be '*vernemen*=learn'.

[202] A *Schaff* is a large bucket, basket, or wash basin, and the name may arise from the method of carrying it (as DIERK pointed out to me). Other translators have read it as '*Schaf*=sheep', perhaps a reference to holding sheep still for shearing. Mayr reads it as *Schopf* and his Latinist renders it *juba*, meaning "horse mane" or "helmet crest" (see, e.g., PMM II: 219[r].3–220[r].1).

...If he falls on you in turn, then grab his right arm with your inverted left hand (in front by the hand) and pull it under your chest, and go over it with your right arm and grab the saddlebow with it, and ride forward so that he must fall.[203]

The 13th figure speaks about this: "The bucket hold defends..."

15 | The is the text of wrestling:
17 Whoever falls on you,
Riding against you, will be joined.
18 Hanging to the ground.
Grab over him correctly with coupling.

Gloss: This is when you ride together with someone; if he comes on your right side with his [right] and falls into your neck in front with the right arm, then go around his with yours like this as well, and go around behind him with the left hand and come in to help with your right, and pull him toward you on the side. Or strike the right arm above over his right and throw him with the bucket hold.[204]

...velt er dich dan wider an, so begreiff im sein rechten armm vorn peÿ der handt mit deiner lincken verkertten handt, vnd ruck den vndter dein prust, vnd far mit deinem rechten armm dar ṽber vnd pegreyff damit den satl pogen, vnd reÿt fũr dich so muß erfallen, Darumb sprich dye xiij figur, der schaff griff wert

Das ist der text von ringen
Wer dich anfellet, wider reÿttens der wirt gesellet, hanngen zw der Erden vber greyff In recht mit geparen

Glosa Das ist, Wen du mit ainem zu samen reyttest, kumbt er dan mit seiner rechten seytten, an|die dein, vnd felt dir mit dem rechten armm vorn In den halß, so far Im auch also vmb dein sein vnd far mit der lincken handt hindten vmb in, vnd kumb da mit deiner rechten zu hilff, vnd ruck in zw dir auff dye seÿtten, Oder schlag den rechten armm oben ṽber sein rechten vnd wirff in mit dem schaffgriff

[203] This is similar to paragraph 116 of the Lew gloss; see JMR: 50v.4–51r.1.

[204] This is a close match to paragraphs 67–68 of the Pseudo-Danzig gloss, and specifically the variant in PKW; see PKW: 116r.5–6.

DRESDEN – SR

GLASGOW – GF

16 |

Another:

Likewise, when you ride together with someone, if he comes on your right side with his left and falls around your neck from behind with his left hand, then go up strongly behind you with the right arm over his left and come with the left hand to help the right, and press his left arm firmly behind the back of your neck. If he deviates with the arm, then grab his left hand with the left and drive the unnamed hold or the obscure hold.[205]

Or likewise, if he grabs around behind you with the left hand and wants to wrestle, then strike strongly down from above with the right arm, outside into the joint of his left arm, and ride forward.[206]

The 26th figure: "Grab over", etc.

Aliud

Item. Wen du mit ainem zu samen reytest, kumb er dan mit seiner lincken seÿtten, an dein rechte, vnd felt er dir mit der lincken hant hyndten vmb den hals, so far mit dem rechten armm hindter dich starck auff, v̄ber sein lincken vnd kumb mit der lincken der rechten hant zu hilff, vnd drück im sein lincken armm vast hindtn in deinen nacken Weicht er dan ab mit dem armm, so begreyff mit der lincken handt sein lincke, vnd treib den vngenantten griff, oder den verporgen grieff.

Item. Oder greÿfft er hindten vmb dich mit der lincken |79r| handt vnd wil ringen, so schlag mit dem rechten armm auswendig starck von oben nÿder, in das gelenck seins lincken arms, vnd reytt fur dich, Dÿ xxvj figur vber greyff etc

MOUNTED FENCING

[205] This is a close match to paragraph 70 of the Pseudo-Danzig gloss; see PD: 46v.2.
[206] This is similar to paragraph 71 of the Pseudo-Danzig gloss; see PD: 46v.3.

17 This is the text of a lesson:
19 Toward both sides
Learn to ride against him.

Gloss: This is that whichever side you approach someone on, you should hold yourself close to him and drive the art like this: if you come on him with your right side, then nimbly drive the pieces that belong to the right side, or if you come on him with the left side, then also drive what belongs to the left side. In this way, he cannot come to his pieces.[207]

18 This is the text to the left side:
20 If you want to ride,
The horse running toward the other side,
21 Cover strongly,
Set upon him with it, distress
22 Whoever defends it.
Catch the sword far, carry near, have the hand.

Gloss: This is if you want to ride toward someone's left side: you should also bind strongly on him with the sword, be it with hewing or with stabbing, and always aim at the exposures with the point, just as you have done toward the right side...

Das ist der text von ainer lerr
Zu payden seytten, du in an ~~ker~~ lere [208] alle wider reÿtten

Glosa. das ist, zw welcher seÿttñ dw zw dem man kumbst, so soltu dich nachent zu Im halten, vnd treyb dye kunst also, kumbst du mit deiner rechten seÿtten an in, so treib die stück behendtlich die zu der rechten seytten gehoren, Oder kumbst du mit der lincken seytten an in, so treib auch waß zu der lincken seÿtten ghort, so mag er zu seinen stucken nit komen.

Das ist der text zw der lincken seÿtten
Ob du wild reÿttn, roß lauffens zu der andern seytten dye sterck schut an setz damit, Not in wer wertt dir das, weyt swert fach trag nachent der handt haß.

Glossa. das ist, ob du ainem zu seiner lincken seytten wilst zu reÿtten, so solstu im auch starck an pinden mit dem swert, es sey mit hawen oder mit stechen, vnd albeg der plõsen remen mit dem ort, als du zu der rechten seytten hast gethon, ...

[207] This is similar to paragraph 73 of the Pseudo-Danzig gloss; see PD: 47ʳ.2.
[208] *lere* inserted above the deletion.

...With this, you constrain him to the pieces that belong to the left side, since there are some pieces that someone drives toward the left side, which they cannot drive toward the right.

The fifth figure speaks about this: "The cover goes before all", etc.

Likewise, this is when you ride toward his left side: hew a strong cut in over him. If he also hews strongly in and wants to set upon you, then you should displace him with the sword while he is far from you, or grab his right hand with your left if he comes close to you.

The 16th figure says, "Catch the weapon", etc.

19 |

23 This is the text of a lesson:
Or turn around,
24 Harry, to hunt the defense
With all art.
He who hunts arranges it favorably.

Gloss: This is if your horse carries you forward away from him, so you can drive no piece on him. Turn yourself toward the side of him where you can drive the best advantage.

The 14th figure speaks about this: "Turn around at length", etc.

...da mit zwingstu in zü den stücken die zu der lincken seytten gehõrn, Wen es seind ettlich stuck die man zu der lincken seytten treibt, die man zu der rechten nit getreiben kan, darumb sprich die funfft figur, das schutten vor geng allen etc

Item das ist, wen du im zu der lincken seytten zu reytst, so haw Im ein oberhaw starck ein, Hawt er dan auch starck ein vnd wil dir ansetzen, so solt du im versetzen, die weil er weit von dir ist mit dem swert, Oder kumbt er nahent an dich so begreif sein rechte hant mit deiner lincken, Die xvj figur spricht, die weer fach etc

79^{v} | Das ist der text von ainer ler
Oder vmker, gerutt zu jagen gewertt, Mit allen kunste der jagt der schick nach gunste.

Glosa. das ist, ob dich dein roß fur In hin weck trug, das dw kain stuck auff In getreiben kunest, So went dich zu im, zu der seÿtten do du den vortail am pesten getreiben magst, Darumb sprich die xiiij figur, In der weit ker vmb etc.

20 | This is the text:

25 If you pass[209]

And go[186] left without thinking,

26 Feel his sword

And wrestle, push, not firmly.

Gloss: This is if your horse carries you away, or however it happens that you must ride to his left side unintentionally: lay your sword on the left arm. If he hews toward your head, go up with the sword and displace with the long edge. If you come close to him with the displaement, then [go] with the left arm over his right hand and press it firmly on your left side and bash him with the pommel under the face.[210]

The 18th figure speaks about this: "If you hunt left, fall upon", etc.

21 | Note:

Or when you press his right arm into his[!] left side and ride away beside him, you take his sword. You can also catch with the reins with the going over in the hand, if you want.[211]

Das ist der text

Ob du vergagst vnd an danck linck gagst, sein schwert auff tast, vnd ring stos nit vast.

Glosa das ist, Ob dich dein roß vertrueg, oder wie das geschach, das dw an danck an sein lincke seÿtten must reÿtten, So leg dein schwert auff den lincken armm, Hawt er dir dan zu dem kopf, so far auff mit dem swert, vnd versetz mit der langen schneid, Kumbst dw dan mit der versatzüng nachent an in, so far mit dem lincken arm vber sein rechte handt vnd trück die also vast in dein lincke seÿtten, vnd stos in mit dem knopf vndter das gesicht, Darumb sprich die xviij figur, Jagst du linck fall aufs etc Nota

Oder wen du im sein rechten armm in sein lincke seytten drückst, vnd so reytt neben im hin wegk, so nymbst du im sein schwert, Auch magst du mit dem ṽberfaren in die handt mit dem zawm fachen ob du wilt.

[209] *Gagen* and *Vergagen* aren't real words. *Gangen* means "go" and *Vergangen* means "pass", but other glosses have '*jagen*=hunt' and '*verjagen*=hunt out'.

[210] The first half (up to "with the long edge") is a close match to paragraph 81 of the Pseudo-Danzig gloss, and specifically the variant in PKW; see PKW: 116^{v}.5. The remainder is a close match to paragraph 129 of the Lew gloss; see JMR: 53^{r}.2 (though the final "under the face" returns to Pseudo-Danzig).

[211] This is similar to paragraphs 131–132 of the Lew gloss; see JMR: 53^{r}.4–5.

Likewise, when you displace his high cut (as is written previously), then hew in freely over above toward the head.

Likewise, when you have your sword on the left arm in the guard,[212] if someone rides at you toward your left side with a lance, go well up with the pommel and let the blade hang to your left side and set aside the lance with it like this, and hew toward his head or set upon him. Or, if he rides at you with the lance toward your right side, then sweep straight up onto his lance with the sword, and wind into the over hanger and set upon him.[213]

22| This is the text:
27 If someone hunts you on the right,
Turn halfway around, await fencing.
28 Catch with the arms,
So harm cannot follow you.

Gloss: This is if you are pressed to flee, and you were armored and have nothing but a sword, and someone sets upon you behind with the lance toward your right side: turn yourself from the thrust against him (on your left side), and turn yourself with the sword against his lance, and set upon him...

Item wen du im den ober haw versetzt als vor geschriben stett, so haw im einen freyen ober haw oben ein zu dem kopff.

Item. wen du dein schwert hast auff dem lincken armm |80^{r}| in der hut, reyt dan ainer auff dich mit ainer gleffen zu deiner glincken seytten, so far wol auff mit dem knopf, vnd laß die klingen hangen zu deiner lincken seytten, vnd setz im die gleffen also da-mit ab, vnd haw Im zu dem kopf, oder setz im an, Oder reytt er dir mit der gleffen zu deiner rechten seÿtten, so streich schlett auff mit dem swert an sein gleffen, vnd wint In das õber gehenng, vnd setz Im an.

Das ist der text
Jagt man rechtens, halb ker vmb wart vechtens,
Mit armm vachen, so mag dir kain schad nachen.

Glosa das ist, ob du gedrungen wirst zu der flucht, vnd du warst gewappent, vnd hest nichtz denn ain swert, vnd setzt dir dan ainer mit der gleffen hindten an dein rechte seytten, so wen dich auß dem stich gegen Im auff dein lincke seyt, vnd wendt dich mit dem swert gegen seiner gleffen, vnd setz im an, ...

[212] This guard is labeled “first guard” in other glosses, but Ainring omits the list of guards.
[213] This is similar to paragraph 88 of the Pseudo-Danzig gloss; see PD: 48^{v}.4.

Dresden – SR

Glasgow – GF

...Or, if he sets upon you behind (on your left side), then turn yourself on your right against him and wind with the sword as before, and set upon him.

The 24th figure speaks about this: "If someone hunts you from both sides", etc.

Likewise, in the flight, you should also know how to nimbly turn yourself around in the saddle from one side to the other, and stab behind you, and set aside sword and lance to both sides.[214]

23 | Remember a nimbleness with the lance:

When you are hunted and have a lance, if someone hunts after you and also has one, hold your lance with your right hand on the right shoulder and when you are sure that he is close behind you, raise the lance over the head onto your left shoulder, and let your point remain behind you, and turn yourself against him on your left side, and with that, strike your lance under the arm, so you come equally with him under the eyes.[215]

...Oder setzt er dir hindte̲n an dein lincke seÿt, so went dich auff dein rechte gegen im, vnd wind mit dem schwert als vor vnd setz im an, Daru̲mb spricht die xxiiij figur, ob ma̲n dich jagt von paid seytte̲n etc Item du solst auch in der flücht behennttigklich dich wissen vmb zu wenden in dem satte̲l von ainer seÿtte̲n zu der andere̲n, vnd hindter dich stechen, vnd absetze̲n zu paide̲n seÿtte̲n schwert vnd gleffen

Merck ein behendigkait mit der gleffen

Wen du vor jagst vnd hast ein gleffen, jagt dir ainer | 80v | nach, vnd hat auch aine, So halt dein gleffe̲n mit der rechte̲n handt auff der rechte̲n achsel, vnd wen du sichst das er schier hindte̲n an dir ist, so heb die gleffe̲n ṽber das haubt, auff dein lincke ~~s~~ achsel, vnd laß dein ort hindter dir pleibe̲n, vnd wendt dich gege̲n im auff dein lincke seÿtte̲n, vnd schlach damit dein gleffe̲n vndter den armm̲, so kumbst du im da mit geleich vnder die augen.

[214] This is a close match to paragraph 143 of the Lew gloss; see JMR: 54v.4.
[215] This is a close match to paragraph 87 of the Pseudo-Danzig gloss; see PD: 48v.3.

24 | 29

This is the text:
Take the knife,
Learn to hold without shame.

Gloss: Here remember how someone should take his sword or his knife: ride toward his right side and seek the exposure however you can, with hewing or stabbing. If he displaces and comes close to you, grab his right arm behind his right hand with your inverted left hand and pull it in front of you and hold him firmly by it, and bend your left arm outward at the hilt of his sword such that he must let his sword fall. And when you hold him by the arm like this, you can strike him with the sword or throw him with the bucket hold.[216]

The 25th figure speaks about this: "Taking the knife", etc.

Das ist der text
Der messer nemen, behalten ler an schemen.

Glosa Hie merck, wie ainem sein swert nemen solt oder sein messer, oder den tegen, Reyt eim zu der rechten seÿtten vnd suech die ploß wie du magst mit hawen oder mit stechen, versetzt er vnd kumbt nachent an dich, so begreiff sein rechten armm hindter seiner rechten handt, mit deiner lincken verkertten hant, vnd rück den fur dich vnd halt in vest da peÿ, vnd peug deinen lincken armm auß wendig an dem pundt seins swertz so muß er das swert fallen lassen, Und wen dw in also peÿ dem armm haltest, so magstu In mit dem schwert schlagen, oder mit dem schaff griff werffen Darumb sprich die xxv figur, der messer nemen etc

[216] This is a close match to paragraphs 144–145 of the Lew gloss; see JMR: 54^{v}.5–55^{r}.2.

25| This is the text of the unnamed hold:
30 The unnamed,
Turn the strong;
31 Their thrust, their strike,
Ruined without any vengeance[!].[217]

Gloss: Remember, this is the unnamed hold. When you come to him with the left side, if he has drawn his weapon and wants to strike you, or grabs your front with the right hand with wrestling, then grab his right arm with your right hand (in front by the hand) and pull it under your chest, and lie yourself on it with the body and ride forward. In this way, you break his arm.[218]

Likewise, if you don't want to break his arm when you have pulled it under your chest, then grab his right elbow with your left hand and shove him away from you with it, and grab his sword by the pommel with your right hand and pull toward yourself with it. In this way, you take his sword.[219]

The 23rd figure speaks about this: "The unnamed hold", etc.

Das ist der text von dem vngenatten griff
Den vngenantten den starcken verwendt Ir stechen,
Ir schlagen verdirbt on alles rechenn

Glosa Merck das ist der vngenant grieff, Wen du mit deiner |81^{r}| lincken seÿtten kumbst an in, hat er dan sein wer gezogen vnd wil dich schlagen, oder greyfft dich mit der rechten handt voren an mit ringen, so begreyff mit deiner rechten handt sein rechten armm vorn pey der hant, vnd ruck den vndter dein prust, vnd leg dich mit dem leib darauff, vnd reÿt fur dich so prichstu im den armm Item wildu im den armm nit prechen, wen du in hast geruck fur dein prust, so greÿff mit deiner lincken hant an sein rechten Elpogen, vnd stoß in damit von dir vnd begreyff mit deiner rechten hant sein swert pey dem knopf vnd ruck da mit an dich so nymbst du im das schwert Darumb sprich die xxiij figur, der vngenant griff etc

217 Should be '*vechten*=fencing'.
218 This is similar to paragraphs 92 of the Pseudo-Danzig gloss and 148 of the Lew gloss; see PD: 49^{r}.5–49^{v}.1 & JMR: 55^{r}.5–55^{v}.1.
219 This is a close match to paragraph 149 of the Lew gloss; see JMR: 55^{v}.2.

26| This is the text of the sun show:[220]
32 If you want to hold,
Do not stop riding beside him.
33 The sun show:
If you want to bend the left arm,
34 Touch the front of the head
Against, press quite firmly after,
35 So that he sinks himself
And rarely rises again.

Gloss: Remember, this is the best wrestling in on horse. When you ride together with him, if you come to his right side with your right, then hold him close to you, and grab around behind with your left hand and grab him by his left arm with it, and pull him firmly around toward you, and grab him underneath on his jaw with your right hand and shove it firmly from you, above itself, against his left side. In this way, you turn his face against the sun. You win his momentum with this so he cannot stop himself.[221]

Or likewise, if you come with the left side on his right, then grab him as before and throw him behind you on your left side, and this wrestle is called 'the sun show'.

The 15th figure speaks about this: "In the After[!], catch the hand", etc.

Das ist der text von dem sunnen zeichen
Wil du an fassen, neben reittens nicht solt du lassen, das sunen zaigen, den lincken armm wil du naigen, das vorhaubt tascht gegen nach druck sere vast das er sich sencket, vnd selten wider auff lencket.

Glosa Merck, das ist der pestn ringen ains, zu roß, wen du mit im zu samen reÿtest, kumbst du den mit deiner rechten seÿtten an sein rechte, so halt dich nachet zu im, vnd greyff mit deiner lincken ~~hinden~~ handt hinten vmb in, vnd pegreyff in do mit pey seinem lincken armm, vnd zeuch in vast her vmb zu dir, vnd mit deiner rechten hant greyff im vndten an sein kindpackn vnd stos in vast an dir, vber sich auff, gegen seiner lincken seÿtten, |81v| so kerstu im sein antlutz gegen der sunen, da mit gewinst du im den schwunck an, das er sich nit mag gehalten.

Item oder kumbstu mit der linckn seytten an sein gerechte, so begreÿff in als vor vnd wirff in hindter dich auff dein lincke seÿtten, vnd das ringen haist das sunen zaichen, Darumb sprich dÿ xv figur In der nach fach die hant etc

[220] This is a piece in which you turn your opponent's face upward to 'show them the sun', but it could also be understood as a reference to the gnomon or 'pointer arm' (*Zeiger*) of a sundial (*Sonnenuhr*).

[221] This is a close match to paragraphs 151–152 of the Lew gloss; see JMR: 55v.4–56r.2.

27|

36 Whoever aims at you,
Grab over, then will he be shamed.[222]
37 Press arm to head,
This grab has often robbed the saddle.

Gloss: Remember here that this is the counter against the sun show. If someone comes on your right side with his right and wants to throw you by showing you the sun, then remember: when he grabs onto your jaw with the right hand, strike the right arm over his right, and pull him firmly onto your chest and lay yourself on it with the body, and ride forward. In this way, you throw him. Or throw him with the bucket hold.[223]

Likewise, if he comes on your right side with his left and grabs around behind you by your left arm with his left hand, then go up behind from underneath, over his left arm, and then press him firmly into the back of the neck. If he yields with the arm, then grab his left hand with the left and throw him with the unnamed hold.

The 9th figure speaks about this: "Whoever defends against the stab, turn, catch", etc.

Wer dir das remet, vber greyff den der wirt beschemet druck armm zu hawbt der griff hat offt sattel beraubt

Glosa hye merck, das ist der pruch wider das sunen zaichen, kumbt ainer mit seiner rechten seytten an dein rechte, vnd wil dich werffen mit dem sunnen zaichen, so merck, wen er dich greyfft mit der rechten handt an, ~~an~~ den kinpacken, so schlag den rechten armm vber sein rechten, vnd ruck In vast an dein prüst, vnd leg dich mit dem leib darauff, vnd reyt fur dich, so wirfstu in, oder wirff in mit dem schaff griff. Item. kumbt er mit seiner lincke seytten an dein rechte, vnd greyft mit seiner lingken hant hindten vmb dich nach deinem lincken armm, so far von vndten auff hinter dich, ṽber sein lincken armm, vnd truck im den vast hindtn in den nacken, weicht er den ab mit dem armm so begreÿff mit der lincken hant sein lincke, vnd wirff in mit dem vngenantten griff, darumb spricht die ix figur, Wer den stich wert dem vach etc

222 This couplet uses similar phrasing to long sword couplet 32, in the section on the wrathful cut, and 80, in the section on overrunning.

223 This is a close match to paragraphs 156–157 of the Lew gloss; see JMR: 56v.3–4.

28 |
38
39

Another text:
If you want to moderate yourself again
In the catching, let him go lightly.
Lead the wrestling,
Caught without any rope.

Gloss: Remember, this is called 'the obscure wrestling'. If you're willing to make it common and let it be obviously seen, then drive it like this: ride with your left side on his right. If he grabs you in front with wrestling, then grab his right arm with the right hand (in front by the hand) and pull it forward, and grab his right elbow with the left hand and shove it upward, and shove his right arm up over his left arm with the right hand, and raise his right arm over itself with the left arm like this. In this way, you have caught and bound him without any bonds. Drive this to both sides.[224]

Ein ander text
|82r| Wil du aber dich massen des fachens liecht von dir lassen Ringens dan fure, gefangen an schünre.
Glosa. Merck, das haist das verporgen ringen, wil du das gemain machen, vnd offenpard sechen lassen, So treib es also, Reÿtt mit deiner lincken seÿtten an sein rechte, greyfft er dich vorn an mit ringen, so begreÿff mit der rechten hant, sein rechten arm vorn pey der hant, vnd ruck den fur dich, vnd greyff mit der lincken hant ~~han~~ sein rechten elpogn, vnd stos den vber sich, vnd mit der rechten hant stos sein rechtn armm oben vber den lincken armm, vnd heb also mit dem lincken armm seinen rechten armm ṽber sich also hastu In gefangen vnd gepunden an alle pandt, vnd das treib zu paiden seÿtten.

[224] This is a close match to paragraphs 99 of the Pseudo-Danzig gloss and 164 of the Lew gloss; see PD: 50v.2 & JMR: 57v.7–58r.1.

29|
40

Remember the Leading Grab,[225]
Which further breaks his strength.

Gloss: This is that you should always come forth with the grappling and wrestling you have learned well ere he does, and foremost with the four main wrestles with which you let him come to no pieces. These are the bucket hold, the unnamed hold, the sun show, and the obscure hold. You should not forget the catches with the reins. And when you can do the wrestling well, no one can throw you hard from the horse without injury.[226]

The end

Der vorgriff merck, die pricht furbaß ~~me~~ sein sterck
Glosa das ist, das albeg soltest vor kumen ee den er, mit den greiffen vnd ringen die du wol vernumen hast, vnd zu vor auß mit den vier haubt ringen, do lest du du in mit zu kainem stuck kumen, das ist der schaffgriff, der vngenant griff, das sünen zaichen, vnd der verporgen griff, vnd des vachens mit dem zaim solstest du mit nicht vergessen, vnd wen du dye ringen wol kanst, so mag dich hart nymat an schaden von dem roß gewerffen

Explicit

[225] *Vorgriffen* is usually read as '*vor*=before' + '*griffen*=grabbing', but Jessica Finley has pointed out that, just as *angriffen* can mean either "grab" + "on" *or* "attack", *vorgriffen* could also mean "attack before" and perhaps even be related to the '*Vorschlag*=leading strike' of N. For this reason, and because the term isn't otherwise clearly defined, I've translated it similarly to that manuscript (see Chidester & Hagedorn 2021). Only the Pseud-Danzig gloss clearly associates this verse with Before and After; see PD: 50v.3–51r.1.
[226] This is a close match to paragraph 165 of the Lew gloss; see JMR: 58r.2.

Paul Kal greets a Bavarian duke armed for mounted combat, presumably Ludwig IX 'the Rich', Duke of Bayern-Landshut (to whom Kal dedicated his book). The banner reads "Gracious lord, I vow to serve you; may my God and his dear Mother help us." (*Genediger herr ich gelob euch den dienst mein gott und sein lieb ein muter sullen unser helfer sein*, PKM: 4r.)

Appendix 1: Hans Medel overflow

Pieces 58–59 of the 'long' sword appear in five witnesses: **GF**, **HS**, **HM**, **JMR**, and **SR**, but there is only space in the reading apparatus for up to four. Since **GF**, **JMR**, and **SR** are the primary witnesses and **HS** has priority of age, **HM**: 32^{v}–33^{r} could not be included there and is instead placed here.

58 | Against displacing:
If you are displaced,
And however it comes to that,
Hear what I advise you:
Sweep or wrench off, hew quickly with haste.

Gloss: This is that however it happens that you are displaced, remember: if someone displaces your upper cut, then in the displacement, go against him with the pommel over his front hand and wrench off downward with it, and with the wrench, strike him on the head with the sword.

32^{v} | Wider die versatzung
Ob dir versetzt ist vnd wie das dar zue kummen ist hör was ich dir rate streich oder reiss ab haw schnell mit drate Glosa

Das ist wies dar zw kummenn ist das dir versetzt wordenn ist | 33^{r} | So merck versetzt er dir ainen oberhaw so var Im in der versatzung mit dem knopf v̈ber sein versetzte handt vnd reiss damit vndersich ab vnd mit dem reissen schlag In mit dem swert auf den kopf...[1]

59 | Again, a piece against displacing:
Remember, when you hew a lower cut from the right side, if he then falls on it with the sword so you cannot come up with it and overwhelm you down on the side, then go nimbly over his sword with the pommel, and strike in with the snapping, with the long edge toward the head.

But if it happens on the left side, then again go with the pommel over his sword and step forward with the right foot and strike in with the short edge.

Ain ander stuck wider die versatzung
Wenn dw hawst ainen vnderhaw von der rechten seitten felt er dir dann mit dem swert darauf das dw damit nit aufkummen magst vnd dringet dich nider in die seitten so var behentlich mit dem knopf v̈ber sein swert vnd schlag yn mit dem schnappen mit der langen schneid zw dem kopf geschicht es aber auf der lincken seitten so var aber mit dem knopf v̈ber sein swert vnd trit mit dem rechten fues für vnd schlag yn mit der kurtzen schneid...

[1] Text in Medel that is not derived from Ainring's gloss is omitted here and represented by ellipses.

Appendix 2: Rhyming *Record* Translation

The unglossed *Record* of the long sword is included before the gloss in SR and may have originally appeared in GF, too (see Chapter 4).

Foreword

1 Young knight, learn first: have love for God,
And honor women—that I laud.
2 So that your honor great may grow,
Practice chivalry and know
3 Arts that adorn you well in play
And bring you fame in war some day.
4 Gain wrestling's artful holds with might;
With lance, spear, sword, and knife do fight,
5 And wield them all with gallant hand,
So that against you none may stand.
6 Cut here, guard there, and close with this;
Charge in whether you hit or miss.
7 The wise ones hate and will disdain
All those who praise themselves in vain;
8 So grasp when all is said and done:
All art needs measure, moderation.

Fencing

9 If you would bring the art to light,
See left advance and strike with right,
10 For you will find that left with right
Is how you may most strongly fight.
11 Who waits and follows after blows,
Will little joy in this art know.
12 Cut close to them; your will revealed,
No changer comes to breach your shield;
13 Toward head and body, forth you race,
The skirmish you should then embrace.
14 And always fight with all your strength,
Your body tight at any length.
15 A simple rule you should not slight:
Fence not from left when you are right.
16 If on your left is how you fight,
You'll be quite clumsy on the right.
17 Before and after, these two things:
The font from which all true art springs.
18 With strong and weak, you'll rule the bind;
The word "within" bear well in mind.
19 Learn onward in this art until
You can defend and work with skill.
20 But if you easily take fright,
You'll find it hard to learn to fight.
21 Five strikes you must now learn to heed,
Perform them with your right hand's speed;
22 Those fencers skilled whose art we vow
To reward well will show us how:
23 The wrathful one hates crook and cross;
So cock your eye and parting cause.
24 A fool will parry all strikes, so
Pursue, run down, and set the blow.
25 Change it through, pull your point back,
Run through, slice off, or hands attack.
26 Then hang and wind, expose them more,
Strike, catch, sweep, and push your point fore.

27 When from above they cut in near,
The point of wrath makes danger clear.
28 If they should sense the threat and shove,
Care not, just take it off above.
29 Or if they're strong, then more strength show
And thrust—if seen, take it below.
30 Bear this in mind once in the fray:
Cut, thrust, then hard or soft you lay,
31 "Within", and then drive after more,
But calmly—do not rush to war;
32 If at the war above they aim,
Go down below and bring them shame.
33 Down every path that you may wind,
Cut, thrust, and slice you'll learn to find;
34 You also must learn to assess
Which one of them would serve you best,
35 So that, whenever steel meets steel,
The masters you'll confound with zeal.

36 Know there are but four exposures;
Clear your path and aim for closure.
37 And caring not how they appear,
You may attack without a fear.
38 Should they attack, revenge is sweet;
Exploit these four exposures neat:
39 To land above you should redouble;
Transmute below to cause them trouble.
40 Now let me make this plain and clear:
No one defends without a fear.
41 And if these truths you'll learn and know,
Then scarcely can they come to blows.

42 Throw a curve with crooked grace;
Onto their hands your point will race.
43 Lure them out well and curve down low,
And step to hinder many blows.
44 Cut crookedly up toward the flat,
Subvert the masters' strength with that.
45 When steel on steel above should spark,
Abscond and I will praise your art.
46 Curve not: cut short, your plan concealed,
Then with it, changing through reveal.
47 Who down a crooked path will lead,
The noble war ensnares indeed:
48 They'll truly have no way to know
Where they'd be safe from any blow.

49 The cross takes what from sky comes down,
Rewarding all with glory's crown.
50 The cross in strength performed its deed,
Your work remains, take careful heed.
51 When to the plow you drive across,
Yoke it hard then to the ox.
52 Cross yourself and take a leap,
And threaten heads while yours you keep.
53 Mislead by missing skillfully;
Go low and harry willfully.
54 Inversion forces and constrains;
Run through and wrestling holds you'll gain:
55 Their elbow take; be sure, then leap;
Done right, their balance you will reap.
56 Now miss them twice, and if you hit,
Then make a classic slice with it.
57 The second time you miss, I say
To step in left, and don't delay.

58 When buffalo will cut or thrust,
The cockeye breaks and enters thus.
59 If, with the changer, threat they lay,
The cockeye robs them anyway.
60 If you see them shortchanging you,
Take your revenge by changing through.
61 Down to their point you cock your eye,
But, fearless, take their neck up high.
62 Or cock your eye up to their part,
If you would spoil their hands with art.
63 Cut from your part to seek your prize
And threaten them under the eyes.
64 Then turn and take it down below,
And threats against their heart bestow.
65 Whatever from your part descends,
Their lofty crown can well defend.
66 Slice through their crown—refuse to kneel—
Its glory broken by your steel.
67 With sweeping cuts press your attack;
Slice through and then pull yourself back.

68 In four lairs only should you lie;
Hold there and vulgar guards decry.
69 The ox that plows, the foolish one,
And from the day you should not shun.

70 Now four displacements learn with care,
Which also flush them from their lair.
71 Be mindful of displacement's game,
It guards you well or brings you shame.
72 Should you become displaced at last,
However this has come to pass,
73 Then listen now to what I say:
Wrench off, cut in, and don't delay.
74 Set on to four extremities,
Learn to remain and end with ease.

75 Learn to pursue, then learn it twice;
Or into their defenses slice.
76 When they're outside, there take them on
In two forms; start what work you want.
77 Then gauge each threat with hand and blade:
If pushing hard or softly laid;
78 For this you must learn how to feel.
The word "within" cuts deep as steel.
79 Pursue them twice, and if you hit,
Then make the same old slice with it.

80 If down below your sword they aim,
Flow over them and bring them shame.
81 When steel on steel above should spark,
Stay strong and I will praise your art.
82 Work onward then with skill and ardor,
Or press them hard and press them harder.

83 Learn well to lure them out at will,
Then hinder cuts and thrusts with skill.
84 Whoever tries to stab at you,
Your point meets theirs and breaks on through.
85 From either side, both left and right,
Your swords will meet if forth you stride.

86 Learn to change through, your sword untied,
Then thrust sharply from either side.
87 Whomever tries to bind on you
You'll quickly find by changing through.

88 Now step in close, engage the bind,
Then pull, and what you seek you'll find.
89 Pull back; if sword you meet, pull more;
Devise a work that hurts them sore.
90 Pull back whenever steel meets steel
And masters you'll confound with zeal.

91 Hold pommel high, let point hang down,
Run through and wrestling abounds.
92 When strength would press and oppress you,
Remember this: just run on through.

93 Whenever hardness blocks your plays,
Slice off from underneath both ways.
94 There are four slices you must know:
Two falling high, two rising low.

95 Now turn your slice with great finesse
And crush their hands as in a press.

96 Two hangers from the ground now guide,
Out from each hand, both left and right:
97 In every threat, in each foray,
Cut, thrust, then hard or soft you lay.
98 Spread windows wide that speech may flow,
And hear their case with joyful glow,
99 But snap the windows shut upon
Whomever tries to cut and run.
100 Now let me make this plain and clear:
No one defends without a fear,
101 And if these truths you'll learn and know,
Then scarcely can they come to blows.

102 If you lead well and break through right,
To this end you may guide the fight,
103 And breaking in with flashing steel,
Three wonders of the sword reveal.
104 So hang your point in straight and true,
And wind your sword to follow through.
105 Now eight winds note with thoughtful mind,
And weigh the paths that each may find:
106 In each and every wind of sword,
Three wonders wait to be explored.
107 They thus expand to twenty-four—
Count one by one, you won't need more.
108 From either side, both left and right,
Learn these eight winds when forth you'd stride.
109 Then gauge each threat with hand and blade:
But pushing hard or softly laid.

DUELING

1 Direct your lance with honor plain;
Against you all will ride in vain.
2 If from the path you'd fall away,
The end will see great disarray.
3 Strike here and there as you close in;
Draw not your blade from scabbard then,
4 But pull them left and seize them right—
You need not fence to win the fight.
5 Learn this to joust or fence with lance:
Break calmly through in your advance.

6 If sudden shift the fight then shows
And swords begin to trade their blows,
7 Then tie the strong up well and good,
And mark them as a cutpurse would.
8 Now learn to cover, strength displayed,
And then whenever blade meets blade,
9 Fear not, but set upon your foe:
Hang in at them from head to toe.
10 Or if you'd harry and distress,
Then hunt them long and grant no rest.
11 Whoever to defense will rise,
Your wind of sword will hurt their
12 eyes. If they will yet defend at bay,
Catch well the reins and don't delay.
13 Consider where they are exposed;
Your knife, not pommel, finds them most.

14 Two sweeps now learn so you may stand
Against all arms with empty hands.
15 And when to wrestling they would hasten,
Learn to hold them like a basin.
16 When underneath their very nose,
Correctly grab as you fly close.
17 Whoever falls on you with might
When riding 'gainst you: hold them tight,
18 And let yourself hang toward the earth
While gripping over—show your worth.
19 On either side, both left and right,
Against them learn to ride and fight.

20 If you should choose to charge and ride
Your horse onto the other side,
21 First cover strongly and suppress,
Then set upon and cause distress.
22 When you defend, their sword catch wide
And carry near, their handle tied.

23 Or wheel about to hunt again,
Prepared to harry and restrain;
24 With every skill that hunters learn,
Seek your advantage as you turn.
25 Should you ride forth without relent
But end up left without intent,
26 Then worry not, but their sword find,
And wrestle well: push hard and bind.
27 If someone hunts you on the right,
Turn halfway 'round, prepare to fight.
28 With arms to catch and drive away,
No harm can reach you in the fray.
29 Take sword away like it's a knife
And learn the holds that cause them strife:
30 Apply the hold that has no name
To move the strong and bring them shame.
31 Spoil strikes and thrusts with this defense;
Have your revenge—no need to fence.

32 If you would grab them as they ride,
Don't fail to ride up alongside.
33 To show the sun and make them bow:
Take first their left arm firmly now,
34 Then grab their head in front with skill,
And press it upward hard until
35 They sink down low and show respect
(And then might never stand erect).
36 If taking you down low's their aim,
Then grab them high and bring them shame;
37 Press arm to head, their grip defeat—
This often robs them of their seat.
38 But should you seek a measured course
To catch and hold them on their horse,
39 Then with this wrestling they'll be found,
And without rope they will be bound.
40 Remember well the grab that leads,
To break through strength and work great deeds.

41 When you dismount or are unhorsed,
The fight on foot must take its course.
42 Take up your spear in steady hands
And face your foe in proper stance;
43 Its sharpened point will serve your needs,
So boldly throw the thrust that leads,
44 Then leap and wind and set upon—
If they defend, pull and you've won.
45 If you would pull and thrust with skill,
Then learn to break through guards at will.
46 Now if your foe fears injury
And backs away and wants to flee,
47 Then close with them and seek your chance
To catch and hold as you advance.

48 And should you wrestle, learn to leap:
Your foot behind their front leg sweep,
49 Or deftly lock the leg you catch
Just like the closing of a latch.
50 From either hand, both left and right,
Perform your art and end the fight.

51 Should it come to pass that here
The sword is drawn against the spear,
52 Observe the way they thrust, then leap
And catch and wrestle as you sweep.
53 What they extend, your left hand beats;
Leap surely in, catch what you meet.
54 Should they again fear harm, and pull
Back from your catch and your control,
55 Then you will find them all exposed,
Your point harassing unopposed.
56 And leather, gauntlets, and the eyes:
Toward these exposures you should rise.

57 Forbidden arts of wrestling learn
And bring them forth in battle's churn;
58 Find ways to lock and take control,
Surmount the strong to reach your goal.
59 In every lesson that you learn,
Your point toward the exposure turn.

60 When each from scabbard draws their sword
And both face off with one accord,
61 Then you should strengthen in the fight
And bear in mind to cover right.
62 Before and after: these two things
Learn well to gauge with backward spring.
63 Pursue whenever steel meets steel,
And you'll confound the strong with zeal.
64 If they defend, pull back and thrust;
If they defend, rush in you must.
65 If they should fight you long and wide,
Then artfully their fate decide.
66 If they attack you fierce and strong,
Defeat them when you shoot in long.
67 The other point can guard and smite,
So meet their sword and don't take fright.
68 With either hand, both left and right,
Turn then your point to piece their sight.
69 With strikes, your forward foot defend,
That you may fight on to the end.

Appendix 3: Concordance

This section includes a concordance of all nine witnesses treated in this book. It's primarily intended to offer a visual overview of the relative coverage and completeness of each witness; since the folio numbers are included in the reading apparatus, it uses a simpler notation of × for a complete piece and ⁄ for a partial piece.

As described in Chapter 4, P only includes a few pieces which can be assigned to Ainring with confidence. In this concordance, however, those illustrations which could be assigned to all the glosses—including Ainring—are noted with a □.

'Long' sword fencing

Piece	GF	JMR	SR	AD	AR	HM	HS	P	WP
1		×	×			×			
2		×	×			×			
3		×	×			×			
4		×	×			×			
5		×	×			×			
6			×			×			
7			×			×			
8		×	×			×			
9			×			×			
10		×	×					□	
11		×	×						
12		×	×					□	
13		×	×					□	
14		×	×						
15		×	×			⁄		□	
16		×	×			⁄			
17		×	×		⁄	×			
18		×	×						
19		⁄	×						
20			×						
21			×		×				
22			×					×	
23			×						

Piece	GF	JMR	SR	AD	AR	HM	HS	P	WP
24			×						
25		/	×						
26			×						
27		×	×		×				
28		×	×					□	
29		×	×						
30		×	×						
31			×			×			
32			×						
33						×	×		
34	×	×	×				×	□	
35	×	×							
36	×		×					□	
37	×		×					□	
38	×		×					□	
39	×	×	/					□	
40	×		×					□	
41	×		×			×		□	
42	×	×				×			
43	×	/	×					□	
44	×		/						
45	×	×	×						
46	×		×			×	×	□	
47	×	×	×					□	
48	×	×	×				×		
49		×	×					□	
50	×	/	×				×	□	
51	×		×			×		□	
52	×		×					□	
53	×		×					□	
54	×		×					□	
55	×		×					□	
56	×		×						
57	/		/				×		
58	×	×	×				×	×	

Piece	GF	JMR	SR	AD	AR	HM	HS	P	WP
59	×		×				×	×	
60	×		×			×		×	
61	×		×			/		×	
62			/			×			
63		/	×			×		☐	
64			×						
65			×						
66	×		×			×			
67	×		×						
68	×		×					☐	
69	×		×						
70	×		×						
71	×		/					☐	
72	×		/			/		☐	
73	×	×	×					☐	
74	×	×	×						
75			×					☐	
76			×					☐	
77	×		×					☐	
78	×		×					☐	
79	×							☐	
80	×		×					☐	
81	×		×					☐	
82	×		×					☐	
83			×						
84		×	×					☐	
85	×		×		×			☐	
86	×		×		×			☐	
87	×		×		×			☐	
88	×		×		×				
89	×	/	×		×				
90	×		×		×				
91	×		×		×				
92	×		×		×				
93	×								

MOUNTED DUELING

Piece	GF	JMR	SR	AD	AR	HM	HS	P	WP
F[1]	×								
1	×		×						
2	×		×						
3	×		×						
4	×		×						
5	×								
6	×								
7	×								
8	×								
9	×								
10	×								
11	×								
12	×								
13	×								
14	×								
15	×								
16	×								
17	×								
18	×								
19	×								
20	×								
21	×								
22	×								
23	×								
24	×								
25	×								
26	×								
27	×								
28	×								

[1] Liechtenauer's 26 figures

DISMOUNTED DUELING

Piece	GF	JMR	SR	AD	AR	HM	HS	P	WP
1		×	×	×					×
2		×	×	×					×
3		×	×	×					×
4		×	×	×					×
5		×	×	×					×
6		×	×	×					×
7		×	×	×					×
8		×	×	×					×
9		/	/	/					/
10		×	×	×					×
11		×	×	×					×
12		×	×	×					×
13		×	×	×					×
14		×	×	×					×
15		×	×	×					×
16		×	×	/					/
17		×	×						
18		×	×	/					/
19		/	×	/					×
20			×	×					×
21		/	×	×					×
22		×	×	×					×
23		×	×	×					×
24		×	×	×					×
25		×	×	×					×

Glossary

Abnemen	Taking away; the second piece of the wrathful cut.
Abschneiden	Slicing off.
Absetzen	Setting aside.
Abzucken	Pulling away/withdrawing; this is extremely rare in the glosses.
Alber	Foolish; a guard. *Albern* is an adjective meaning "foolish" whereas *Alber* is a noun meaning "poplar tree"; Joachim Meyer interprets it as the former, which he justifies with the explanation that only a fool would use such a worthless guard; Paul Hektor Mayr reads it as the latter, and his Latin translator renders it '*populus arbor*=poplar tree' without any explanation (PMM: 89r.2–90r.3). There's no way to be sure which one Ainring intended, but PD: 2r renders it *Alwer*, a MHG word meaning "simple or worthless", and Hans Medel uses '*alberlich*=foolishly' in relation to it (HM: 32r.5), suggesting a shared understanding early in the tradition.
Alten Schnitt	The old slice; mentioned in verse 79 (in the pursuit) but not described anywhere. Possibly meant to mean "the previous slice", referring to the similar verse and gloss in the double misser.
Ansetzen	Setting upon; putting your point on your opponent and pushing. Primarily appears in the armored fencing section.
Auser myn/nym	Outer form/taking; a piece of the pursuit. The second word in the name of this piece is sometimes spelled '*Myn*=form' and sometimes '*Nym*=take'. *Myn* tends to appear more commonly in the 15th century and *Nym* in the 16th, but it's hard to say which one is the original term and both are used in this passage. Either term would capture the essence of the piece, which is actions from a bind on the "outside" (left) instead of the "inside" (right).
Binden	Binding.
Bloss	Exposure.
Buffel	Buffalo; a fencer who relies on strength rather than feeling.
Dringen	Crowding; wedging your point in your opponent's armor to move them around.

Duplieren Redoubling; a piece of the four exposures. This would literally just be "doubling", but "redoubling" has the same meaning in English and "doubling" has taken on the very different meaning of "simultaneously hitting and being hit" in our parlance.

Durchlauffen Running through; the umbrella for most wrestling pieces.

Durchwechsel Changing through.

Einlauffen Running in; approaching close to an opponent's body, presumably to wrestle with them.

Fechten Fencing or, more generically, fighting.

Fechtschule A public training day or competitive event in early modern German towns.

Feler Misser; a *Fehler* is the opposite of a *Treffer* ("hitter") and signifies something that *doesn't hit* whatever target it was aimed at. In this text, it's clearly an intentional miss.

Gloss Word used in both English and German that means commentary.

Halbes Schwert Half sword; equivalent to short sword.

Hauptstucke Main pieces; the seventeen sections of Liechtenauer's teachings on the long sword.

Hengen Angling or hanging; "to hang" is the cognate for *hengen*, but "angle" is the translation that will make the most sense to readers not steeped in the German terminology. (Mayr's Latin translator agrees and uses '*inclinatio*=inclination or angle'; see PMM: 82^{r}.3, 93^{v}.3, 96^{r}.3, & 97^{r}.1–2.)

Hut Guard; see *Leger*.

Indes Within; the meaning of the word changes significantly from Middle High German to Early New High German. It could be translated "within" or "inside" in both languages, but in MHG, *Indes* was a spatial adverb (i.e., within a place or location) and in ENHG it became a temporal adverb (i.e., within a time or event). Liechtenauer's *Record* seems to have been written during this transition and straddles these senses: the word 'Within' is used to describe actions the instant (time) when you have felt pressure in the bind (place) and must choose a response.

Kampff Duel or struggle.

Krieg War; a situation in which you remain in the bind and try to win through turning your sword. When used against the crooked cut, it's labeled the '*edel Krieg*=noble (or straightforward) war'.

Kron Crown; a guard.

Krumphaw Curved or crooked cut; *krumm* specifically describes a thing that is usually straight but has *unnatural* curvature. These translations use "curved" as the adjective form and "crookedly" as the adverb.

Kurtzes Schwert Short sword; the sword held with the dominant hand on the grip and the nondominant hand on the blade. This is the literal translation, but since it isn't the sword that's short, rather it's the way the sword is held that 'shortens' it, "retracted sword" seems clearer. Compare '*langes Schwert*=long sword' in the unarmored lessons, which refers to placing both hands on the grip.

Kurtzhaw Retracted cut; a piece of the curved cut.

Langen Ort Extended point; a guard. Possibly the same as *Sprechfenster*.

Langes Schwert 'Long' sword; the sword held with both hands on the grip. This is the literal translation, but since it isn't the sword that's long and instead it's the way the sword is held that 'lengthens' it, "extended sword" seems clearer. Compare '*kurtzes Schwert*=short sword' in the armored lessons, which refers to placing one hand on the blade.

Leger Lair; a *Leger* is a lair or place to lie down, and also a military encampment (especially a siege camp). A *Hut* is a defensive position or guard. It's possible that at one time *Leger* was a guard used offensively and *Hut* was one used defensively, but by the time this text was written, the terms were used interchangeably.

Mutieren Transmutation; a piece of the four exposures.

Nach After; when your opponent constrains you to parry.

Nachreisen Pursuit.

Oberhaw A cut that comes in over the target.

Ort	Point or thrust; can be used as "point" both in the sense of the sharp end of a sword and the sense of a location.
Ortlauffen	Running point; this isn't described but is referred to once in HM.
Ochs	Ox; a guard. While 'oxen' refers to bovines trained as draft animals—usually castrated males, but sometimes bulls or cows—the term *Ochs* also encompassed the aurochs, the wild ancestor of domestic cows that was prized for its leather and hunted to extinction by the 17th century. Unlike domestic cattle, the aurochs had large horns that pointed inward.
Pflug	Plow; a guard.
Ringen	Wrestling.
Ruren	Harrying; it also means is "to stir up", "to cause something to move", and "to touch or hit" (including "to land a blow in fencing").
Schaff Griff	Bucket hold; an armored wrestling piece.
Schaitelhaw/ Schaitler	Part cut/parter; *Scheitel* refers to the top of the front of the head, the hairline, and the part of the hair. The verb *scheiden* means to divide or separate, and the noun *Schaiden* refers to damage or harm. Thus, while Ainring is clear that the name refers to starting from the part of *your* hair (i.e., over your head) rather than targeting *their* hairline, the other potential meaning of "to part" (divide the opponent in half) may also be intended.
Schilhaw/ Schiller	Cockeyed cut/cockeye; *schielen* is a tricky verb because it has a double meaning of "aslant/askew" and "cross-eyed" (once upon a time called "squint-eyed"), and the text relies on both meanings. "Cockeyed" is the closest we can get to both meanings in one English word.
Schirmaister	A *Schirmmeister* is a fencing teacher, using the late Medieval term for fencing (*schirmen* rather than *fechten*). A *Schirrmeister* is a lord's stable-master, or a logistics officer in military setting in charge of animals and anything pulled by animals (wagons, cannon, etc.). '*Schirmeister*' could be a spelling of either one; Hans Medel reads it as the former.
Schlagenden Ort	Smashing point; a name for the pommel in armored fencing.

Schrankhut	Lattice fence guard; *Schränke* are the fences around a dueling or tournament field; *schranken* can mean "to cross or form a lattice" and can also describe the crisscrossed outward tilt of saw teeth, so this definition produces "crossed guard". Mayr's Latin translator renders it '*custodia cancellata*=lattice guard' (PMM: 84^{v}.1).
Schwertnemen	Disarm; literally "sword-taking".
Schwache	Weakness; the part of the sword from the middle to the point.
Schwert	Sword.
Sprechfenster	Spreading window; often interpreted as *Sprachfenster*, literally "window for speaking" and the term for the screen through which cloistered monastics could communicate with the outside world (but not, as far as I can tell, a small window inside a door that can be used to talk without opening the door). However, Paul Hector Mayr's Latin translator renders it '*fenestra patula*=spread-open window' (PMM: 96^{v}.1–4), which seems to be based on reading it as *Spreichfenster* from '*spreien*=spread'. It's also possible that both definitions are intended.
Stercke	Strength; the part of the sword from the guard to the middle.
Stuck	Piece; a fencing technique, maneuver, or sequence.
Sunnen zeichen	Sun show; an armored wrestling piece.
Taschenhaw	Pouch cut.
Twerhaw	Crosswise cut; *twer* is translated as both "crosswise" and "across" in these translations.
Uberlauffen	Overrunning; this is also the term used when water overflows a container.
Ungenatten Griff	Unnamed hold; an armored wrestling piece.
Unterhaw	A cut that comes in under the target.

Vorgriffen — Leading grab; a concept introduced in the mounted fencing. *Vorgriffen* is usually read as '*vor*=before' + '*griffen*=grabbing', but just as *angriffen* can mean either "grab" + "on" *or* "attack", *vorgriffen* could also mean "attack before" and perhaps even be related to the '*Vorschlag*=leading strike' of manuscript N. (Thanks to JESSICA FINLEY for pointing this out.) Because of this possibility, and since this term isn't clearly defined, it's translated similarly.

Verhawen — Mis-hew. The prefix *ver-* can indicate a thing done preemptively or in front, but also a thing done poorly. *Verhawen* seems to refer to the second pursuit in which the opponent hews to the ground in front of you; this could be rendered 'fore-hewing' (whether meaning hewing before you or hewing in front of you), but 'mis-hewing' emphasizes that it's an error. Mayr's Latin translator has '*ictus longius*=long strike' (PMM: 91^{v}.3–92^{r}.1 & 92^{v}.4), describing the nature of the error.

Verkerer — Inverter.

Verporgen ringen — Obscure wrestling; an armored wrestling piece.

Verpoten ringen — Forbidden wrestling; pieces banned in friendly competition.

Versetzen — Displacement or parrying; *Versetzen* doesn't exactly mean "to parry", but in the 15th century it encompasses both '*versetzen*=to move or shift in location' and '*vorsetzen*=to place in front', and in this text it usually refers to a defensive action we'd call a parry. Also used to describe four of the five strikes (curved, crosswise, cockeye, and parter).

Vom Tag — From the day; could also be '*von Dach*=from the roof'.

Vor — Before; when you constrain your opponent to parry.

Winden — Turning or winding; the sense isn't turning a key or coiling a rope, but rather moving back and forth (like 'a winding road').

Wunder — Wonder; sometimes translated as "wounder", but this isn't supported in any dictionary or other source. *Wunder* is pretty much always "miracle" or similar.

Zecken Skirmish; this is commonly translated with a variety of words suggesting minor attacking actions (and *Zeck* also means "tick", leading people to read it as "bug bites"), but the *Lexer* gives an alternative reading of "skirmishing" or harassing actions, which contrasts nicely with '*Krieg*=war' in the wrathful cut section and which I interpret as the term for actions that leave the bind (such as the take-away, many crosswise pieces including the misser, and the gloss of this section of the *Record* in Pseudo-Danzig and Lew).

Zeckrur Harrying strikes; this word doesn't appear in *Grimm* or *Lexer*, so I read it as a compound of *zecken* and *ruren*.

Zedel Record; the name of Liechtenauer's teaching poem.

Zornhaw Wrathful cut.

Zucken Pulling.

Zufechten Describes approaching an opponent with the intent to fence with them.

Zugangen Describes approaching, and in this gloss may be a reference to the structure of *Gänge* (or 'rounds' of counted blows) used in *Fechtschule* competitions. Unlike other glossators, Ainring adds a strike toward the head at the end of the sequence where this term appears followed by a very rare instruction to pull back/withdraw (*abzucken*). This would line up well with fencing a '*Gang* of five' blows at the *Fechtschule*.

Bibliography

Primary sources

This manuscript reference system was developed by Dierk Hagedorn. It encompasses all known manuscript fight books from the German language area, as well as those from surrounding areas that are typically grouped with or compared to the German texts. These manuscripts are all profiled on Wiktenauer <*www.wiktenauer.com*>.

A *Codex Amberger*
Towson, private collection

AD *Albrecht Dürer*
Wien, Albertina (Graphische Sammlung), Hs. 26-232

ADB *Albrecht Dürer-Breslau*
Lost (formerly Breslau, Universitätsbibliothek, Cod. 1246)

ADL *Albrecht Dürer-London*
London, British Library, Sloane MS No. 5229

AR *Anton Rast*
Augsburg, Stadtarchiv, Reichsstadt Nr. 82

BB *Verzeichnis etlicher Stücke des Fechtens im Rapier*
Berlin, Staatsbibliothek, Ms. germ. fol. 1476

BF *Bauman's fight book*
Universitätsbibliothek Augsburg, Cod.I.6.4°.2

BKW1 *Blume des Kampfes-Wien #1*
Wien, Österreichische Nationalbibliothek, Cod. 5278

BKW2 *Blume des Kampfes-Wien #2*
Wien, Österreichische Nationalbibliothek, Cod. 10799

BS *Berlin picture book*
Berlin, Staatsbibliothek, Libr. pict. A 83

FA *Fabian von Auerswald*
Göttingen, Niedersächsische Staats- und Universitätsbibliothek, 2° Col. Ms. philos. 62

FB *Fechtbüchlein*
Wolfenbüttel, Herzog August-Bibliothek, Cod. Guelf. 1074 Novi

FLG *Fior di Battaglia* or *Fiore de'i Liberi-Getty*
Los Angeles, J. Paul Getty Museum, Ms. Ludwig XV 13

FLM *Fiore de'i Liberi-Morgan*
New York, Morgan Library & Museum, Ms. M.383

FLP *Florius de Arte Luctandi* or *Fiore de'i Liberi-Paris*
Paris, Bibliothèque nationale de France, Mss. Latin 11269

FLPD *Flos Duellatorum* or *Fiore de'i Liberi-Pisani Dossi*
Corbetta, Museo archeologico Villa Pisani Dossi

FLZ *Fiore de'i Liberi-Zeno*
Biblioteca Guarneriana, Ms. XXIV

G *Goliath*
Kraków, Biblioteka Jagiellonska, Berol Ms. Germ. Qu. 2020

GE *Gregor Erhart*
Glasgow, Kelvingrove Museum, E.1939.65.354

GF *Glasgow Fechtbuch* or *Sigmund Emring*
Glasgow, Kelvingrove Museum, E.1939.65.341

GK *Gladiatoria-Kraków*
Kraków, Biblioteka Jagiellonska, Berol Ms. Germ. Qu. 16

GN *Gladiatoria-New Haven*
New Haven, Yale Center for British Art, U860 .F46 1450

GW *Gladiatoria-Wien*
Wien, Kunsthistorisches Museum, KK 5013

GWo *Gladiatoria-Wolfenbüttel*
Wolfenbüttel, Herzog August-Bibliothek, Cod Guelf. 78.2 Aug. 2°

HB *Hugold Behr*
Rostock, Universitätsbibliothek, Mss. var. 83

HC *Hans Czynner*
Graz, Universitätsbibliothek, Ms. 963

HE *Hans Entner*
Lost (formerly Nürnberg, Germanisches Nationalmuseum)

HF *Hans Folz*
Weimar, Herzogin Anna Amalia Bibliothek, Q.566

HG *Heinrich von Günterode*
Dresden, Sächsische Landesbibliothek, Mscr. Dresd. C.15

HM *Hans Medel* or *Sigmund Schining*
Augsburg, Universitätsbibliothek, Cod.I.6.2°.5

HS *Hans von Speyer*
Salzburg, Universitätsbibliothek, M. I. 29

HTA *Hans Talhoffer-Augsburg*
Augsburg, Universitätsbibliothek, Cod.I.6.2°.1

HTB *Hans Talhoffer-Berlin*
Berlin, Stiftung Preußischer Kulturbesitz, 78 A 1

HTC *Hans Talhoffer-Coburg*
Coburg, Kunstsammlungen der Veste Coburg, Hz. 014

HTG *Hans Talhoffer-Gotha*
Gotha, Forschungsbibliothek Schloss Friedenstein, Chart. A. 558

HTGö *Hans Talhoffer-Göttingen*
Göttingen, Niedersächsische Staats- und Universitätsbibliothek, 2° Col. Ms. philos. 61

HTK *Hans Talhoffer-Königsegg*
Königseggwald, Königsegg-Aulendorf Collection, Hs. XIX 17-3

HTKa *Hans Talhoffer-Kassel*
Kassel, Landesbibliothek, 2° Ms. iurid. 29

HTKo *Hans Talhoffer-København*
København, Det Kongelige Bibliotek, Thott 0290 2°

HTM *Hans Talhoffer-München #1*
München, Bayerische Staatsbibliothek, Cod. icon. 394a

HTM2 *Hans Talhoffer-München #2*
München, Bayerische Staatsbibliothek, Cod. icon. 394

HTM3 *Hans Talhoffer-München #3*
München, Bayerische Staatsbibliothek, Cod. icon. 395

HTN *Hans Talhoffer-New York*
New York, Metropolitan Museum of Art, 26.236

HTP *Hans Talhoffer-Private*
Private collection (formerly Wien, Österreichische Nationalbibliothek, Cod. ser. nov. 2978)

HTW *Hans Talhoffer-Wien*
Wien, Kunsthistorisches Museum, KK 5342

HTWo *Hans Talhoffer-Wolfenbüttel*
Wolfenbüttel, Herzog August Bibliothek, Cod. Guelf. 125. 16. Extrav.

HW *Hugo Wittenwiler*
München, Bayerische Staatsbibliothek, Cgm 558

J *Beringer-Jena*
Jena, Thüringer Universitäts- und Landesbibliothek, Ms. G. B. f. 18a

JH *Le jeu de la hache*
Paris, Bibliothèque nationale de France, Mss. Français 1996

JML *Joachim Meyer-Lund*
Lund, Universitetsbibliotek, Msc. A. 4° 2

JMM *Joachim Meyer-München*
München, Bayerisches Nationalmuseum, Bibl. 2465

JMR *Joachim Meyer-Rostock* or *Sigmund Einring*
Rostock, Universitätsbibliothek, Mss. var. 82

JSN *Jeremias Schemel-New York*
New York, Metropolitan Museum of Art, 23.279

JSW *Jeremias Schemel-Wien*
Wien, Kunsthistorisches Museum, KK 5247

JSW01 *Jeremias Schemel-Wolfenbüttel #1*
Wolfenbüttel, Herzog August Bibliothek, Cod. Guelf. 1.6.3 Aug. 2°

JSW02 *Jeremias Schemel-Wolfenbüttel #2*
Wolfenbüttel, Herzog August Bibliothek, Cod. Guelf .5 Blank

JWA2 *Jörg Wilhalm-Augsburg #2*
Augsburg, Universitätsbibliothek, Cod.I.6.2°.3

JWA3 *Jörg Wilhalm-Augsburg #3*
Augsburg, Universitätsbibliothek, Cod.I.6.4°.5

JWM1 *Jörg Wilhalm-München #1*
München, Bayerische Staatsbibliothek, Cgm 3711

K *Kölner Fechtregeln*
Köln, Historisches Archiv, Best. 7020 (W*)150

Ka *Kasseler picture book*
Kassel, Universitätsbibliothek, 2° Ms. Math. 7

KH *Krakauer Hausbuch*
Kraków, Biblioteka Jagiellonska, Rkp. Przyb. 35/64

L *Lew fight book*
Augsburg, Universitätsbibliothek, Cod.I.6.4°.3

LEE *Ludwig von Eyb-Erlangen*
Erlangen, Universitätsbibliothek, Ms. B.26

LEM *Ludwig von Eyb-München*
München, Bayerische Staatsbibliothek, Cgm 961

LF *Lutegerus fight book* or *Walpurgis fight book*
Leeds, Royal Armouries Museum, FECHT 1 (formerly MS I.33)

LH *Hans Leckküchner-Heidelberg*
Heidelberg, Universitätsbibliothek, Cod. Pal. germ. 430

LM *Hans Leckküchner-München*
München, Bayerische Staatsbibliothek, Cgm 582

LQ *Liber Quodlibetarius*
Erlangen, Universitätsbibliothek, B.200

LSA *Lienhart Sollinger-Augsburg* or *Jörg Wilhalm-Augsburg #1*
Augsburg, Universitätsbibliothek, Cod.I.6.2°.2

LSM *Lienhart Sollinger-München* or *Jörg Wilhalm-München #2*
München, Bayerische Staatsbibliothek, Cgm 3712

LSWo *Lienhart Sollinger-Wolfenbüttel* or *Jörg Wilhalm-Wolfenbüttel*
Wolfenbüttel, Herzog August-Bibliothek, Cod. Guelf. 38.21 Aug. 2°

N *Nicolas Pol Hausbuch* or *Nürnberger Hausbuch*
Nürnberg, Germanisches Nationalmuseum, Cod. Hs. 3227[a]

O *Ortenburg fight book*
Private collection (formerly Tambach, Ortenburg Hofbibliothek)

P *Paris fight book* or *Cluny fight book*
Paris, Musée National du Moyen Age, CL23842

PD *Peter von Danzig fight book*
Roma, Accademia Nazionale dei Lincèi, Cod. 44.A.8

PFS *Peter Falkner-Strasbourg*
Destroyed (formerly Strasbourg, Stadtarchiv)

PFW *Peter Falkner-Wien*
Wien, Kunsthistorisches Museum, KK 5012

PKB *Paul Kal-Bologna*
Bologna, Universitätsbibliothek, Ms. 1825

PKG *Paul Kal-Gotha*
Gotha, Forschungsbibliothek Schloss Friedenstein, Chart. B. 1021

PKG1 *Paul Kal-Botticino*
Botticino, Museo dell'Arte Marziale, M.A.M. Kal MS

PKM *Paul Kal-München*
München, Bayerische Staatsbibliothek, Cgm 1507

PKP *Paul Kal-Private*
Private collection

PKS *Solothurner Fechtbuch*
Solothurn, Zentralbibliothek, S. 554

PKW *Paul Kal-Wien*
Wien, Kunsthistorisches Museum, KK 5126

PMA *Paul Hektor Mair-Augsburg*
Augsburg, Universitätsbibliothek, Cod.I.6.2°.4

PMD *Paul Hektor Mair-Dresden*
Dresden, Sächsische Landesbibliothek, Mscr. Dresd. C.93/94

PMM *Paul Hektor Mair-München*
München, Bayerische Staatsbibliothek, Cod. icon. 393

PMW *Paul Hektor Mair-Wien*
Wien, Österreichische Nationalbibliothek, Cod. 10825/6

PVR *Philippo di Vadi-Roma*
Roma, Biblioteca Nazionale Centrale, Vitt. Em. 1324

SK *Shermkunst*
Chicago, Newberry Library, VAULT Case MS folio U 423 .792

SR *Sigmund ain ringeck fight book*
Dresden, Sächsische Landesbibliothek, Mscr. Dresd. C.487

V *Vechtboek*
Leiden, Universiteitsbibliotheek Leiden, BPL 3281

W *Wiener fight book*
Wien, Österreichische Nationalbibliothek, Cod. 11093

WF *Wassmannsdorff's fight book*
Lost

WP Willibald Pirckheimer
Nürnberg, Stadtbibliothek, PP. 353

W01 *Wolfenbüttel fight book #1*
Wolfenbüttel, Herzog August-Bibliothek, Cod. Guelf. 83.4 Aug. 8°

SECONDARY SOURCES

ACUTT 2019

JAY ACUTT. *Swords, Science, and Society: German Martial Arts in the Middle Ages*. Glasgow: Fallen Rook Publishing.

ALDERSON 2010

KEITH ALDERSON. "On the Art of Reading: An Introduction to Using the Medieval German 'Fightbooks'." *In the Service of Mars: Proceedings from the Western Martial Arts Workshop, 1999–2009*: 251–286. Wheaton: Freelance Academy Press.

ALDERSON 2014

———. "Arts and Crafts of War: *die Kunst des Schwerts* in its Manuscript Context." *Can These Bones Come to Life? Insights from Reconstruction, Reenactment, and Re-creation* **1**: 24–29. Wheaton: Freelance Academy Press.

ALTMANN 2006

LOTHAR ALTMANN. "Die Schlacht bei Alling und die Wittelsbacher Memorialkapelle Hoflach". *Amperland* **46**: 223–230.

ANGLO 2000

SYDNEY ANGLO. *The Martial Arts of Renaissance Europe*. New Haven & London: Yale University Press.

BAS 2013

PIERRE-HENRY BAS. The true edge: a comparison between self-defense fighting from German "*fight-books*" (*Fechtbücher*) and the reality of judicial sources (1400–1550). *Acta Periodica Duellatorum* **1**(1): 179-195.

BAUER 2014A

MATTHIAS JOHANNES BAUER. "Ein Zedel Fechter ich mich ruem/Im Schwert un Messer ungestuem. Fechtmeister als protagonisten und als (fach)literarisches Motiv in den deutschsprachigen Fechtlehren des ittelalters und der frühen Neuzeit." *Das Mittelalter* **19**(2): 302–325. Ed. by CHRISTIAN JASER and UWE ISRAEL. doi:10.1515/mial-2014-0018.

BAUER 2014B

———. "Fechten lehren 'mitt verborgen vnd verdeckten worten'. Fachsprache, Dialekt, Verballhornung und Geheimsprache in frühneuhochdeutschen Zweikampftraktaten." *Das Schwert – Symbol und Waffe*: 163–172. Ed. by LISA DEUTSCHER, MIRJAM KAISER, and SIXT WETZLER. Rahden: Verlag Marie Leidorf GmbH.

BECKER 2020

PAUL BECKER. "Hans Talhoffer: Insights into the Life of a Fencing Master in the 15th Century." *Alte Armature und Ringkunst: The Royal Danish Library Ms. Thott 290 2°*: 47–68. Ed. Michael Chidester. Somerville: HEMA Bookshelf.

BERGNER & GIEßAUF 2006

UTE BERGNER and JOHANNES GIEßAUF. *Würgegriff und Mordschlag. Die Fecht- und Ringlehre des Hans Czynner (1538).* Graz: Akademische Druck- u. Verlagsanstalt.

BODEMER 2008

HEIDEMARIE BODEMER. *Das Fechtbuch. Untersuchungen zur Entwick-lungsgeschichte der bildkünstlerischen Darstellung der Fechtkunst in den Fechtbüchern des mediterranen und westeuropäischen Raumes vom Mittelalter bis Ende des 18. Jahrhunderts* [Unpublished doctoral dissertation]. Universität Stuttgart, 2008.

BURGER 2021

DANIEL BURGER. "Hans Lecküchner of Nuremberg and His Fencing Treatise in the Long Knife." *Kunst und Zettel im Messer: Bavarian State Library Cgm 582*: 121–140. Ed. MICHAEL CHIDESTER. Somerville: HEMA Bookshelf.

BURKART 2016

ERIC BURKART. "The Autograph of an Erudite Martial Artist: A Close Reading of Nuremberg, Germanisches Nationalmuseum, Hs. 3227a." *Late Medieval and Early Modern Fight Books. Transmission and Tradition of Martial Arts in Europe*: 451–480. Ed. by DANIEL JAQUET, KARIN VERELST, and TIMOTHY DAWSON. Boston: Brill. doi:10.1163/9789004324725_017.

BURKART 2018

———. "Informationsverarbeitung durch autographe Notizen: Die ältesten Aufzeichnungen zur Kampfkunst des Johannes Liechtenauer als Spuren einer Aneignung praktischen Wissens." *Mittelalter. Interdisziplinäre Forschung und Rezeptionsgeschichte* **S2**: 117–158. doi:10.26012/mittelalter-25866.

CAMARGO 2021

Sigmund ain Ringeck. *El Arte del Combate con Espada Larga.* Trans. by ARTURO CAMARGO. Self-published.

CAVINA 2016

MARCO CAVINA. "Science of Duel and Science of Honour in the Modern Age: The Construction of a New Science between Customs, Juris-

prudence, Literature and Philosophy". *Late Medieval and Early Modern Fight Books. Transmission and Tradition of Martial Arts in Europe*: 571–593. Ed. by DANIEL JAQUET, KARIN VERELST, and TIMOTHY DAWSON. Boston: Brill, 2016.
doi: 10.1163/9789004324725_020

CHENEY 2020A
Sigmund ain Ringeck, et al. *Ringeck • Danzig • Lew Long Sword.* Trans. by STEPHEN CHENEY. Self-published.

CHENEY 2020B
"Sigmund ain Ringeck." Trans. by STEPHEN CHENEY. *Wiktenauer.* <*wiktenauer.com/wiki/Sigmund_ain_Ringeck*>

CHIDESTER, ET AL. 2016
Sigmund Ainring, Peter von Danzig, Paulus Kal, et al. *The Recital of the Chivalric Art of Fencing of the Grand Master Johannes Liechtenauer.* Trans. by MICHAEL CHIDESTER, MIKE RASMUSSON, DAVID RAWLINGS, THOMAS STOEPPLER, CHRISTIAN HENRY TOBLER, CHRISTIAN TROSCLAIR, and CORY WINSLOW. Ed. by MICHAEL CHIDESTER and DIERK HAGEDORN. Wiktenauer (self-published).

CHIDESTER, ET AL. 2020
Hans Talhoffer. *Alte Armature und Ringkunst: The Royal Danish Library Ms. Thott 290 2°.* Trans. by REBECCA L. R. GARBER. Ed. by MICHAEL CHIDESTER and DIETER BACHMANN. Somerville: HEMA Bookshelf.

CHIDESTER 2021
Fiore de'i Liberi. *The Flower of Battle: MS M 383.* Trans. by MICHAEL CHIDESTER. Somerville: HEMA Bookshelf.

CHIDESTER & HAGEDORN 2021
The Foundation and Core of All the Arts of Fighting': The Long Sword Gloss of GNM Manuscript 3227a. Trans. by MICHAEL CHIDESTER. Ed. by MICHAEL CHIDESTER and DIERK HAGEDORN. Somerville: HEMA Bookshelf.

CHIDESTER, ET AL. 2021
Johannes Lecküchner. *Kunst und Zettel im Messer: Bavarian State Library Cgm 582.* Ed. by MICHAEL CHIDESTER. Somerville: HEMA Bookshelf.

CHIDESTER, ET AL. 2023
Bauman's Fight Book: Augsburg University Library I.6.4° 2. Trans. by CHRISTIAN HENRY TOBLER and JESSICA FINLEY. Ed. by MICHAEL CHIDESTER. Medford: HEMA Bookshelf.

CHIDESTER 2024

MICHAEL CHIDESTER. "Reception of Joachim Meyer's *Gründtliche Beschreibung der... Kunst des Fechtens*, 1570–1686." *Acta Periodica Duellatorum* (forthcoming).

CHIDESTER, ET AL. 2024

Various authors. "Sigmund ain Ringeck." Ed. by MICHAEL CHIDESTER. *Wiktenauer.*
<*wiktenauer.com/wiki/Sigmund_ain_Ringeck*>

CONAN & LEBLANC 2024

BENJAMIN CONAN and HÉLÈNE LEBLANC. *Comme le dit Maître Hans... L'art de l'escrime à l'épée longue.* Les éditions Solstices.

CÓRDOBA 2013

Craft Treatises and Handbooks: The Dissemination of Technical Knowledge in the Middle Ages. Ed. by RICARDO CÓRDOBA. Turnhout: Brepols.

DANE 2012

DANE, JOSEPH A. *What Is a Book? The Study of Early Printed Books.* Notre Dame: University of Notre Dame.

DEACON 2016

JACOB HENRY DEACON. "Prologues, Poetry, Prose and Portrayals: The Purposes of Fifteenth Century Fight Books According to the Diplomatic Evidence." *Acta Periodica Duellatorum* **4**(2): 69–90. doi:10.36950/apd-2016-014.

DÖRNHÖFFER 1910

FRIEDRICH DÖRNHÖFFER. *Albrecht Dürers Fechtbuch.* Wien: F. Tempsky; Leipzig: G. Freytag.

DOROFEEVA 2019

ANNA DOROFEEVA. "Visualizing Codicologically and Textually Complex Manuscripts." *Manuscript Studies* **4**(2): 334–360.

DUPUIS 2009

"Transkription der Handschrift I.6.2°.5." Ed. by OLIVIER DUPUIS. *Gesellshaft für Pragmatische Schriftlichkeit.*
<*www.pragmatische-schriftlichkeit.de/Cod.I.6.20.5.html*>

DUPUIS 2013

DUPUIS, OLIVIER. "A fifteenth-century fencing tournament in Strasburg." *Acta Periodica Duellatorum* **1**(1): 67–79. 2013.
doi:10.36950/apd-2013-004

ELEMA 2012
ARIELLA ELEMA. *Trial by Battle in France and England* [Unpublished dissertation]. University of Toronto.

ELEMA 2019
———. "Tradition, Innovation, Re-enactment: Hans Talhoffer's Unusual Weapons." *Acta Periodica Duellatorum* 7(1): 3–25. doi:10.2478/apd–2019–0001.

ENGSTRÖM 2009
"Codex I.6.2.5 (Library of the University of Augsburg)." Ed. by ANDREAS ENGSTRÖM. *HROARR*.
<*www.hroarr.com/manuals/liechtenauer/Codex_I.6.2.5_transcription_1.0.pdf*>

FABIAN 2003
BERNHARD FABIAN. *Handbuch der historischen Buchbestände in Deutschland, Österreich und Europa.* Niedersächische Saats- und Universitätsbibliothek.
<*fabian.sub.uni-goettingen.de/fabian*>

FABIAN 2024
MARTIN FABIAN. *Fechtbuch Fabian.* Self-published.

FINLEY 2019
JESSICA FINLEY. "Visualizing Liechtenauer: Using Structure to Understand Verse" [Lecture]. *Longpoint*, 28 Mar 2019.

FINLEY 2021
JESSICA FINLEY. "'What's in a Name?' A Comparative Analysis of the Nomenclature of Johannes Leckü̈chner and Johannes Liechtenauer." *Kunst und Zettel im Messer: Bavarian State Library Cgm 582*: 157–174. Ed. by MICHAEL CHIDESTER. Somerville: HEMA Bookshelf.

FINLEY & TOBLER 2022
——— and CHRISTIAN HENRY TOBLER. "Warp and Weft: The *Bauman Fight Book*'s Place in the Tapestry of German *Fechtkunst*." *Bauman's Fight Book: Augsburg University Library I.6.4° 2*: 85–102. Ed. by MICHAEL CHIDESTER. Medford: HEMA Bookshelf.

FORGENG 2012
FORGENG, JEFFREY L. "Owning the Art: The German Fechtbuch Tradition." *The Noble Art of the Sword: Fashion and Fencing in Renaissance Europe 1520–1630*: 167–168. Ed. by TOBIAS CAPWELL. London: Paul Holberton Publishing.

FORGENG 2015

Johannes Leckùchner. *The Art of Swordsmanship by Hans Leckùchner.* Trans. by JEFFREY L. FORGENG. Woodbridge: Boydell Press.

FORGENG 2016

Joachim Meyer. *The Art of Sword Combat: A 1568 German Treatise on Swordsmanship.* Trans. by JEFFREY L. FORGENG. London: Frontline Books.

GAITE 2018

PIERRE GAITE. "Exercises in Arms: the Physical and Mental Combat Training of Men-at-Arms in the Fourteenth and Fifteenth Centuries." *Journal of Medieval Military History* **XVI**. Ed. by KELLY DEVRIES, JOHN FRANCE, and CLIFFORD J. ROGERS. Woodbridge: Boydell Press.

GARBER 2023

Joachim Meyer. *Foundational Description of the Art of Fencing: The 1570 Treatise of Joachim Meyer (Reference Edition)* (2 vols). Illus. by Hans Christoph Stimmer. Trans. by REBECCA L. R. GARBER. Ed. by MICHAEL CHIDESTER. Medford: HEMA Bookshelf.

GRIMM

JACOB GRIMM and WILHELM GRIMM. *Deutsches Wörterbuch von Jacob Grimm und Wilhelm Grimm*, Version 01/23. Wörterbuchnetz des Trier Center for Digital Humanities.
<*www.woerterbuchnetz.de/DWB*>

HAGEDORN 2008A

Transkription und Übersetzung der Handschrift 44 A 8. Ed. by DIERK HAGEDORN. Herne: VS-Books.

HAGEDORN 2008B

"Sigmund Ringeck." Ed. by DIERK HAGEDORN. *Hammaborg Historischer Schwertkampf.*
<*www.hammaborg.de/de/transkriptionen/sigmund_ringeck/*>

HAGEDORN 2009A

"Hans von Speyer." Ed. by DIERK HAGEDORN. *Hammaborg Historischer Schwertkampf.*
<*www.hammaborg.de/de/transkriptionen/hans_von_speyer/*>

HAGEDORN 2009B

"Glasgow: E.1939.65.341." Ed. by DIERK HAGEDORN. *Hammaborg Historischer Schwertkampf.*
<*www.hammaborg.de/de/transkriptionen/emring_glasgow/*>

HAGEDORN 2015A
DIERK HAGEDORN. "The Adventures of the Gladiatoria Manuscript from New Haven." *Acta Periodica Duellatorum* **3**(1): 183–208. doi:10.1515/apd-2015-0006.

HAGEDORN 2015B
"Joachim Meyer, Rostock." Ed. by DIERK HAGEDORN. *Hammaborg Historischer Schwertkampf.* <www.hammaborg.de/de/transkriptionen/meyer_rostock/>

HAGEDORN & WALCZAK 2015
DIERK HAGEDORN and BARTŁOMIEJ WALCZAK. *Gladiatoria. New Haven – MS U860.F46 1450.* Herne: VS-Books.

HAGEDORN 2016A
DIERK HAGEDORN. "German Fechtbücher from the Middle Ages to the Renaissance." *Late Medieval and Early Modern Fight Books. Transmission and Tradition of Martial Arts in Europe*: 247–279. Ed. by DANIEL JAQUET, KARIN VERELST, and TIMOTHY DAWSON. Boston: Brill. doi: 10.1163/9789004324725_011.

HAGEDORN 2016B
"Albrecht Dürer." Ed. by DIERK HAGEDORN. *Hammaborg Historischer Schwertkampf.* <www.hammaborg.de/de/transkriptionen/albrecht_duerer/>

HAGEDORN 2017
DIERK HAGEDORN. "Spezialeinsatz mit Mordschlag und Straße der Glieder: Der Kampf im Harnisch in den Gladiatoria-Handschriften." *'Kunst dye dich zyret' – Fechten als Mittel persönlicher und institutioneller Repräsentation.* Ed. by THORE WILKENS and UWE FIEDLER. Dresden: Sandstein Verlag.

HAGEDORN, ET AL. 2017
Jude Lew: Das Fechtbuch. Ed. by DIERK HAGEDORN. Herne: VS-Books.

HAGEDORN 2018
Hans Talhoffer. *Medieval Combat in Colour: Hans Talhoffer's Illustrated Manual of Swordfighting and Close-Quarter Combat from 1467.* Ed. by DIERK HAGEDORN. Barnsley: Greenhill Books.

HAGEDORN 2020
DIERK HAGEDORN. "Talhoffer Galore." *Alte Armature und Ringkunst: The Royal Danish Library Ms. Thott 290 2°*: 69–88. Ed. by MICHAEL CHIDESTER. Somerville: HEMA Bookshelf.

HAGEDORN & AMBERGER 2020
——— and CHRISTOPH AMBERGER. *Codex Amberger.* Herne: VS-Books.

HAGEDORN 2021A
Albrecht Dürer. *Albrecht Dürer. Das Fechtbuch.* Ed. by DIERK HAGEDORN. Herne: VS-Books.

HAGEDORN 2021A
DIERK HAGEDORN. "Many Magnificent Messer Manuscripts (Plus Plentiful Picturesque Prints)." *Kunst und Zettel im Messer: Bavarian State Library Cgm 582*: 175–190. Ed. by MICHAEL CHIDESTER. Somerville: HEMA Bookshelf.

HAGEDORN, ET AL. 2021
Jörg Wilhalm. *Renaissance Combat. Jörg Wilhalm's Fightbook, 1522–1523.* Trans. by DIERK HAGEDORN, HELEN HAGEDORN, HENRI HAGEDORN. Barnsley: Greenhill Books.

HAGEDORN & TOBLER 2021
The Peter von Danzig Fight Book. Trans. by CHRISTIAN HENRY TOBLER. Ed. by DIERK HAGEDORN. Wheaton: Freelance Academy Press.

HAGEDORN 2022
DIERK HAGEDORN. "Joining the Dots: Additional Comments on the Connection Between Bauman's Fight Book, the Gladiatoria Group, and Fiore de'i Liberi." *Bauman's Fight Book: Augsburg University Library I.6.4° 2*: 195–210. Ed. by MICHAEL CHIDESTER. Medford: HEMA Bookshelf.

HAGEDORN & JAQUET 2022
Albrecht Dürer. *Dürer's Fight Book: The Genius of the German Renaissance and his Combat Treatise.* Ed. by DIERK HAGEDORN and DANIEL JAQUET. Barnsley: Greenhill Books.

HAGEDORN 2023
Das Ortenburger Fechtbuch. Ed. by DIERK HAGEDORN. Herne: VS-Books.

HAGEDORN 2024
Das Spiel mit der Axt – Le Jeu de la Hache. Ed. by DIERK HAGEDORN. Herne: VS-Books.

HERGSELL 1887
Hans Talhoffer. *Talhoffers Fechtbuch aus dem jahre 1467: Gerichtliche und andere zweikämpfe darstellend.* Ed. by GUSTAV HERGSELL. Praha: J. G. Calve.

HERGSELL 1889A
———. *Talhoffers Fechtbuch (Gothaer Codex) aus dem Jahre 1443: Gerichtliche und andere Zweikämpfe darstellend.* Ed. by GUSTAV HERGSELL. Praha: Self-published.

HERGSELL 1889B
———. *Talhoffers Fechtbuch (Ambraser Codex) aus dem Jahre 1459: Gerichtliche und andere Zweikämpfe darstellend.* Ed. by GUSTAV HERGSELL. Praha: Self-published.

HERGSELL 1894
———. *Livre d'escrime de Talhoffer de l'an 1467.* Ed. by GUSTAV HERGSELL. Praha: Chez L'Auteur.

HERMANN 2017
HANS-GEORG HERMANN. "Die prozessuale Konfiguration des spatmittelalterlichen Zweikampfes als Deutungskontext von Fechtbüchern." *Die Kunst des Fechtens*: 23–76. Ed. by MATTHIAS JOHANNES BAUER and ELISABETH VAVRA. Heidelberg: Universitätsverlag Winter GmbH.

HILS 1985
HANS-PETER HILS. *Meister Johann Liechtenauers Kunst des langen Schwertes.* Frankfurt am Main: P. Lang.

HIRSCH 1933
Valuable Manuscripts of the Middle-Ages Mostly Illuminated. München: Emil Hirsch Antiquariat.

HOEPLI 1933
Autographes, manuscrits enluminés, incunables, livres illustrés du XVIe au XVIIIe siècle, éditions originales françaises du XIXe siècle, ouvrages d'intérêt musical, éditions de luxe modernes, gravures, Helvetica. Milano: Librairie Ancienne Ulrico Hoepli.

HOFFMANN 2022
WERNER J. HOFFMANN. *Die Mittelalterlichen Deutschen und Niederländischen Handschriften der Sächsischen Landesbibliothek – Staats- Und Universitätsbibliothek Dresden.* Die Deutsche Nationalbibliothek.

HUGHES 2007
STEVEN HUGHES. "Soldiers and Gentlemen: The Rise of the Duel in Renaissance Italy". *Journal of Medieval Military History* **V**: 99–152.

HULL, ET AL. 2007
———, GRZEGORZ ŻABIŃSKI, MONIKA MAZIARZ. *Knightly Dueling: The Fighting Arts of German Chivalry.* Boulder: Paladin Press.

HULL 2008

JEFFREY HULL. "The Longsword Fight Lore of Mertin Siber." *Masters of Medieval and Renaissance Martial Arts*: 223–238. Ed. by JOHN CLEMENTS. Boulder: Paladin Press.

JAQUET & WALCZAK 2014

DANIEL JAQUET and BARTŁOMIEJ WALCZAK. "Liegnitzer, Hundsfeld or Lew? The question of authorship of popular Medieval fighting teachings." *Acta Periodica Duellatorum* **2**(1): 105–148. doi:10.1515/apd-2015-0015.

JAQUET 2017

DANIEL JAQUET. "The collection of Lew the Jew in the lineage of German Fight Books corpus." *Acta Periodica Duellatorum* **5**(1): 151–191. doi:10.1515/apd-2017-0004.

JAQUET 2018

———. "Six Weeks to Prepare for Combat: Instruction and Practices from the Fight Books at the End of the Middle Ages, a Note on Ritualised Single Combats: Perspectives on Fighters in the Middle Ages." *Killing and Being Killed: Bodies in Battle*: 131–164. Bielefeld: Transcript-Verlag. doi:10.14361/9783839437834-009.

JUNGREITHMAYR 1988

"M I 29. Fechtbuch." *Die deutschen Handschriften des Mittelalters der Universitätsbibliothek Salzburg* **2**: 5–7. Ed. by ANNA JUNGREITH-MAYR. Wien: Verlag der Österreichischen Akademie der Wissenschaften. doi.org/10.1553/0x00058838d

KELLETT 2012

RACHEL E. KELLETT. "Royal Armouries MS I.33: The Judicial Combat and the Art of Fencing in Thirteenth- and Fourteenth- Century German Literature." *Oxford German Studies* **41**(1): 32–56. doi:10.1179/0078719112Z.0000000003.

KELLETT 2015

RACHEL E. KELLETT. "The Placement and Consequences of Swordblows in Sigmund Ringeck's Fifteenth-Century Fencing Manual." *Wounds and wound repair in Medieval culture*: 128–150. Ed. by LARISSA TRACY and KELLY DEVRIES. Leiden: Brill. doi:10.1163/9789004306455_007.

KURRAS 1980

LOTTE KURRAS. *Die deutschen mittelalterlichen Handschriften. 2. Teil: Die naturkundlichen und historischen Handschriften, Rechtshandschriften, Varia.* Wiesbaden: Otto Harrassowitz, 1980.

KWAKKEL 2012

ERIK KWAKKEL. "Late Medieval Text Collections: A Codicological Typology Based on Single-Author Manuscripts." *Reader, Author, Book: Medieval Authorship in Theory and Practice*: 56–79. Ed. STEPHEN PARTRIDGE and ERIK KWAKKEL. Toronto, Buffalo, & London: University of Toronto Press, 2012.

LEE 2022

LUCIEN LEE. "*Messerfechten* from the Augsburg Tradition." *Bauman's Fight Book: Augsburg University Library I.6.4° 2*: 129–156. Ed. by MICHAEL CHIDESTER. Medford: HEMA Bookshelf.

LENG 2008

RAINER LENG. *Katalog der deutschsprachigen illustrierten Handschriften des Mittelalters (KdiH), begonnen von Hella Frühmorgen-Voss, fortgeführt von Norbert H. Ott zusammen mit Ulrike Bodemann und Christine Stöllinger-Löser, Bd. 4/2 (Lfg. 1/2) (Fecht- und Ringbücher).* München: C. H. Beck, 2008.

LEXER

MATTHIAS LEXER. *Mittelhochdeutsches Handwörterbuch von Matthias Lexer*, Version 01/23. Wörterbuchnetz des Trier Center for Digital Humanities.
<*www.woerterbuchnetz.de/Lexer*>

LINDHOLM & SVÄRD 2003

Sigmund ain Ringeck, et al. *Sigmund Ringeck's Knightly Art of the Longsword.* Illus. by PETER SVÄRD. Trans. by DAVID LINDHOLM. Boulder: Paladin Press.

LINDHOLM & SVÄRD 2006

———. *Sigmund Ringeck's Knightly Arts of Combat.* Illus. by PETER SVÄRD. Trans. by DAVID LINDHOLM. Boulder: Paladin Press.

MITTERMÜLLER 1867

RUPERT MITTERMÜLLER. *Albert der Dritte, Herzog von München-Straubing* (Vol. 1). Landshut: Thomann.

MITTERMÜLLER 1869

RUPERT MITTERMÜLLER. *Albert der Dritte, Herzog von München-Straubing* (Vol. 2). Landshut: Thomann.

MUHLBERGER 2005

STEVEN MUHLBERGER. *Deeds of Arms*. Highland Village: Chivalry Bookshelf.

MUHLBERGER 2019

STEVEN MUHLBERGER. *Murder, Rape, and Treason: Judicial Combat in the Late Middle Ages*. Wheaton: Freelance Academy Press.

MÜLLER 1992

JAN-DIRK MÜLLER. "Bild – Verse – Prosakommentar am Beispiel von Fechtbüchern. Probleme der Verschriftlichung einer schriftlosen Praxis." *Pragmatische Schriftlichkeit im Mittelalter. Erscheinungsfor-men und Entwicklungsstufen*: 251–282. Ed. by HAGEN KELLER and KLAUS GRUBMÜLLER, NIKOLAUS STAUBACH. München: Fink.

RECTOR 2000

Hans Talhoffer. *Medieval Combat: A Fifteenth-Century Illustrated Manual of Swordfighting and Close-Quarter Combat*. Trans. by MARK RECTOR. Barnsley: Greenhill Books.

RIDGEWAY 2019

Peter von Danzig. Trans. by HARRY RIDGEWAY. Self-published.

ROTH 1917

Die Chroniken der deutschen Städte vom 14. bis ins 16. Jahrhundert **32**. Ed. by FRIEDRICH ROTH. Leipzig: S. Hirzel.

SCHEIBLE & HOLZBERG 2012

HELGA SCHEIBLE and NIKLAS HOLZBERG. *Provisorischer Katalog der Pirckheimer-Papiere (PP) in der Stadtbibliothek Nürnberg* [unpublished]. München: Stadtbibliothek Nürnberg.

SCHNEIDER 1978

KARIN SCHNEIDER. *Die deutschen Handschriften der Bayerischen Staatsbibliothek München: Cgm 501–690*. Wiesbaden: Harrassowitz.

SHARP 2024

TRISTAN W. SHARP. "Putting the Violence Back in the Late Medieval German Feud." *Austrian History Yearbook* **55**: 1–12.

SMALLRIDGE & CHENEY 2022

The Fencing Art. Trans. by MIKE SMALLRIDGE. Ed. by STEPHEN CHENEY. Self-published.

SMITH 2010

PAMELA H. SMITH. "Why Write a Book? From Lived Experience to the Written Word in Early Modern Europe." *Bulletin of the German Historical Institute* **47**: 25–50.

SMITH 2021
BENJAMIN J. SMITH. *The Art of Longsword Fighting: Teaching the Foundations of Sigmund Ringeck's Style.* Barnsley & Philadelphia: Frontline Books.

SMITH 2022
PAMELA H. SMITH. *From Lived Experience to the Written Word: Reconstructing Practical Knowledge in the Early Modern World.* Chicago & London: The University of Chicago Press.

STRAUSS 1974
WALTER L. STRAUSS. *The complete drawings of Albrecht Dürer.* New York: Abaris Books.

TALAGA & TALAGA 2018
MACIEJ TALAGA and SZYMON TALAGA. "Do you even Zornhaw? A set-theoretic approach to HEMA reconstruction." *Acta Periodica Duellatorum* **6**(1): 151–181.

TALAGA 2020
MACIEJ TALAGA. "Affordances theory as an operational framework for interpretation of past material culture and practices. Praxiography of things, bodies, and motions." *AVANT* **XI** (2): 1–22. doi:10.26913/avant.2020.02.11.

TALAGA 2022
———. "'Have the Highest Righteous Fencer in Your Mind's Eye': Medieval Martial Ethic as a Conceptual Repository for Just War Theory." *Martial Arts Studies* **12**: 8–18. doi:10.18573/mas.154.

TOBLER 2001
Sigmund ain Ringeck, et al. *Secrets of German Medieval Swordsmanship.* Trans. by CHRISTIAN HENRY TOBLER. Union City: Chivalry Bookshelf.

TOBLER 2006
Paulus Kal. *In Service of the Duke: The 15th Century Fighting Treatise of Paulus Kal.* Trans. by CHRISTIAN HENRY TOBLER. Highland Village: Chivalry Bookshelf.

TOBLER 2010A
CHRISTIAN HENRY TOBLER. *In Saint George's Name: An Anthology of Medieval German Fighting Arts.* Wheaton: Freelance Academy Press.

TOBLER 2010B
———. *Messer Fighting from the Glasgow Fechtbuch*. Wheaton: Freelance Academy Press.

TOBLER 2011
Peter Falkner. *Captain of the Guild: Master Peter Falkner's Art of Knightly Defense*. Trans. by CHRISTIAN HENRY TOBLER. Wheaton: Freelance Academy Press.

TOBLER 2022
CHRISTIAN HENRY TOBLER. *Lance, Spear, Sword, & Messer: A German Medieval Martial Arts Miscellany*. Wheaton: Freelance Academy Press.

TROSCLAIR 2015A
"Sigmund ain Ringeck." Trans. by CHRISTIAN TROSCLAIR. *Wiktenauer*. <*wiktenauer.com/wiki/Sigmund_ain_Ringeck*>

TROSCLAIR 2015B
"Hans Medel." Trans. by CHRISTIAN TROSCLAIR. *Wiktenauer*. <*wiktenauer.com/wiki/Hans_Medel*>

UEBERSCHÄR 2018
WERNER UEBERSCHÄR, DANIEL BURGER. "Transkription (vorläufige Version) Anthony Rast (Anthoni Rasch) 1553. Schätze 82 Reichsstadt, Stadtarchiv Augsburg." *Schwertbund Nurmberg e.V.* <*www.schwertbund-nurmberg.de/Transkription%20Anthony%20Rast.pdf*>

VANSLAMBROUCK 2023
CHRIS VANSLAMBROUCK. "Introduction: The Brief Life and Times of Joachim Meÿer." *Foundational Description of the Art of Fencing: The 1570 Treatise of Joachim Meyer (Reference Edition)* **1**. Ed. by MICHAEL CHIDESTER. Medford: HEMA Bookshelf.

VERELST 2016
KARIN VERELST. "Finding a Way through the Labyrinth: Some Methodological Remarks on Critically Editing the Fight Book Corpus." *Late Medieval and Early Modern Fight Books. Transmission and Tradition of Martial Arts in Europe*: 117–188. Ed. by DANIEL JAQUET, KARIN VERELST, and TIMOTHY DAWSON. Boston: Brill. doi:10.1163/9789004324725_008.

VODIČKA 2019
ONDŘEJ VODIČKA. "Origin of the oldest German Fencing Manual Compilation (GNM Hs. 3227a)." *Waffen- und Kostümkunde* **61**(1): 87–108.

WALCZAK & STARKO 2018
BARTŁOMIEJ WALCZAK and BARTOSZ STARKO. "Additional Transmissions of Hundsfeld and Lignitzer Dagger Teachings." *Acta Periodica Duellatorum* **6**(1): 137–149.
doi:10.2478/apd-2018-0005

WALCZAK 2022
BARTŁOMIEJ WALCZAK. "Bauman Dagger Techniques and the Augsburg Group." *Bauman's Fight Book: Augsburg University Library I.6.4° 2*: 103–128. Ed. by MICHAEL CHIDESTER. Medford: HEMA Bookshelf.

WALDMANN 2013
SZABOLCS WALDMANN. "Mertein Hündsfelder: Fechtlehre mit dem Kurzen Schwert, circa 1491 AD, Fight-Teaching with the Shortened Sword from Codex Speyer (137r–141r)." *Acta Periodica Duellatorum* **1**(2): 93–105.
doi:10.36950/apd-2013-006.

WASSMANNSDORFF 1870
KARL WASSMANNSDORFF. *Die Ringkunst des deutschen Mittelalters.* Liepzig: Priber.

WASSMANNSDORFF 1888
KARL WASSMANNSDORFF. "Nachtrag über das sog. Lebkhommer'sche Fechtbuch und dessen Verhältnis zu der Breslauer Dürer Handschrift von 1512." *Monatsschrift für das Turnwesen, mit besonderer Berücksichtigung des Schulturnens und der Gesundheitspflege* **7**: 138–145.

WELLE 1993
RAINER WELLE. *'...und wisse das alle höbischeit kompt von deme ringen'. Der Ringkampf als adelige Kunst im 15. und 16. Jahrhundert.* Pfaffenweiler: Centaurus-Verlagsgesellschaft.

WELLE 2009
RAINER WELLE. "Ordnung als Prinzip. Eine Besprechung von Bd. 4/2, Lieferung 1/2, Nr. 38: 'Fecht- und Ringbücher' des Katalogs der deutschsprachigen illustrierten Handschriften des Mittelalters." *Medium Aevum Quotidianum* **59**: 37–59. Ed. by GERHARD JARITZ.

WELLE 2014
———. *"...vnd mit der rechten faust ein mordstuck". Baumanns Fecht- und Ringkampfhandschrift. Edition und Kommentierung der anonymen Fecht- und Ringkampfhandschrift Cod. I.6.4° 2 der UB Augsburg aus den Beständen der ehemaligen Öttingen-Wallersteinschen Bibliothek* (2 vols). München: Herbert Utz Verlag.

WELLE 2017
———. "Ein unvollendetes Meisterwerk der Fecht- und Ringkampfliteratur des 16. Jahrhunderts sucht seinen Autor: der Landshuter Holzschneider und Maler Georg Lemberger als Fecht- und Ringbuchillustrator?" *Codices manuscripti & impressi* **S12**. Purkersdorf: Verlag Brüder Hollinek.

WELLE 2020
———. "Mit Dürer auf dem langen Weg zum 'Führer'. Eine Rezeptionsgeschichte über Albrecht Dürers Zweikampfhandschrift Ms 26232 der Albertina Wien." *SportZeiten* **20**(3): 8–42. Bielefeld: Die Werkstatt.

WELLE 2021
———. *Albrecht Dürer und seine Kunst des Zweikampfes: auf den Spuren der Handschrift 26232 in der Albertina Wien* (2 vols). Kumberg: Sublilium Schaffer, Verlag für Geschichte, Kunst & Buchkultur.

WIDAUER 2017
HEINZ WIDAUER. "Das Ring- und Fechtbuch der Albertina. Eine Handschrift mit vielen Rätseln." *Die Kunst des Fechtens*: 267–284. Ed. by MATTHIAS JOHANNES BAUER and ELISABETH VAVRA. Heidelberg: Universitätsverlag Winter GmbH.

WIERSCHIN 1965
MARTIN WIERSCHIN. *Meister Johann Liechtenauers Kunst des Fechtens*. München: C. H. Beck.

WILKENS 2015
THORE WILKENS. "Ringeck, Siegmund." *Deutsches Literatur-Lexicon. Das Mittelalter* **7**: 603–605. Ed. by WOLFGANG ACHNITZ. Berlin & Boston.

ŻABIŃSKI 2008
"Unarmored Longsword Combat by Master Liechtenauer via Priest Döbringer." *Masters of Medieval and Renaissance Martial Arts*: 59–116. Trans by. GRZEGORZ ŻABIŃSKI. Ed. by JOHN CLEMENTS. Boulder: Paladin Press.

ŻABIŃSKI 2010
GRZEGORZ ŻABIŃSKI. *The Longsword Teachings of Master Liechtenauer. The Early Sixteenth Century Swordsmanship Comments in the 'Goliath' Manuscript*. Toruń: Wydawnictwo Adam Marshall.

ZMORA 2011
HILLAY ZMORA. *The Feud in Early Modern Germany*. New York: Cambridge University Press.

INDEX

About the Authors

MICHAEL CHIDESTER has been studying historical European martial arts since 2001, and has been Editor-in-Chief of Wiktenauer since 2011. In 2019, he started a publishing company called HEMA Bookshelf to produce facsimiles of fencing manuals and publish translations and historical research. He has lectured on historical martial arts across North America and Europe and has written several books. He currently trains at Athena School of Arms in Cambridge, MA.

DIERK HAGEDORN is instructor for the long sword at Hammaborg, a club for historical fencing in Hamburg; he is also a passionate wearer of armor. He has transcribed 30 of the approximately 80 surviving fight books in German language and made them available on the website of his club or published them as an edition in book form. He works for the IT subsidiary of a large German airline for numerous creative issues such as user experience, interface, and corporate design.

* 9 7 8 1 9 5 3 6 8 3 4 7 2 *
ND - #0193 - 150926 - C500 - 246/189/23 [25] - CB - 9781953683472 - Matt Lamination